Promoting Sustainable Management Through Technological Innovation

Supriya Lamba Sahdev
ISBR Business School, India

Chitra Krishnan
Amity University, Noida, India

Ahdi Hassan
Global Institute for Research Education and Scholarship, India

IGI Global
PUBLISHER of TIMELY KNOWLEDGE

A volume in the Advances in Environmental Engineering and Green Technologies (AEEGT) Book Series

Published in the United States of America by
IGI Global
Engineering Science Reference (an imprint of IGI Global)
701 E. Chocolate Avenue
Hershey PA, USA 17033
Tel: 717-533-8845
Fax: 717-533-8661
E-mail: cust@igi-global.com
Web site: http://www.igi-global.com

Library of Congress Cataloging-in-Publication Data

Names: Sahdev, Supriya Lamba, 1984- editor. | Krishnan, Chitra, 1983- editor. | Hassan, Ahdi, 1991- editor.
Title: Promoting sustainable management through technological innovation / edited by Supriya Lamba Sahdev, Chitra Krishnan, Ahdi Hassan.
Description: Hershey, PA : Engineering Science Reference, [2024] | Includes bibliographical references and index. | Summary: "The primary objective of the book is to explore the intersection of technology and sustainability, with a focus on how technological innovations can be used to promote sustainable management practices in various industries"-- Provided by publisher.
Identifiers: LCCN 2023034578 (print) | LCCN 2023034579 (ebook) | ISBN 9781668499795 (hardcover) | ISBN 9781668499801 (ebook)
Subjects: LCSH: Management--Technological innovations. | Sustainable development--Technological innovations. | Information technology--Management.
Classification: LCC HD30.2 .P765 2024 (print) | LCC HD30.2 (ebook) | DDC 658.4/038--dc23/eng/20230807
LC record available at https://lccn.loc.gov/2023034578
LC ebook record available at https://lccn.loc.gov/2023034579

This book is published in the IGI Global book series Advances in Environmental Engineering and Green Technologies (AEEGT) (ISSN: 2326-9162; eISSN: 2326-9170)

British Cataloguing in Publication Data
A Cataloguing in Publication record for this book is available from the British Library.

For electronic access to this publication, please contact: eresources@igi-global.com.

Advances in Environmental Engineering and Green Technologies (AEEGT) Book Series

Sang-Bing Tsai
Zhongshan Institute, University of Electronic Science and Technology of China, China & Wuyi University, China
Ming-Lang Tseng
Lunghwa University of Science and Technology, Taiwan
Yuchi Wang
University of Electronic Science and Technology of China Zhongshan Institute, China

ISSN:2326-9162
EISSN:2326-9170

Mission

Growing awareness and an increased focus on environmental issues such as climate change, energy use, and loss of non-renewable resources have brought about a greater need for research that provides potential solutions to these problems. Research in environmental science and engineering continues to play a vital role in uncovering new opportunities for a "green" future.

The **Advances in Environmental Engineering and Green Technologies (AEEGT)** book series is a mouthpiece for research in all aspects of environmental science, earth science, and green initiatives. This series supports the ongoing research in this field through publishing books that discuss topics within environmental engineering or that deal with the interdisciplinary field of green technologies.

Coverage

- Green Technology
- Alternative Power Sources
- Air Quality
- Electric Vehicles
- Water Supply and Treatment
- Pollution Management
- Renewable Energy
- Contaminated Site Remediation
- Industrial Waste Management and Minimization
- Policies Involving Green Technologies and Environmental Engineering

IGI Global is currently accepting manuscripts for publication within this series. To submit a proposal for a volume in this series, please contact our Acquisition Editors at Acquisitions@igi-global.com or visit: http://www.igi-global.com/publish/.

The Advances in Environmental Engineering and Green Technologies (AEEGT) Book Series (ISSN 2326-9162) is published by IGI Global, 701 E. Chocolate Avenue, Hershey, PA 17033-1240, USA, www.igi-global.com. This series is composed of titles available for purchase individually; each title is edited to be contextually exclusive from any other title within the series. For pricing and ordering information please visit http://www.igi-global.com/book-series/advances-environmental-engineering-green-technologies/73679. Postmaster: Send all address changes to above address.

Titles in this Series

For a list of additional titles in this series, please visit: www.igi-global.com/book-series

Perspectives on Global Biodiversity Scenarios and Environmental Services in the 21st Century
Naveen Kumar Chourasia (Government College, Bichhua, India) and Kavita Chahal (Government College, Bichhua, Chhindwara, India)
Engineering Science Reference • copyright 2023 • 300pp • H/C (ISBN: 9781668490341) • US $250.00 (our price)

Environmental and Technological Aspects of Redox Processes
Gheorghe Duca (Institute of Chemistry, Moldova State University, Moldova) and Ashok Vaseashta (International Clean Water Institute, USA)
Engineering Science Reference • copyright 2023 • 435pp • H/C (ISBN: 9798369305126) • US $295.00 (our price)

Climate Change Mitigation With Hydrogel in Agriculture
Elsayed Gamal Zaki (Egyptian Petroleum Research Institute, Egypt) and Shimaa Mohamed Elsaeed (Egyptian Petroleum Research Institute, Egypt)
Engineering Science Reference • copyright 2024 • 350pp • H/C (ISBN: 9781668498774) • US $250.00 (our price)

Machine Learning and Deep Learning for Smart Agriculture and Applications
Mohamamd Farukh Hashmi (National Institute of Technology, Warangal, India) and Avinash G. Kesakr (Visvesvaraya National Institute of Technology, India)
Engineering Science Reference • copyright 2023 • 257pp • H/C (ISBN: 9781668499757) • US $265.00 (our price)

Artificial Intelligence Applications in Water Treatment and Water Resource Management
Victor Shikuku (Kaimosi Friends University, Kenya)
Engineering Science Reference • copyright 2023 • 270pp • H/C (ISBN: 9781668467916) • US $225.00 (our price)

Impactful Technologies Transforming the Food Industry
Şule Aydın (Nevsehir Haci Bektas Veli University, Turkey) Eda Özgül Katlav (Nevşehir Haci Bektaş Veli University, Turkey) Koray Çamlıca (Nevşehir Haci Bektaş Veli University, Turkey) and Firdevs Yönet Eren (Nevşehir Haci Bektaş Veli University, Turkey)
Engineering Science Reference • copyright 2023 • 343pp • H/C (ISBN: 9781668490945) • US $240.00 (our price)

Convergence of Cloud Computing, AI, and Agricultural Science
Avinash Kumar Sharma (ABES Institute of Technology, India) Nitin Chanderwal (University of Cincinnati, USA) and Rijwan Khan (ABES Institute of Technology, India)
Engineering Science Reference • copyright 2023 • 375pp • H/C (ISBN: 9798369302002) • US $265.00 (our price)

IGI Global
PUBLISHER of TIMELY KNOWLEDGE

701 East Chocolate Avenue, Hershey, PA 17033, USA
Tel: 717-533-8845 x100 • Fax: 717-533-8661
E-Mail: cust@igi-global.com • www.igi-global.com

Table of Contents

Detailed Table of Contents

In recent years, the integration of the internet of things (IoT) and sustainable development has gained significant attention across industries. This chapter explores the role of IoT in driving sustainable development and examines the future trends that are expected to shape industries. By leveraging IoT technologies, industries can enhance resource efficiency, reduce environmental impacts, and create new opportunities for sustainable growth. It further delves into the future trends in industries influenced by IoT-driven sustainable practices. By analyzing existing literature, case studies, and expert opinions, this chapter aims to provide insights into the transformative potential of IoT in fostering sustainable development and shaping the future of industries.

The changing dynamics of the business and growing pressure from the stakeholders has forced the organisations to take another important aspect into their boardroom – sustainability. Investment in technological interventions is a must to enable sustainability within the business. The problem faced by the businesses is to find the right balance between their profits and technological investment to achieve sustainability. The current study focuses on understanding the factors which affect investment in sustainability using the multi-criteria approach of fuzzy-DEMATEL approach. The methodology is utilised to depict the context-specific link between the factors to aid the business managers in making right decisions. The results demonstrate that the factors such as commitment of top management (TS1), cost of training and development (TS2), cost of regulatory audits and fines (TS5), cost of subsidies and incentives (TS6), and cost of searching for suitable partners (TS9) are crucial for the incorporation of sustainability.

Concerning gender diversity, the IT industry faces both a challenge and an opportunity. As in many other disciplines, women are underrepresented in the leadership of the technology sector. However, this situation also presents an opportunity: tech companies that can expand their ranks of female executives stand to gain significantly. It is critical to promote and support more women and girls to enter STEM (science, technology, engineering, and mathematics) sectors for the sake of global economies and the global transition to sustainability. There were many ways in which technological advancements aided women's capability and resource development. Women's adoption, usage, and pervasiveness of ICTs span the broad areas of outreach (e.g., health promotion), education (e.g., health literacy opportunities), lifestyle (e.g., peer coaching and planning), prevention (e.g., screening opportunities), health challenges (e.g., intimate partner violence apps), and perceptions of barriers.

The aim of this study is to explain the concept of digital economy and to examine the effect of digital technology on economic growth. In the study, the causality relationship between digital technology and economic growth in the Turkish economy for the period 2004-2021 was investigated with the Toda-Yamamoto test. As a result of the study, according to 10% significance level, bidirectional causality between internet usage-economic growth, computer usage-economic growth, and one-way causality between mobile line usage and economic growth from mobile line usage to economic growth was determined. No causal relationship was found between website ownership and economic growth in enterprises. According to the 5% significance level, a one-way causality relationship was determined from only computer and internet use to economic growth.

The study tries to investigate the long run and short run relationship between trade openness (TO), political stability (PO), and FDI on sustainable development of select developed and developing nations. Time series data from 1995 to 2021 of about 25 economies—10 developed economies and 15 developing economies—was collected and analyzed using Phillips Perron Fisher panel unit root test, panel auto regressive distributed lag (PARDL) model, and panel fully modified least squares/fully modified OLS.

From the result, it found that FDI and TO are positively contributing to sustainability development index (SDI) in developing countries rather than the developed countries in the long run. In addition to this, changes in the SDI score is significantly influenced by the present and past import and export activities in developed as well as developing economies in the short run.

This chapter delves into the factors influencing the implementation of business analytics in the healthcare sector of MENA (Middle East and North Africa) countries. A comprehensive theoretical model has been developed to encompass various organizational, technological, human, and external factors. The data analysis plan involves both descriptive and inferential statistical methods to analyze the collected data. Regression analysis, factor analysis, and correlation analysis will be used to examine the relationships between the identified factors and the success of business analytics implementation. Additionally, qualitative analysis will be conducted to identify common strategies and approaches that have been effective in overcoming barriers to the adoption of business analytics in healthcare. The findings of this study will contribute to the existing body of knowledge on business analytics adoption in healthcare and provide insights for managers and policymakers in MENA countries to overcome barriers and promote the effective utilization of business analytics in the healthcare sector.

The causal relationship between the consumption due to increased international trade and resultant CO2 emissions has been a subject of debate. This study examines the relationship between CO2 emissions and energy consumption along with global trade indicators for India using data from World Bank for the period 1982-2013. The study employed Augmented Dickey Fuller unit root test for stationarity among the variables, Johansen Cointegration technique to determine the order of the cointegration equation, and then the Vector Error Correction mechanism and Wald tests to check for short-run or long-run relationship among the variables. The results indicate that the variables are cointegrated. The VECM suggested no long-run causality from all the variables to CO2 emissions. Wald tests indicate no short-run causality also from FDI, exports, imports, and energy consumption to CO2 emissions but existence of short-run causality from trade openness to CO2 emissions. No long-run or short-run causality from FDI to CO2 emissions indicate the non-existence of Pollution Haven Hypothesis in India.

Preface

In a world grappling with unprecedented environmental and social crises, the urgency to secure a sustainable future for all has never been more pronounced. The relentless march of climate change, the perilous depletion of vital resources, and the persistent specter of social inequality compel us to act decisively. In this defining moment, technology emerges as a potent ally—a beacon of hope capable of reshaping our world for the better.

Yet, we must tread cautiously, for technology's promise is twofold—it holds the potential for both salvation and peril. Its profound impact on society, the environment, and the economy demands astute management and ethical consideration. As editors, we embarked on a compelling journey to unravel the complex relationship between technology and sustainability, fully aware of its immense potential for positive transformation and its capacity to sow new seeds of discord.

Promoting Sustainable Management Through Technological Innovation embodies our collective endeavor to dissect this intricate alliance. Within its pages, we dive into the tantalizing promise of technological innovation across diverse industries, spanning energy, transportation, and agriculture. Yet, we do not shy away from scrutinizing the shadows it may cast, dissecting the ethical dilemmas and potential pitfalls that lurk within.

This book aspires to be more than a mere compendium—it is a rallying cry for change, a source of inspiration, and a catalyst for action. It showcases instances where technology has successfully ushered in sustainable solutions, offering invaluable insights and recommendations to policymakers, visionary business leaders, and dedicated researchers. Our ambition is to cultivate a deep-seated understanding of how technological innovation can be harnessed to forge a path towards a more sustainable future for all.

The central mission of *Promoting Sustainable Management Through Technological Innovation* is to illuminate the nexus of technology and sustainability, with a sharp focus on the promotion of sustainable management practices across a spectrum of industries. In pursuit of this grand objective, the book endeavors to:

1. Provide a Sweeping Overview: By presenting a holistic perspective on technology's capacity to confront sustainability challenges head-on and facilitate sustainable management practices.
2. Probe Benefits and Risks: By meticulously dissecting the potential advantages and risks associated with a spectrum of technologies, navigating the ethical considerations that underscore their development and deployment.
3. Spotlight Success Stories: By bringing to light cases of sustainable technology innovation that offer priceless lessons for policymakers, forward-thinking business leaders, and relentless researchers.
4. Contribute to a Global Dialogue: By actively engaging in the ongoing discourse on sustainable development and fervently shaping a more sustainable future for all.

In our quest to achieve these monumental objectives, this book is poised to make a series of meaningful contributions:

- It will offer a comprehensive and up-to-date synthesis of existing research at the intersection of technology and sustainability, pinpointing areas ripe for further exploration.
- It seeks to foster a more nuanced and sophisticated understanding of the potential advantages and risks linked to diverse technologies, laying the foundation for future groundbreaking inquiry in this critical field.
- By furnishing invaluable insights and actionable recommendations, it aims to inform the creation of policies and strategies designed to fortify sustainable management practices and harness technological innovation to realize sustainability objectives.

Promoting Sustainable Management Through Technological Innovation is not merely a book; it is a manifesto for a brighter, more sustainable future. It is tailored for a diverse audience united by a shared passion for the synergy of technology and sustainability. Whether you are a policymaker envisioning a better world, a business leader daring to innovate, a researcher on the cusp of breakthroughs, an educator nurturing future change-makers, or a student aspiring to shape a brighter tomorrow, this book offers relevance and transformative power.

For:

- Policymakers: It provides a roadmap—an arsenal of insights and recommendations to fortify policies aimed at sustainable management and the potent leveraging of technological innovation for sustainability.
- Business Leaders: It offers more than guidance—it offers a call to action, encouraging the fostering of sustainable management practices within organizations and the strategic leveraging of technological innovations to propel sustainability goals to new heights.
- Researchers: It is a clarion call for collaboration and exploration, a definitive repository of current research at the nexus of technology and sustainability, identifying the untrodden paths that beckon us toward progress.
- Educators: It stands as an invaluable resource, a beacon of knowledge for molding the minds of future leaders, offering a comprehensive and up-to-date overview of research in sustainability, technology, and responsible management.
- Students: It is your gateway to enlightenment, a roadmap to navigate the intricate terrain of sustainability, technology, and responsible management, propelling you toward transformative action.

CHAPTER OVERVIEW

Chapter 1: IoT-Driven Sustainable Development and Future Trends in Industries

Authors: Seema Garg, Namrata Pancholi

This chapter delves into the dynamic realm where the Internet of Things (IoT) converges with sustainable development across industries. It unveils the pivotal role played by IoT technologies in driving sustainable practices, illuminating how they enhance resource efficiency, curtail environmental impacts, and open new avenues for sustainable growth. Furthermore, the chapter peers into the horizon, offering insights into the future trends that promise to reshape industries through IoT-driven sustainability. Drawing from a rich tapestry of literature, case studies, and expert opinions, this chapter unravels the transformative potential of IoT, providing a roadmap for sustainable development and the evolution of industries.

Chapter 2: Analysis of Factors Affecting Investment in Technology for Sustainability

Authors: Vernika Agarwal, Pragya Arya

In today's fast-evolving business landscape, the imperative of sustainability is increasingly apparent. This chapter addresses the pressing challenge of balancing profitability with sustainability, focusing on the critical role of technological investments. Employing a multi-criteria approach, the chapter investigates the factors influencing investment in sustainability. By peering into the complexities of decision-making, it provides valuable insights for business leaders striving to strike the right equilibrium. The findings elucidate key factors, such as top management commitment, training costs, regulatory pressures, subsidies, and partner search costs, that are instrumental in fostering sustainability within organizations.

Chapter 3: Sustainability Threats to the Information Society – Gap of Sustainability Knowledge and Practice

Authors: Prabhat Ranjan, Bhupendra Kumar Singh

The ubiquity of electronic communication and the vast reservoir of scholarly knowledge have propelled our society into the information age. However, amid the technological marvels, sustainability issues have emerged, raising questions about the vulnerability of our information society. This chapter embarks on a journey to uncover the underlying factors contributing to this vulnerability. Drawing on extensive research and analysis, it underscores the critical role of ethics in knowledge deployment and the necessity for adept political structures to frame policies conducive to a sustainable society. It advocates a shift from a consumer-driven society to one founded on values, emphasizing the importance of environmental education over mere environmental knowledge.

Chapter 4: Innovation, Sustainability, and Business Models

Authors: Namrata Pancholi, Mona Chaudhary, Seema Garg

The world is undergoing a transformative phase fueled by technological innovation, presenting new challenges and opportunities. One of the noteworthy shifts is the evolving perspective of customers who increasingly demand sustainable business models. This chapter explores this changing landscape, delving into the intersection of innovation and sustainability. It scrutinizes the reasons behind the failure of many business model innovations and proposes a novel conceptual framework that bridges the gap between innovation and sustainability. By forging sustainable business models that create value for stakeholders, this chapter aims to usher organizations toward a sustainable competitive advantage.

Chapter 5: Green Technology Innovations

Authors: Ingrid Cuevas Zuñiga, Jessica Aimeé Gómez Quiroz, María del Rocío Soto Flores, Samantha Cuevas Antunez

Amidst growing concerns about environmental crises and global warming, markets are increasingly demanding rational and efficient processes that mitigate the environmental impact of daily operations. This chapter embarks on a quest to identify green technological innovations as the catalysts for organizations to fulfill their commitment to sustainable development. It unveils how these innovations, ranging from energy and water conservation to waste reduction and biotechnology, enable economic, social, and environmental benefits when integrated into production processes. The chapter showcases how organizations can uphold their pledge to sustainable development by incorporating green technological innovations.

Chapter 6: Creating a Sustainable Win-Win Situation – Addressing Gender Disparity in Sustainable Technological Innovation

Author: Maria Mansab

The tech industry presents both a challenge and an opportunity in the realm of gender diversity. This chapter sheds light on the underrepresentation of women in technology leadership roles while highlighting the potential benefits of increasing female representation. It underscores the importance of promoting and supporting women and girls in STEM fields, emphasizing the profound impact they can have on global economies and the transition to sustainability. The chapter explores how technological advancements empower women and foster resource development, showcasing their adoption, usage, and influence across various areas, from health to education, and beyond.

Chapter 7: Digital Economy, the Correlation Between Digital Technology and Economic Growth – An Empirical Application

Authors: Dilek Dinç, Aytaç Gökmen

This chapter embarks on an exploration of the digital economy, unraveling the intricate relationship between digital technology and economic growth. Drawing upon empirical evidence from the Turkish economy, it investigates causality relationships between digital technology usage and economic growth. The findings provide valuable insights into the bidirectional causality between internet usage and economic growth, highlighting the significant impact of digital technology on economic development. Moreover, the chapter reveals how computer usage and internet use are intertwined with economic growth, paving the way for informed decision-making in the digital age.

Chapter 8: Assessing the Role of Trade Openness, FDI, and Political Stability on Sustainable Development

Authors: Sathish Pachiyappan, Jabez J. P., Shylaja H. N., John Paul Raj, Saravanan Vellayan

This chapter embarks on a comprehensive investigation of the factors influencing sustainable development in both developed and developing economies. With a focus on trade openness, foreign direct investment (FDI), and political stability, it explores their long-run and short-run relationships with sustainability. Utilizing advanced econometric methods, the chapter unveils the varying impacts of these factors on the sustainable development index (SDI). Notably, it reveals how FDI and trade openness play a pivotal role in enhancing sustainability in developing nations. This chapter offers critical insights into the complex interplay between these factors and sustainable development, empowering policymakers and business leaders to drive progress.

Chapter 9: Factors Affecting the Implementation of Business Analytics in the Healthcare Sector in MENA Countries

Author: Aftab Ara

Delving into the complex realm of healthcare analytics in the Middle East and North Africa (MENA) region, this chapter examines the factors influencing the implementation of business analytics in healthcare. Through the lens of a theoretical model, it navigates through organizational, technological, human, and external factors shaping the landscape. The chapter employs a blend of quantitative and qualitative analyses, from regression to factor and correlation analysis, to uncover the relationships between these factors and the success of business analytics in healthcare. The findings offer crucial insights for healthcare managers and policymakers in the MENA region seeking to overcome barriers and harness the potential of business analytics in healthcare.

Chapter 10: On the Causality Between FDI, International Trade Indicators, Energy Consumption, and CO2 Emissions in India

Authors: Ritu Rana, Manoj Sharma

This chapter embarks on an empirical exploration of the intricate relationships between foreign direct investment (FDI), international trade indicators, energy consumption, and CO2 emissions in the context of India. Drawing from extensive data spanning nearly three decades, it employs a rigorous econometric framework to decipher causality patterns. The chapter unveils the nuanced dynamics, highlighting how FDI and trade indicators positively contribute to sustainability development in developing countries. It also dispels the notion of a Pollution Haven Hypothesis in India, emphasizing the multifaceted nature of the relationship between economic activities and CO2 emissions.

Chapter 11: Mapping the Future of Sustainable Development Through Cloud-Based Solutions – A Case Study of OpenStreetMap

Authors: Munir Ahmad, Asmat Ali

OpenStreetMap (OSM) is at the forefront of advancing sustainability across economic, environmental, and societal domains. This chapter paints a vivid portrait of OSM's pivotal role in bolstering economic sustainability through infrastructure planning, disaster response, and energy efficiency. It also underscores OSM's contributions to environmental sustainability through resource management, climate change mitigation, and biodiversity protection. Moreover, the chapter illuminates how OSM promotes social sustainability by empowering communities and supporting healthcare, education, and gender equality. Through spatial analysis, OSM's potential to identify areas in need of essential services and address social inequalities is unveiled. The chapter concludes by highlighting the symbiotic relationship between OSM and cloud computing, which promises to enhance OSM's role in advancing sustainability further.

Chapter 12: Sustainability in Upstream Oil and Gas – Integrating Energy and Ecology

Authors: Kaustubh Laturkar, Kasturi Laturkar

The chapter casts a discerning eye on the upstream oil and gas industry, illuminating its pivotal role in meeting global energy demands while acknowledging its environmental and social implications. It outlines the imperative for sustainability improvements in key areas such as transitioning to cleaner energy sources, reducing methane emissions, and adopting innovative carbon capture and storage (CCS) technologies. This chapter offers an in-depth overview of the challenges and opportunities in implementing sustainable practices within the upstream oil and gas sector. It envisions an equitable future where the industry thrives through sustainable initiatives, fostering a culture committed to ecological and societal well-being.

Chapter 13: Eco-Friendly Colors – A Review of Recent Advances in Natural Dyes

Author: Rupayan Roy

Exploring the world of fashion through an eco-friendly lens, this chapter offers a comprehensive review of recent developments in natural dyes. It shines a spotlight on the advancements in natural dye research, showcasing their benefits, the colors they produce, and the methods employed to extract and process these dyes. Moreover, the chapter accentuates the sustainability and environmental impact of natural dyes, drawing a sharp contrast with synthetic alternatives. It underscores the significance of adopting eco-friendly dyes in the fashion industry and contemplates the potential future developments that could revolutionize fashion's ecological footprint.

Chapter 14: Optimization of Parameters and Modeling for Breakthrough Curve for Chromium Removal in Fixed Bed

Authors: Sunil Kulkarni, Saima Parkar, Rutuja Mulukh, Gautami Narhari

Wastewater tainted with heavy metals poses a significant environmental challenge. This chapter presents an exhaustive exploration into the removal of hexavalent chromium (Cr VI) from simulated wastewater using activated carbon. The experiments encompass both batch and continuous modes, with a focus on fixed bed adsorption. It scrutinizes the impact of varying parameters such as bed height, concentration, and flow rate on breakthrough curves. The findings illuminate the dynamics of batch and fixed bed adsorption and offer insights into optimizing the process. By delving into complex scientific methodologies, this chapter provides valuable guidance for addressing environmental challenges associated with heavy metal contamination.

IN SUMMARY

As we conclude this preface for *Promoting Sustainable Management Through Technological Innovation*, we reflect on the profound journey we have embarked upon. The book, authored by a distinguished group of experts, serves as a beacon guiding us through the intricate interplay between technology and sustainability.

In an era where our world faces unprecedented environmental and social challenges, the integration of technology and sustainable management practices emerges as a promising path forward. Climate change, resource depletion, and social inequality demand our immediate attention, and technological innovation offers a formidable tool to address these pressing issues.

This edited reference book, comprising diverse chapters authored by experts in their respective fields, traverses the spectrum of challenges and opportunities that lie at the intersection of technology and sustainability. From IoT-driven sustainable development to the analysis of factors affecting technology investments for sustainability, from the threats posed to the information society to the intricacies of green technological innovations, each chapter unravels a unique facet of our evolving landscape.

We have explored the significance of innovation, sustainability, and business models, delved into the implications of digital technology on economic growth, and assessed the role of trade openness, FDI, and political stability in driving sustainable development. We have mapped the future of sustainable development through cloud-based solutions, delved into the complexities of the upstream oil and gas industry, and reviewed recent advances in eco-friendly natural dyes. Additionally, we have scrutinized the optimization of parameters for heavy metal removal in wastewater, showcasing the diversity and depth of knowledge presented within this volume.

Through these chapters, this book accomplishes several critical objectives. It provides a comprehensive overview of the potential of technology to address sustainability challenges and promote sustainable management practices. It examines the benefits and risks associated with various technologies while navigating the ethical considerations that must guide their development and deployment. The book spotlights successful sustainable technology innovations and offers insights and recommendations for policymakers, business leaders, and researchers on harnessing technological innovation for a more sustainable future.

Furthermore, this volume contributes to the ongoing conversation surrounding sustainable development and endeavors to shape a future where sustainability is at the forefront of our global agenda. By highlighting areas for further research, providing a nuanced understanding of the complexities involved, and offering actionable insights for decision-makers, this book leaves an indelible mark on the field of technological innovation for sustainability.

We envision a world where technology becomes an enabler of sustainable practices across industries, where businesses thrive while safeguarding the environment, where social equity is paramount, and where innovation and responsibility coexist harmoniously. *Promoting Sustainable Management Through Technological Innovation* is our collective effort to move closer to this vision.

This book is intended for a diverse audience – policymakers, business leaders, researchers, educators, students, and anyone with an interest in sustainable resource management and the transformative power of technology. It offers something for everyone, from the seasoned expert seeking cutting-edge insights to the novice eager to explore this dynamic intersection.

In closing, we extend our gratitude to the dedicated authors who have contributed their expertise and knowledge to make this book a reality. We also express our appreciation to the readers who embark on this intellectual journey with us, as we collectively strive to promote a sustainable future through technological innovation. We hope the pages that follow provide inspiration, spark innovation, and catalyze meaningful change on the path toward a more sustainable world for all.

Warmly,

Supriya Lamba Sahdev
ISBR Business School, India

Chitra Krishna
Symbiosis International University, India

Ahdi Hassan
Global Institute for Research Education and Scholarship, India

Chapter 1
IoT-Driven Sustainable Development and Future Trends in Industries

Seema Garg
https://orcid.org/0000-0002-4750-026X
Amity University, Noida, India

Namrata Pancholi
Amity University, Noida, India

ABSTRACT

In recent years, the integration of the internet of things (IoT) and sustainable development has gained significant attention across industries. This chapter explores the role of IoT in driving sustainable development and examines the future trends that are expected to shape industries. By leveraging IoT technologies, industries can enhance resource efficiency, reduce environmental impacts, and create new opportunities for sustainable growth. It further delves into the future trends in industries influenced by IoT-driven sustainable practices. By analyzing existing literature, case studies, and expert opinions, this chapter aims to provide insights into the transformative potential of IoT in fostering sustainable development and shaping the future of industries.

1. INTRODUCTION

The Internet of Things (IoT) has emerged as a powerful technology paradigm with the potential to drive sustainable development across various industries. The Internet of Things (IoT) describes the network of physical objects—"things"—that are embedded with sensors, software, and other technologies for the purpose of connecting and exchanging data with other devices and systems over the internet. These devices range from ordinary household objects to sophisticated industrial tools. The IoT has had a significant influence on the development of enterprises too. Internet of Things helps to integrate digital and physical systems, which ultimately result in an increase in productivity, a reduction in expenses, and

DOI: 10.4018/978-1-6684-9979-5.ch001

an enhancement in end users' lifestyle quality of life. As a direct consequence of this, it is anticipated that the Internet of Things will cause widespread disruption across a wide variety of distinct business sectors in the years to come. These emerging technologies have revolutionized the way we live, work and interact with the world around us. With the use of IOT, industries are becoming more efficient and sustainable, while streamlining their operations to increase productivity and profitability.

The history of IOT can be traced back to the 1990s, when the concept of machine-to-machine communication was first introduced. Since then, the technology has evolved to include a wide range of devices and sensors that can communicate with each other and collect data in real-time. It is a technological advancement that enables the connection of numerous distinct machines to gather data in real-time and respond to certain issues. It provides a wealth of opportunities and is crucial for sustainability while working to increase energy efficiency. The role of technology and internet of things in sustainable development is involving new types of product innovations resulting in sustainability and value chain solutions. IOT is increasingly playing a crucial role in enabling sustainable development, both from a consumer and an industrial standpoint. The UN Sustainable Development Goals for 2030 can be achieved with the aid of IoT. Blockchain and the Internet of Things (IoT) are two dimensions of industry 4.0 that play a crucial role in sustainable supply chain management (SSCM) and reverse logistics. (Dutta et al, 2022). Industries can utilise IoT to reduce food waste, conserve water, or boost energy efficiency to encourage sustainable innovation and profitability. Implementing I.O.T in various industries is not without its challenges. Inadequate infrastructure, a lack of skilled workers, and legal and regulatory difficulties are major obstacles to the widespread use of IoT in India (Agarwal et al., 2021). There are Design challenges of Scalability, Interoperability and Power Consumption. There are certain development challenges of Software Complexity, Connectivity and data management. Also Security challenges include data privacy, security, Authentication, Authorization and Standardization.

The term "green" typically refers to "planning and investing in a technology network that serves both current and future needs while preserving energy and reducing costs." The Green Internet of Things is primarily concerned with enhancing energy efficiency and the validity of IoT concepts. The term "Green IoT" refers to an IoT strategy that uses energy efficiently to either reduce or remove the greenhouse effect brought on by current uses. The term "green IoT" refers to a procedure that minimizes the negative effects on society and the environment when producing, developing, utilizing, and disposing of computers, servers, and associated subsystems (such as printers, displays, communications equipment, and storage devices). In order to lessen the detrimental effects of IoT on human health and the environment, new resources are needed.

Examining the Impact of IoT Implementation on Industries and sustainable development. There has been a lot of focus in recent years on the use of the IoT businesses. The Internet of Things (IoT) has the potential to dramatically alter how we live, work, and interacts with the world around us. The Internet of Things allows for greater resource efficiency, higher levels of public safety, and a higher quality of life for city residents. The industrial sector, meanwhile, may use IoT to boost production, save expenses, and streamline operations. The purpose of this study is to investigate how widespread use of the Internet of Things will affect sustainable development and future trends in industry.

Overall, IoT-driven sustainable development relies on the efficient gathering and analysis of data from connected devices to enable smarter decision-making and resource management. By leveraging IoT technology, we can make significant strides towards achieving a more sustainable and environmentally conscious future.

2.1 IoT Applications for Sustainable Development

2.1.1 Energy Management

IoT-enabled smart grids and energy monitoring systems enable efficient energy distribution, load balancing, and demand response, leading to reduced energy consumption and greenhouse gas emissions. Smart buildings equipped with IoT sensors and automation systems optimize energy usage through intelligent lighting, HVAC (heating, ventilation, and air conditioning), and occupancy management.

2.1.2 Water and Waste Management

IoT-based water quality monitoring systems detect contaminants in real-time, enabling prompt interventions for water treatment and ensuring clean water supply. Smart waste management solutions utilizing IoT sensors optimize waste collection routes, reduce overflowing bins, and promote recycling and waste reduction.

2.1.3 Supply Chain and Logistics

IoT-powered supply chain management enhances visibility and transparency, enabling real-time tracking and monitoring of goods, optimizing transportation routes, and reducing fuel consumption and emissions. Cold chain management systems utilizing IoT sensors ensure proper temperature control and minimize food spoilage during transportation and storage.

2.1.4 Agriculture and Farming

IoT applications in precision agriculture enable efficient use of resources such as water, fertilizers, and pesticides, resulting in increased crop yield, minimized environmental impact, and reduced costs. Smart farming systems equipped with IoT sensors and data analytics provide real-time insights into crop health, weather conditions, and pest infestation, enabling proactive interventions and optimizing farming practices.

2.1.5 Environmental Monitoring

IoT sensors can be deployed to monitor air quality, water quality, and other environmental parameters. This data can help identify pollution sources, assess environmental risks, and support evidence-based decision-making for sustainable development.

2.1.6 Smart Transportation

IoT can contribute to sustainable transportation by enabling smart traffic management, intelligent parking systems, and efficient public transportation systems. By reducing congestion and optimizing traffic flow, IoT can minimize fuel consumption and air pollution.

2.1.7 Smart Buildings

IoT can enable smart building management systems that optimize energy usage, monitor indoor air quality, and automate lighting and HVAC (heating, ventilation, and air conditioning) systems. This leads to energy savings, improved occupant comfort, and reduced environmental impact.

21.8 Conservation and Wildlife Protection

IoT-based solutions such as wildlife tracking devices and connected sensors can aid in monitoring and protecting endangered species, preventing illegal activities like poaching, and managing protected areas more effectively.

2.2 Future Trends in IoT-Driven Sustainable Industries

2.2.1 Edge Computing and AI Integration

The integration of edge computing and AI algorithms with IoT devices enables real-time data processing, analysis, and decision-making at the edge, reducing latency, enhancing efficiency, and enabling autonomous systems.

2.2.2 Blockchain and Distributed Ledger Technologies

Blockchain technology ensures transparency, traceability, and security in supply chains, enabling reliable tracking of sustainable practices, certifications, and product origins. Decentralized energy systems utilizing blockchain facilitate peer-to-peer energy trading, promoting renewable energy generation and fostering a more sustainable energy ecosystem.

2.2.3 Circular Economy and Waste Management

IoT-enabled systems facilitate the tracking, sorting, and recycling of waste materials, supporting the transition to a circular economy where resources are reused, repurposed, and recycled. Collaborative consumption platforms powered by IoT promote the sharing and utilization of underutilized resources, reducing waste and promoting sustainable consumption patterns.

2.2.4 5G and Connectivity Enhancements

The deployment of 5G networks and improved connectivity infrastructure supports the scalability and reliability of IoT systems, enabling faster data transfer, lower latency, and more extensive IoT deployments across industries.

2.2.5 Augmented Reality (AR) and Virtual Reality (VR) Applications

AR and VR technologies integrated with IoT devices offer immersive and interactive experiences for training, simulation, and remote collaboration, reducing the need for physical travel and enhancing sustainable business practices.

3. LITERATURE REVIEW

The phrase "Internet of Things" was first used by Kevin Ashton in 1999. (Gubbi et al, 2013) Ashton characterised it as "a system of ubiquitous sensors, actuators, and networks that enable the seamless integration of physical items into digital systems."

IoT has also revolutionized the manufacturing industry, leading to the concept of Industry 4.0. According to McKinsey Global Institute (2018), "IoT-enabled predictive maintenance and real-time monitoring can reduce downtime by up to 50%." Moreover, IoT can optimize the supply chain and production processes by providing real-time data on inventory levels and machine performance. However, Bandyopadhyay and Sen (2011) note that the integration of legacy systems with IoT technologies is a major challenge in the manufacturing industry. In addition, the lack of skilled workforce to handle IoT systems is a major barrier to adoption (National Association of Software and Service Companies, 2019). Khan et al. (2018) highlight the challenges of implementing IoT in smart cities, including the high cost of deployment, lack of standardization, and privacy and security concerns. Marelli et al. (2021) Page 9 recommends the use of edge computing and block chain technologies to enhance safety and confidentiality of IoT systems in smart cities. Khan et al. (2018) argue that "IoT has the ability to optimise resource use, reduce energy consumption, boost mobility and accessibility, improve public safety, and enable better healthcare outcomes." Although there are many benefits to adopting IoT, there are also risks and difficulties associated with doing so. The Internet of Things (IoT) has many uses in the transportation, healthcare, energy management, and public safety sectors of smart cities and companies. McKinsey Global Institute (2018) estimates that by 2025, IoT will have created $1.2 trillion in economic value in smart cities. Predictive maintenance, streamlined supply chains, and increased worker safety and productivity are just Page 10 some of the industrial benefits that are anticipated from IoT. (Marelli et al., 2021). Inadequate infrastructure, a lack of skilled workers, and legal and regulatory difficulties are major obstacles to the widespread use of IoT in India (Agarwal et al., 2021).

4. RESEARCH QUESTIONS

1. What are the key benefits of IoT implementation in sustainable development?
2. How can IoT be leveraged to improve the efficiency of the industrial sector?
3. What are the challenges and barriers to IoT implementation in industries?
4. What are the future trends or prospects for IoT in industries?

5. RESEARCH METHODOLOGY

This includes searching for and analysing scholarly articles, reports, and other sources of information on the issue. The steps taken were as follows: Identification of relevant keywords, Literature search: To locate relevant papers and articles on the issue, a thorough search was undertaken utilising major academic scholarly archive sites such include Google Scholar, Scopus, and Web of Science.

6. ENVIRONMENTAL BENEFITS OF IOT IN INDUSTRIES

Sustainable Development Goals (SDGs) are a set of 17 global goals established by the United Nations in 2015 as part of the 2030 Agenda for Sustainable Development. These goals aim to address various social, economic, and environmental challenges faced by the world and provide a blueprint for a more sustainable future. The Internet of Things (IoT) has the potential to contribute significantly to the achievement of these SDGs by enabling innovative and efficient solutions. The integration of the Internet of Things (IoT) in industries offers several environmental benefits by enabling enhanced resource management, energy efficiency, and reduced environmental impacts. Some Key environmental benefits of IoT in industries includes Energy Efficiency, Resource Optimization, Waste Reduction and Recycling Emissions Reduction: Environmental Monitoring: Sustainable Agriculture.

The pursuit of sustainable development aims to protect environmental preservation, social equity, and justice. Although these three factors can work together harmoniously, it is frequently observed that they conflict. Throughout the second half of the 20th century, economic growth for a better quality of living has played a significant role in degrading the climate.The planet is currently being polluted by waste products and we are using more energy than ever before. More recently, society has developed the understanding that we cannot live in a stable society or economy if there is so much poverty and environmental degradation. The mainstay of human progress will continue to be economic expansion, but it will need to change to be less damaging to the environment. So as to bring

7. FUTURE TRENDS IN IOT-DRIVEN SUSTAINABLE INDUSTRIES:

The integration of the Internet of Things (IoT) in industries is continually evolving, and several future trends are expected to shape IoT-driven sustainable industries. These trends encompass technological advancements and innovative approaches that can further enhance sustainability and promote a more efficient and responsible use of resources. Here are some key future trends in IoT-driven sustainable industries:

7.1 Edge Computing and AI Integration

Edge computing involves processing and analyzing data at the edge devices or gateways, reducing latency, and enabling real-time decision-making. By integrating AI algorithms with IoT devices at the edge, industries can achieve more intelligent and autonomous systems that optimize resource usage, energy efficiency, and overall sustainability.

7.2 Blockchain and Distributed Ledger Technologies

Blockchain technology provides a decentralized and transparent way of recording and verifying transactions, ensuring trust, security, and traceability. In sustainable industries, blockchain can be used to track and authenticate products, certify sustainable practices, and facilitate peer-to-peer energy trading, fostering a more transparent and accountable supply chain.

7.3 Circular Economy and Waste Management

The transition to a circular economy aims to minimize waste generation, promote recycling, and maximize resource efficiency. IoT can play a crucial role by enabling smart waste management systems that optimize waste collection, sorting, and recycling processes, reducing waste and promoting the circular flow of materials.

7.4 5G and Enhanced Connectivity

The deployment of 5G networks and improved connectivity infrastructure will significantly impact IoT-driven industries. Faster data transfer, lower latency, and increased network capacity provided by 5G will support the scalability and reliability of IoT systems, enabling more extensive deployments and real-time data analytics.

7.5 Augmented Reality (AR) and Virtual Reality (VR) Applications

The integration of AR and VR technologies with IoT devices offers immersive and interactive experiences. In sustainable industries, AR and VR can be used for remote training, maintenance, and collaboration, reducing the need for physical travel, saving costs, and minimizing environmental impacts.

7.6 Energy Harvesting and Self-Powered IoT Devices

The development of energy harvesting technologies allows IoT devices to generate and store their energy from renewable sources or ambient surroundings. Self-powered IoT devices reduce the dependence on batteries and extend the lifespan of devices, contributing to sustainable operations and reducing electronic waste.

7.7 Data Analytics and Predictive Insights

Advanced data analytics techniques, including machine learning and predictive modeling, enable industries to derive valuable insights from the vast amounts of data collected by IoT devices. Predictive analytics can optimize resource management, predict maintenance needs, and enhance energy efficiency, supporting sustainable practices and reducing costs.

7.8 Collaboration and Ecosystem Partnerships

Collaboration among stakeholders, including industries, government bodies, technology providers, and research institutions, is crucial for the success of IoT-driven sustainable industries. Partnerships can facilitate knowledge sharing, resource pooling, and co-creation of innovative solutions, fostering sustainable development on a broader scale.

These future trends in IoT-driven sustainable industries hold immense potential to further optimize operations, promote resource efficiency, and address environmental challenges. By embracing these trends and leveraging IoT technologies, industries can contribute to a more sustainable future while driving innovation and economic growth.

Projections show that IoT spending will increase annually until it reaches $1.1 trillion in 2024. Businesses must learn the best ways to integrate IoT into the issues they are now facing as the number of connected devices increases. By 2025, 75% of enterprise-generated data will be produced and processed at the edge, predicts a recent Gartner estimate. The rising use of artificial intelligence (AI) and machine learning (ML) to the analysis of data from connected devices is another IoT trend. IoT in India is predicted to continue to grow and be adopted across a variety of industries, with a focus on cybersecurity and data protection. Technology is anticipated to play a big influence in a variety of businesses and sectors when it comes to AI. IoT has the potential to have an infinite future. Increased network agility, integrated artificial intelligence (AI), and the ability to install, automate, coordinate, and protect a variety of cases for hyperscale and 5G will expedite the development of the industrial internet.

CONCLUSION

The integration of IoT and sustainable development presents immense potential for industries to transform their operations, enhance efficiency, and reduce environmental impacts. By leveraging IoT technologies and embracing future trends such as edge computing, blockchain, circular economy practices, and enhanced connectivity, industries can drive sustainable development and create a more sustainable future. By embracing the future trends, industries can further optimize their operations, enhance efficiency, reduce environmental footprints, and contribute to the achievement of sustainable development goals. The transformative power of IoT lies in its ability to collect and analyze real-time data, enabling informed decision-making and facilitating the adoption of sustainable practices across industries. It is crucial for businesses, policymakers, and stakeholders to embrace these trends, invest in IoT infrastructure, and foster collaborations to unlock the full potential of IoT. The real-time monitoring, data analytics, and automation capabilities of IoT enable industries to make informed decisions, optimize processes, and adopt sustainable practices, leading to a more sustainable and environmentally conscious future.

By harnessing the transformative power of IoT, we can not only address pressing environmental issues but also create economic opportunities and improve the quality of life for communities worldwide. With continued innovation, collaboration, and responsible deployment, IoT-driven sustainable development has the potential to create a greener, more resilient, and inclusive future for generations to come.As industries move forward, it is essential to prioritize investments in IoT infrastructure, address challenges such as privacy and security concerns, and establish robust policies and regulations to ensure the responsible and ethical use of IoT technologies. By doing so, industries can unlock the full potential of IoT-driven sustainable development and pave the way for a more sustainable and prosperous future.

REFERENCES

Adiono, T., Fathany, M. Y., Putra, R. V. W., Afifah, K., Santriaji, M. H., Lawu, B. L., & Fuada, S. (2016). Live demonstration: MINDS - Meshed and internet networked devices system for smart home: Track selection: Embedded systems. *Proceedings of the 2016 IEEE Asia Pacific Conference on Circuits and Systems, APCCAS*, 736–737.

Agarwal, A., Singh, P., & Kumar, A. (2021). Internet of Things (IoT) for smart cities: challenges and opportunities in India. *Journal of Ambient Intelligence*.

Al-Turjman, F. (2018). Internet of things for smart cities: Technologies and applications. *Intelligent Buildings International, 10*(4), 223–239.

Al-Turjman, F.; Malekloo, A. (2019). Smart parking in IoT-enabled cities: A survey. *Sustain. Cities Soc., 49*.

Andreão, R. V., Athayde, M., Boudy, J., Aguilar, P., de Araujo, I., & Andrade, R. (2018). Raspcare: A Telemedicine Platform for the Treatment and Monitoring of Patients with Chronic Diseases. In Assistive Technologies in Smart Cities. IntechOpen.

Bandyopadhyay, D., & Sen, J. (2011). Internet of things: Applications and challenges in technology and standardization. *Wireless Personal Communications, 58*(1), 49–69. doi:10.100711277-011-0288-5

Dutta, J., Chowdhury, C., Roy, S., Middya, A. I., & Gazi, F. (2017). Towards smart city: Sensing air quality in city based on opportunistic crowd-sensing. *ACM Int. Conf. Proc. Ser.*

Dutta, P., Chavhan, R., Gowtham, P., & Singh, A. (2023). The individual and integrated impact of Blockchain and IoT on sustainable supply chains: A systematic review, Supply Chain Forum. *International Journal (Toronto, Ont.), 24*(1), 103–126. doi:10.1080/16258312.2022.2082851

El-Sayed, H., Sankar, S., Prasad, M., Puthal, D., Gupta, A., Mohanty, M., & Lin, C.T. (2017). Edge of Things: The Big Picture on the Integration of Edge, IoT and the Cloud in a Distributed Computing Environment. *IEEE Access, 6*, 1706–1717.

Farag, S. G. (2019). Application of Smart Structural System for Smart Sustainable Cities. In *Proceedings of the 2019 4th MEC International Conference on Big*. 10.1109/ICBDSC.2019.8645582

Ghosh, A., & Chakraborty, N. (2016). Design of smart grid in an University Campus using ZigBee mesh networks. *Proceedings of the 1st IEEE International Conference on Power Electronics, Intelligent Control and Energy Systems*, 1–6. 10.1109/ICPEICES.2016.7853432

Gubbi, J., Buyya, R., Marusic, S., & Palaniswami, M. (2013). Internet of Things (IoT): A vision, architectural elements, and future directions. *Future Generation Computer Systems, 29*(7), 1645–1660. doi:10.1016/j.future.2013.01.010

Haverkort, B. R., & Zimmermann, A. (2017). Smart Industry: How ICT Will Change the Game! *IEEE Internet Computing, 21*(1), 8–10. doi:10.1109/MIC.2017.22

Huang, Y., Dang, Z., Choi, Y., Andrade, J., & Bar-ilan, A. (2018). High-precision smart system on accelerometers and inclinometers for Structural Health Monitoring: Development and applications. *Proceedings of the 2018 12th France-Japan and 10th Europe-Asia Congress on Mechatronics*, 52–57.

India IoT Market Overview 2020-2025. (2020). *MarketsandMarkets*. www.marketsandmarkets.com/Market-Reports/india-iotmarket-277424354.html

Kang, L., Poslad, S., Wang, W., Li, X., Zhang, Y., & Wang, C. (2016). A public transport bus as a flexible mobile smart environment sensing platform for IoT. *Proceedings of the 12th International Conference on Intelligent Environments*, 1–8. 10.1109/IE.2016.10

Keane, P. A., & Topol, E. J. (2018). With an eye to AI and autonomous diagnosis. *NPJ Digital Medicine*, *1*(1), 10–12. doi:10.103841746-018-0048-y PMID:31304321

Khan, Z., Anjum, A., Soomro, K., & Tahir, M. A. (2015). Towards cloud based big data analytics for smart future cities. *Journal of Cloud Computing (Heidelberg, Germany)*, *4*(1), 4. doi:10.118613677-015-0026-8

Koubaa, A., Aldawood, A., Saeed, B., Hadid, A., Ahmed, M., Saad, A., Alkhouja, H., Ammar, A., & Alkanhal, M. (2020). Smart palm: An IoT framework for red palm weevil early detection. *Agronomy (Basel)*, *10*(7), 987. doi:10.3390/agronomy10070987

Kuzlu, M., & Pipattanasomporn, M. (2013). Assessment of communication technologies and network requirements for different smart grid applications. *Proceedings of the 2013 IEEE PES Innovative Smart Grid*, 1–6. 10.1109/ISGT.2013.6497873

Lele, A. (2019). Internet of things (IoT). *Smart Innov. Syst. Technol.*, *132*, 187–195. doi:10.1007/978-981-13-3384-2_11

O'Grady, M., Langton, D., & O'Hare, G. (2019). Edge computing: A tractable model for smart agriculture? *Artif. Intell. Agric.*, *3*, 42–51. doi:10.1016/j.aiia.2019.12.001

Pardini, K., Rodrigues, J. J., Kozlov, S. A., Kumar, N., & Furtado, V. (2019). IoT-based solid waste management solutions: A survey. *J. Sens. Actuator Netw.*, *8*(1), 5. doi:10.3390/jsan8010005

Rahman, A., Hossain, M. S., Alrajeh, N. A., & Guizani, N. (2020). B5G and Explainable Deep Learning Assisted Healthcare Vertical at the Edge: COVID-19 Perspective. *IEEE Network*, *34*(4), 98–105. doi:10.1109/MNET.011.2000353

Rayan, Z., Alfonse, M., & Salem, A. B. M. (2018). Machine Learning Approaches in Smart Health. *Procedia Computer Science*, *154*, 361–368. doi:10.1016/j.procs.2019.06.052

Rojek, I., & Studzinski, J. (2019). Detection and localization of water leaks in water nets supported by an ICT system with artificial intelligence methods as away forward for smart cities. *Sustainability (Basel)*, *11*(2), 518. doi:10.3390u11020518

Saini, A., & Malik, A. (2017). Routing in internet of things: A survey. *Commun. Comput. Syst.*, *I*, 855–860.

Sakhardande, P., Hanagal, S., & Kulkarni, S. (2016). Design of disaster management system using IoT based interconnected network with smart city monitoring. *Proceedings of the 2016 International Conference on Internet of Things and Applications*, 185–190. 10.1109/IOTA.2016.7562719

Sharma, N., Solanki, V. K., & Davim, J. P. (2019). Basics of the Internet of Things (IoT) and Its Future. In Handbook of IoT and Big Data. CRC Press.

Talari, S., Shafie-Khah, M., Siano, P., Loia, V., Tommasetti, A., & Catalão, J. P. (2017). A review of smart cities based on the internet of things concept. *Energies, 10*(4), 421. doi:10.3390/en10040421

Tao, F., Cheng, J., & Qi, Q. (2018). IIHub: An industrial internet-of-things hub toward smart manufacturing based on cyber-physical system. *IEEE Transactions on Industrial Informatics, 14*(5), 2271–2280. doi:10.1109/TII.2017.2759178

Trakadas, P., Simoens, P., Gkonis, P., Sarakis, L., Angelopoulos, A., Ramallo-González, A. P., Skarmeta, A., Trochoutsos, C., Calvo, D., Pariente, T., & .... (2020). An artificial intelligence-based collaboration approach in industrial iot manufacturing: Key concepts, architectural extensions and potential applications. *Sensors (Basel), 20*(19), 5480. doi:10.339020195480 PMID:32987911

Trencher, G., & Karvonen, A. (2019). Stretching "smart": Advancing health and wellbeing through the smart city agenda. *Local Environment, 24*(7), 610–627. doi:10.1080/13549839.2017.1360264

Wan, J., Yang, J., Wang, Z., & Hua, Q. (2018). Artificial intelligence for cloud-assisted smart factory. *IEEE Access, 6*, 55419–55430.

Wang, Y., Ram, S., Currim, F., Dantas, E., & Sabóia, L. A. (2016). A big data approach for smart transportation management on bus network. *Proceedings of the IEEE 2nd International Smart Cities Conference: Improving the Citizens Quality of Life*, 1–6. 10.1109/ISC2.2016.7580839

Yan, Y., Qian, Y., & Sharif, H. (2011). A secure data aggregation and dispatch scheme for home area networks in smart grid. *Proceedings of the GLOBECOM —IEEE Global Telecommunications Conference,* 1–6. 10.1109/GLOCOM.2011.6133655

Yousefpour, A., Fung, C., Nguyen, T., Kadiyala, K., Jalali, F., Niakanlahiji, A., Kong, J., & Jue, J. P. (2019). All one needs to know about fog computing and related edge computing paradigms: A complete survey. *Journal of Systems Architecture, 98*, 289–330. doi:10.1016/j.sysarc.2019.02.009

Chapter 2
Analysis of Factors Affecting Investment in Technology for Sustainability

Vernika Agarwal
Apeejay School of Management, India

Pragya Arya
Apeejay School of Management, India

ABSTRACT

The changing dynamics of the business and growing pressure from the stakeholders has forced the organisations to take another important aspect into their boardroom – sustainability. Investment in technological interventions is a must to enable sustainability within the business. The problem faced by the businesses is to find the right balance between their profits and technological investment to achieve sustainability. The current study focuses on understanding the factors which affect investment in sustainability using the multi-criteria approach of fuzzy-DEMATEL approach. The methodology is utilised to depict the context-specific link between the factors to aid the business managers in making right decisions. The results demonstrate that the factors such as commitment of top management (TS1), cost of training and development (TS2), cost of regulatory audits and fines (TS5), cost of subsidies and incentives (TS6), and cost of searching for suitable partners (TS9) are crucial for the incorporation of sustainability.

1. INTRODUCTION

The world has been seeing an upsurge in technological interventions over past few decades. While the industry has led to this upsurge through research and development, it also provides avenues for large scale applications of various technological interventions. Mature organisations, start-ups with funding and entrepreneurial ventures without external funding, all exploit technology for some benefit depending upon their financial capabilities. However, the role technology plays in helping organisations become truly sustainable on environmental and social parameters is not very apparent, as organisations do not

DOI: 10.4018/978-1-6684-9979-5.ch002

invest in technology solely for the purpose of achieving sustainability. Economic performance is the purpose of an organisation's existence. Conceptually and practically, an organisation exists to maximise the benefits for its stakeholders. However, in the present context, the stakeholder pressure has forced the organisations to take another important aspect into their boardroom – Sustainability. In 1987, World Commission on Environment and Development, in its report *Our Common Future also known as the Brundtland Report* defined sustainable development as *'development that meets the needs of the present without compromising the ability of future generations to meet their own needs'*. Since then, many efforts to conceptualise and achieve the objective of sustainability were taken. The most widely accepted conceptualisation is the Triple Bottom Line (TBL) which says that true sustainability is achieved only when efforts are directed toward environmental, social and economic development (Melville, 2010; Seuring & Muller, 2008). In the year 2016, United Nations also brought forward the twelve Sustainable Development Goals (SDGs) to provide an objective and direction for the nations to foray on the road to sustainability. Businesses are the pillars of development in any economy. Hence, they must also take the onus of making their operations sustainable on the three pillars of TBL. Today organisations are not left with any choice but to take initiatives to enhance their environmental and social sustainability.

Despite the fact that industrial development and advancement of technology has degraded the ecological conditions surrounding us, businesses alone cannot be held responsible and accountable to bring about a change in our surrounding ecological conditions (Van den Berg & De Langen, 2017). This is an era of sustainable collaborative advantage. The organisations are dependent on many external organisations like outsourcing partners, vendors, logistics service providers, etc. to carry out their operations effectively and efficiently. At present the efforts by organisations to achieve environmental sustainability of their operations are isolated. This is probably the reason why the results of these efforts are not visible. It is high time that various organisations collaborate to see the results of their efforts towards achieving environmental sustainability. Collaboration will not only help in easily materialising the results of efforts towards sustainability, it will also reduce the financial pressure on an individual organisation to do so.

Another aspect that must be taken into account are the ways and means by which the results of efforts implemented to achieve widespread benefits of sustainability efforts while not forcing the organisations to compromise with their economic sustainability. The chapter started with a discussion on technological interventions and that is the answer (Lee & Wu, 2014). Technological interventions specific to industries are a must to enable environmental sustainability of their operations without compromising with the profits. Research is abundant with literature that supports this belief. However, still organisations struggle to find the right balance between their profits and technological investment to achieve sustainability. The present investigation aimed to tackle two primary research inquiries:

1. What are the major factors that affect an organisations decision to invest in technology for sustainability?
2. What are the interrelationship between the factors that affect an organisations decision to invest in technology for sustainability?

This chapter is structured to discuss the various reasons or factors that affect an organisations decision to invest in technology for sustainability. In the second section, we discuss these factors. In the third section, Fuzzy DEMATEL approach is applied to analyse which of these factors are the most essential to motivate organisations to invest in sustainability of their operations. Section four, elaborates the results and conclusion is given in section five.

2. FACTORS AFFECTING INVESTMENT IN TECHNOLOGY FOR SUSTAINABILITY

As discussed achieving true sustainability cannot be the responsibility of a single business organisation. Sustainability efforts need a systemic approach of implementation with corporates, vendors, service providers and regulators all coming together to play a role. The factors affecting sustainability can be broadly classified into three categories (Peshkov, 2020):

a) Business Organisation and Planning Factors
b) Regulatory or State Level Factors
c) Supply Chain Factors

Business organisation and planning factors include internal management organisation and planning that a firm must engage in for enhancing sustainability of its operations on triple bottom line and to create a conscious culture (Kumar & Chandrakar, 2012) with consistent focus on sustainability. This category further includes commitment of top management towards sustainability, training and development and internal audits to assess sustainability.

Regulatory or state level factors include policies to promote investment by businesses towards sustainable operations, policies to keep a check on non-adherence to minimum required standards for sustainability, regulatory audits to assess the sustainability efforts of organisations, developing infrastructure that will ease implementation of sustainability efforts and developing legal infrastructure to address issues that may arise during collaborative implementation of sustainability efforts.

Supply chain factors include collaboration (Riandita, 2020) with vendors and contract manufacturers in the upstream section and with logistics service providers in the downstream section. The main aim of these factors is to create a pressure on these partners to develop and provide eco-friendly and socially sustainable materials and services.

The above factors are discussed in detail below. Many factors are termed and discussed in terms of the associated cost because in business everything is evaluated from an economic perspective.

Business Organisation and Planning Factors:

i) **Commitment of Top Management**: Commitment of top management towards achieving business sustainability is of prime importance in pushing the drive throughout the organisation. This commitment may surface in the form of regular communication by the top management to all levels of management to ensure sustainability at the minutest level of operations, setting aside of funds to specifically invest in technology for sustainable business operations and giving guidelines to senior managers to incorporate sustainability within their own departments. The purpose is to create a culture that fosters sustainability (Gandhi et. al, 2015) and ingrains it in the bloodstream of the organisation.
ii) **Cost of Training and Development:** There are many facets of business operations which create different types of social and environmental impacts. Training the employees about specific environmental and social impacts of their immediate work related responsibilities and how these can be mitigated or minimised to the largest possible extent, will help in enhancing sustainable performance. Training and development should also be directed towards use of technology that may enhance sustainable performance.

iii) **Cost of Internal Audits:** The audits play an important role in ensuring that the guidelines coming from top management are strictly adhered and implemented at all levels. For this the top management must conduct regular audits to check if all that has been planned to make business operations sustainable is being done and done in correct manner, within the defined organisational and regulatory framework. The internal audits will also prepare the organisation to identify loopholes and rectify them prior to an external audit by the regulatory bodies.

Regulatory or State Level Factors

iv) **Cost of Infrastructure Development:** Many industries like logistics and shipping, air travel, manufacturing, etc. require specific infrastructure for smooth and efficient operations. Sometimes these infrastructures are not in the hands of the businesses to develop and maintain unless directed by the regulatory bodies. Such state-owned infrastructures must be developed and maintained by the government to ensure that the efforts of the corporate sector can lead to desired level of sustainable performance.

v) **Cost of Regulatory Audits and Fines:** Government bodies vested with the responsibility of promoting and implementing sustainability in all spheres of economic development of the country, must conduct regular audits to check sustainable performance of business organisations. This will help in creating a pressure on business organisations to work hard and stick to their triple bottom line objectives and strive to consistently improve on this parameter. Imposing fines on those organisations that fail to stick to their triple bottom line objectives must be fined by the regulatory bodies on account of non-performance and non-seriousness. Like regulatory audits, this step will also create a pressure that an organisation cannot reap the same level of benefits and recognition without investing for sustainability as an organisation that seriously strives to work consistently in that direction.

vi) **Cost of Subsidies and Incentives:** Imposing fines and conducting audits will only ensure that organisations do the minimum required bit to escape regulatory fine and non-certification as a responsible organisation (Varsei et. al, 2014). While negative reinforcement is important, it is also equally important to reward the efforts of those organisations that not only achieve the targets of sustainable performance but also exceed them.

vii) **Cost of Developing Legal and Regulatory Framework:** It has been highlighted above that collaboration is the key to successful attainment of sustainability objectives. However, when there is collaboration, rifts are bound to occur. Proper legal and regulatory framework to address issues arising due to collaboration initiatives must be developed to quickly resolve them and set the positive tone for further action. This may also result in smooth implementation of carbon trading guidelines for the businesses.

Supply Chain Factors

viii) **Cost of Collaboration:** Focal firms or anchor firms play a significant role in driving the traditional supply chain operations. From a sustainability perspective, the supply chains are also the implementation zones of sustainable performance strategies of the higher management. Hence, the supply chains must focus on developing successful collaborative relationships with its upstream and downstream partners (El Bz & Laguir, 2017) to ensure that its own sustainable performance objectives are aligned with those of its partners (Ralston, 2014). This will help in easy attainment of these objectives by all the collaborating parties.

ix) **Cost of Searching for Suitable Partners:** Just the way regulatory bodies use the carrot and stick approach with the industry to ensure consistent efforts towards sustainability, similarly, anchor firms must ensure they give contracts to or partner only with those vendors and service providers that are engaged in incorporating sustainability into their routine operations and those that provide sustainable material and service offerings along with some added economic benefits like real time stock visibility and location (Montabon et. al, 2016). Such partners must also be rewarded by anchor firms by giving them repeat business, investing into their operations to enhance their sustainable performance and aiming to build long-term collaborative relationships with them.

3. FUZZY DEMATEL APPROACH

We have coupled fuzzy set theory with DEMATEL to analyse the assessments of specific impediments to technological empowerment. The integrated fuzzy DEMATEL approach's phases are as follows:

Step 1: Identifying the decision-making structure and shortlisting factors-

Step2: Adopting the fuzzy linguistic scale-

The adopted fuzzy scale is shown in Table 1.

Step3: Determining initial direct matrix (IDM).

Step 4: Defuzzification of the IDM matrix (K)-

The linguistic assessments by the DMs were de-fuzzified into crisp value by using CFCS (Converting Fuzzy data into Crisp Scores) adopted from Opricovic & Tzeng, 2003

Step 5: Determining the Normalized direct relation matrix. The DRM is normalized as:

$$K = \left[k_{ij}\right]_{nxn} = \frac{A}{s} \quad where \quad s = \max\left(\max_{1\le i\le n}\sum_{j=1}^{n} k_{ij}, \max_{1\le j\le n}\sum_{i=1}^{n} k_{ij}\right)$$

Step 6: Calculate the Total relation matrix T.

$$T = \left[t_{ij}\right]_{nxn} = \lim_{m\to\infty}\left(K + K^2 + K^3 \ldots\ldots + K^m\right) = K\left(I - K\right)^{-1}$$

Table 1. Fuzzy linguistic scale

Value	Representation	Fuzzy Value
VH	Very high influence	(0.75,1,1)
H	High influence	(0.5,0.75,1)
L	Low influence	(0.25,0.5,0.75)
VL	Very low influence	(0,0.25,0.5)
No	No influence	(0,0,0.25)

Step 7: Constructing the causal diagram. Let Ri and Cj denote n x1 and 1 x n matrices, respectively, representing the row-sum and column-sum of T. The variable Ri denotes the aggregate impact that factor i exerts on other factors, whereas Cj denotes the total impact that factor j receives. The expression (Ri+Ci) denotes the degree of importance of the ith factor in the process of arriving at decisions. Conversely, the notation (Ri-Ci) denotes the resultant impact of the i-th factor. In order to minimise the impact of insignificant variables on other variables, it may be advisable for decision makers to establish a predetermined threshold limit. The variables that exceed the threshold limit will be selected for mapping in the dataset, represented as a causal diagram in the format of (Ri+Ci, Ri-Ci).

4. ILLUSTRATION

This segment identifies 9 factors (TS1- TS9) which affect an organisations decision to invest in technology for sustainability. The factors are presented in Table 2 and the consolidated assessment data of decision makers is given in Table 3. The initial direct relation matrix (IDM) is created for the identified barriers as given by the decision makers and is given in Table 4.

After calculating the normalized direct matrix, the total relation matrix is calculated following the step 6 as discussed above. The total relation matrix is given below in Table 5.

Table 2. List of factors

Factors	Sub-Factors	Notation
Business Organisation and Planning Factors	Commitment of Top Management	TS1
	Cost of Training and Development	TS2
	Cost of Internal Audits	TS3
Regulatory or State Level Factors	Cost of Infrastructure Development	TS4
	Cost of Regulatory Audits and Fines	TS5
	Cost of Subsidies and Incentives	TS6
	Cost of Developing Legal and Regulatory Framework	TS7
Supply Chain Factors	Cost of Collaboration	TS8
	Cost of Searching for Suitable Partners	TS9

Table 3. The assessment data of decision makers

Factors	TS1	TS2	TS3	TS4	TS5	TS6	TS7	TS8	TS9
TS1	No	VL	L	L	VL	VL	H	VH	VL
TS2	L	No	VL	VH	L	VL	VH	L	VL
TS3	VL	L	No	H	L	VL	L	L	VL
TS4	VL	H	L	No	L	VL	L	VL	VL
TS5	L	L	H	L	No	VL	L	H	L
TS6	VL	L	H	L	VL	No	L	L	VL
TS7	L	VL	VL	L	H	L	No	L	H
TS8	L	H	L	VL	L	VL	L	No	VL
TS9	VL	L	VL	L	H	L	L	VL	No

Table 4. The initial direct relation matrix

Factors	TS1	TS2	TS3	TS4	TS5	TS6	TS7	TS8	TS9
TS1	0.334	0.267	0.517	0.517	0.267	0.267	0.767	0.962	0.267
TS2	0.517	0.334	0.267	0.962	0.517	0.267	0.962	0.517	0.267
TS3	0.267	0.517	0.334	0.767	0.517	0.267	0.517	0.517	0.267
TS4	0.267	0.767	0.517	0.334	0.517	0.267	0.517	0.267	0.267
TS5	0.517	0.517	0.767	0.517	0.334	0.267	0.517	0.767	0.517
TS6	0.267	0.517	0.767	0.517	0.267	0.334	0.517	0.517	0.267
TS7	0.517	0.267	0.267	0.517	0.767	0.517	0.334	0.517	0.767
TS8	0.517	0.767	0.517	0.267	0.517	0.267	0.517	0.334	0.267
TS9	0.267	0.517	0.267	0.517	0.767	0.517	0.517	0.267	0.334

Table 5. The total relation matrix

Factors	TS1	TS2	TS3	TS4	TS5	TS6	TS7	TS8	TS9
TS1	0.393	0.466	0.476	0.539	0.471	0.321	0.610	0.602	0.357
TS2	0.464	0.522	0.473	0.675	0.566	0.353	0.701	0.568	0.386
TS3	0.366	0.498	0.429	0.575	0.500	0.309	0.549	0.501	0.344
TS4	0.351	0.519	0.442	0.480	0.481	0.297	0.530	0.438	0.330
TS5	0.468	0.567	0.573	0.606	0.538	0.356	0.630	0.623	0.441
TS6	0.361	0.493	0.506	0.529	0.450	0.321	0.545	0.498	0.340
TS7	0.449	0.497	0.466	0.575	0.594	0.391	0.566	0.554	0.474
TS8	0.420	0.542	0.466	0.493	0.504	0.313	0.560	0.480	0.348
TS9	0.369	0.497	0.424	0.534	0.549	0.361	0.553	0.462	0.362

Then, using formula as given in step 7, the final matrix representing Identification of cause and effect segments is calculated and shown in Table 6.

Table 5 highlights the cause-and-effect factors. The influence of one metric over other can be given by identifying threshold level in the total relation matrix which in this case comes out to be as 0.476. R + C) gives the index of strength of influences exerted or received as shown in Table 7. More positive the value of (R + C), greater is the degree of influence on other attributes. On the other hand, if (R-C)

Table 6. Identification of cause and effect

Sub-Factors		Ri	Ci	Ri+Ci	Ri-Ci	Cause/Effect
Commitment of Top Management	TS1	4.235	3.641	7.875	0.594	Cause
Cost of Training and Development	TS2	4.718	4.601	9.319	0.118	Cause
Cost of Internal Audits	TS3	4.069	4.254	8.323	-0.185	Effect
Cost of Infrastructure Development	TS4	3.869	5.005	8.874	-1.136	Effect
Cost of Regulatory Audits and Fines	TS5	4.802	4.653	9.455	0.149	Cause
Cost of Subsidies and Incentives	TS6	4.043	3.023	7.066	1.020	Cause
Cost of Developing Legal and Regulatory Framework	TS7	4.567	5.244	9.811	-0.678	Effect
Cost of Collaboration	TS8	4.126	4.726	8.851	-0.600	Effect
Cost of Searching for Suitable Partners	TS9	4.111	3.392	7.502	0.719	Cause

is positive, then attribute A is affecting other attributes acting as a cause, and if (R-C) is negative, then attribute A is being influenced by other attributes, hence acting as the effect. Figure 1 gives the graphical representation of the cause-effect relation.

It can be seen that the cause factors include Commitment of Top Management (TS1), Cost of Training and Development (TS2), Cost of Regulatory Audits and Fines (TS5), Cost of Subsidies and Incentives (TS6) and Cost of Searching for Suitable Partners (TS9). The topmost management in the organisation's hierarchy, must make sustainability a permanent agenda with specific KPIs defined. Positive enforcement in the form of subsidies and incentives for green efforts and efforts to address social issues will go a longer way in inculcating the sustainability oriented culture in the industry. The industry needs to have strong channel partners to ensure consistent efforts towards sustainability. The effect group includes Cost of Internal Audits (TS3), Cost of Infrastructure Development (TS4), Cost of Developing Legal and Regulatory Framework (TS7) and Cost of Collaboration (TS8).

5. CONCLUSION

The organisations have been addressing sustainability to some extent as part of their CSR initiatives, the efforts to improve the environmental and social performance of business operations are still seen as a cost head. This changing mindset of the business and growing pressure from the stakeholders had forced the organisations to rethink their investment in sustainability. Investing in corporate social responsibil-

Figure 1. Graphical representation of cause-effect relation

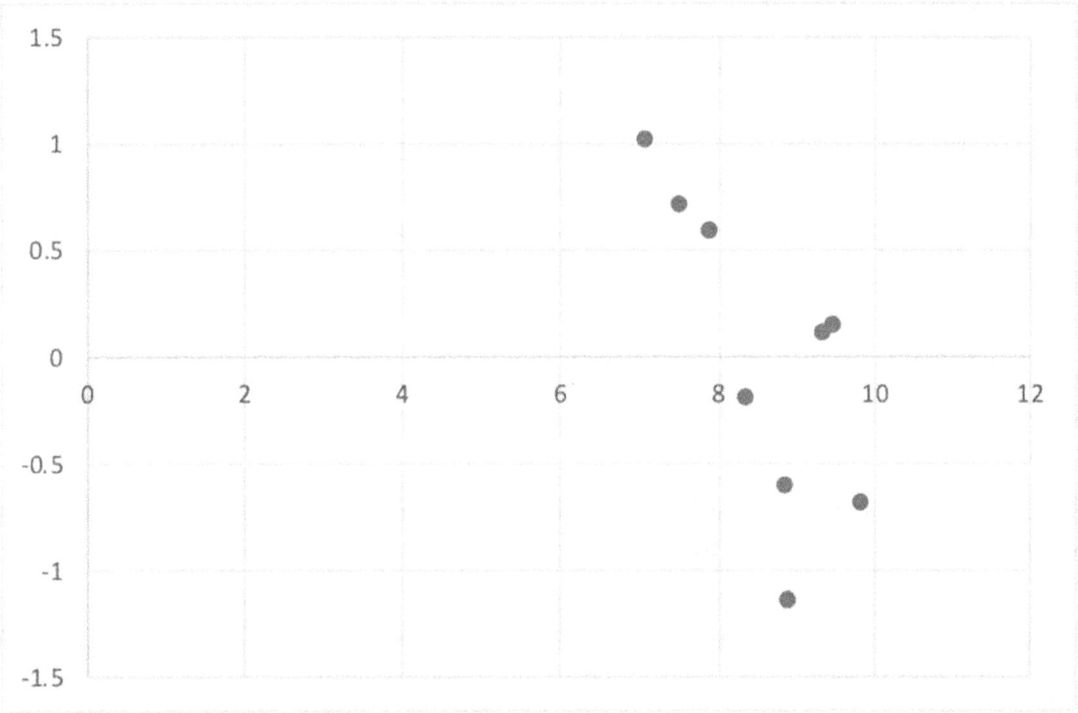

ity is not adding these businesses in enhancing the sustainability within their business. The problem faced by the businesses is to understand the investment in technological aspects and strategic issues to achieve sustainability. The current study focuses on understanding the factors which affect investment in sustainability using the multi-criteria approach of Fuzzy-DEMATEL approach. The methodology is utilised to depict the context-specific link between the factors, to aid the business managers in making right decisions. The study analyses nine factors affect sustainability.

The results demonstrate that the factors such as Commitment of Top Management (TS1), Cost of Training and Development (TS2), Cost of Regulatory Audits and Fines (TS5), Cost of Subsidies and Incentives (TS6) and Cost of Searching for Suitable Partners (TS9) are crucial for the incorporation of sustainability. Technological interventions specific to industries will enable environmental sustainability of their operations without compromising with the profits.

REFERENCES

El Baz, J., & Laguir, I. (2017). Third-party logistics providers (TPLs) and environmental sustainability practices in developing countries: The case of Morocco. *International Journal of Operations & Production Management, 37*(10), 1451–1474. doi:10.1108/IJOPM-07-2015-0405

Gandhi, S., Mangla, S. K., Kumar, P., & Kumar, D. (2015). Evaluating factors in implementation of successful green supply chain management using DEMATEL: A case study. *International Strategic Management Review, 3*(1-2), 96-109.

Kumar, R., & Chandrakar, R. (2012). Overview of green supply chain management: Operation and environmental impact at different stages of the supply chain. *International Journal of Engineering and Advanced Technology, 1*(3), 1–6.

Lee, K. H., & Wu, Y. (2014). Integrating sustainability performance measurement into logistics and supply networks: A multi-methodological approach. *The British Accounting Review, 46*(4), 361–378. doi:10.1016/j.bar.2014.10.005

Melville, N. P. (2010). Information systems innovation for environmental sustainability. *Management Information Systems Quarterly, 34*(1), 1–21. doi:10.2307/20721412

Montabon, F., Pagell, M., & Wu, Z. (2016). Making sustainability sustainable. *The Journal of Supply Chain Management, 52*(2), 11–27. doi:10.1111/jscm.12103

Opricovic, S., & Tzeng, G. H. (2003). Defuzzification within a multicriteria decision model. *International Journal of Uncertainty, Fuzziness and Knowledge-based Systems, 11*(05), 635–652. doi:10.1142/S0218488503002387

Ralston, P. (2014). *Supply chain collaboration: A literature review and empirical analysis to investigate uncertainty and collaborative benefits in regards to their practical impact on collaboration and performance*. Academic Press.

Riandita, A. (2020). *Inter-firm relationships for sustainability: Incumbent firms and sustainability ventures*. Academic Press.

Seuring, S., & Müller, M. (2008). From a literature review to a conceptual framework for sustainable supply chain management. *Journal of Cleaner Production, 16*(15), 1699–1710. doi:10.1016/j.jclepro.2008.04.020

Van den Berg, R., & De Langen, P. W. (2017). Environmental sustainability in container transport: The attitudes of shippers and forwarders. *International Journal of Logistics, 20*(2), 146–162. doi:10.1080/13675567.2016.1164838

Varsei, M., Soosay, C., Fahimnia, B., & Sarkis, J. (2014). Framing sustainability performance of supply chains with multidimensional indicators. *Supply Chain Management, 19*(3), 242–257. doi:10.1108/SCM-12-2013-0436

World Commission on Environment and Development. (1987). *Our Common Future*. Oxford University Press.

Chapter 3
Sustainability Threats to the Information Society:
Gap of Sustainability Knowledge and Practice

Prabhat Ranjan
https://orcid.org/0000-0002-8139-6902
Jagat Taran Girls Degree College, India

Bhupendra Kumar Singh
University of Allahabad, India

ABSTRACT

The electronic communication mechanism of information has empowered human society to develop as an information society and dominate the planet. An immense amount of scholarly information has been generated in recent decades, and human technological development is far more than what was expected earlier. However, sustainability issues arose in the past decades that show the vulnerability of information society in its present form. There is a big question of why humanity is under threat despite such immensity of knowledge within it. In this chapter, probable factors behind it have been discussed. The chapter is based on an extensive literature search and analysis. The study clarifies that ethics is necessary for knowledge deployment and needs a competent political structure to make proper decisions and frame proper policies for society. The present consumer society should be replaced by a value-based society. The chapter is an original study and elucidates that environmental education is important, rather than merely environmental knowledge in society.

1. INTRODUCTION

The gradual proliferation of knowledge in human society gave rise to the present information society. The primary, secondary, and tertiary industrial sectors constitute the current information society. Primary sectors of industries are involved in the extraction and production of raw materials in society. Secondary

DOI: 10.4018/978-1-6684-9979-5.ch003

sectors are manufacturing industries supporting the activities of the primary sectors. The tertiary services are service industries that include services rather than the production of physical materials. The problems due to the pollution had started in industrial society i.e. start of mass manufacturing using fossil fuels; however, sustainability was not an issue till some decades before.

1.1. Sustainability Threats to the Information Society

The ecological footprint per capita has increased tremendously after the entry of the electronic age. As a result, an immense deficit of non-renewable resources has been created along with high-scale pollution. The problem has so much intensified that it has posed questions about the future of human society. The enormity of biodiversity loss reflects the dreadfulness of the emerging problems to the whole life system on the planet. The number of mammals, reptiles, and birds has decreased tremendously in a few decades. Not only large-sized vertebrates are being lost on the earth today, but the heavy population of small-sized insects is also reducing at the rate of 1% per year (Wagner et al., 2021), which signals the deterioration of life-sustaining conditions on the earth. The changes in the environment due to the mechanical ingenious of the human species are so huge that compositions of organisms in any part of the globe have been changed enormously including the rain forests and ice zones on the globe (Attenborough, 2021). The recent 100 years of changes have been so gigantic that a new term 'Anthropocene' has been started to denote this epoch on the geological timescale (Gibbard& Walker, 2013; Lewis & Maslin, 2015). Such changes are steepest in any specified period by a single species on the earth (Dasgupta, 2021).

The situations are devastating to humans themselves too. The condition of the environment is worsening and the human population is being heavily affected. 90% of the human population in the world has no option now but to live in unsafe air (World Health Organization, 2022). However, there are no climate policies yet in some cities in the world (Hsu, 2020). Human society has posed a danger to it by its uncontrolled growth and now needs to save itself from itself (Williams, 2021). The environmental cost of high-level living is very high and disgusting for the future. The use of natural resources by human society is more than the replenishment of such resources. The ecological footprint per person is very high in developed countries; and not less than 1 gha in developing countries also (Ewing et al., 2010).

The study by Miller et al. (2011) underlines limitations of the economic growth. The researchers found an inverted u-shaped relation between ecological footprint and economic growth clarifying that increasing ecological footprints has negative impacts on economic growth after a certain time; and underlines the need to reduce ecological footprints. Several essentialities on the earth for life processes have reached the zone of uncertainty (risk) and beyond uncertainty (high risk) due to disturbed intrinsic biophysical processes on the earth. It has reached such a position that external impacts on the Earth due to other solar planets may prove crucial to the degeneration of the life system on the earth (Stephen et al., 2015).

The agrarian societies did not have pollution as a by-product (Dasgupta, 2021). Most of such problems started with the utilization of fossil fuels as energy sources for mass production during industrial revolutions in industrial society. The post-industrial era could not solve the issues of pollution caused by burning such fuels but added many more pollutants including plastic and electronic wastes in this period with an immense increase in ecological footprints per capita. The depletion of the environment and natural resources intensified severely in this period. The situations have turned so problematic that now our outfits will prove insufficient to reverse the deteriorations made by us if proper steps are not taken within the present decade (United Nations, 2019). The United Nations has set seventeen goals to achieve by 2030, which include poverty eradication, women empowerment, and climate change (United

Nations, 2023). Hence, the present time has a great need to think seriously on the issues of environmental preservation and sustainability.

1.2. Gaps Between Sustainability Knowledge and Practice

Substantial information is available on sustainability issues. However, the environmental threats are increasing day by day in each country. It confirms that sustainability threats are increasing even after vast technological knowledge of sustainability issues. Societies need economic growth, and hence governments also work for the same. It creates a question if fulfilling such needs is scientific and economical in the long run.

1.3. Objectives of Study

The main objectives of this study are to seek answers about sustainability threats to present information to society and the reasons behind the gap between sustainability knowledge and sustainability practices and remedial activities and measures to maintain and regenerate the environment. The most underlying fact is that sustainability issues exist after vast environmental knowledge within society.

1.4. Scopes and Limitations

The present discussions are interdisciplinary; however, it does not penetrate on seeking a technical solution to the sustenance problems, but it seeks societal solutions to the threats towards sustainability.

1.5. Definitions of Terms Used

Some terms, definitions, and contexts need to be clarified.

1.5.1. Societal Knowledge

There is no word available 'societal knowledge' in the Oxford Learner's Dictionary. However, meaning of the word 'societal' explains meaning of the present term also: Meaning of 'societal' is "connected with society and the way it is organized" (Oxford Learner's Dictionary, 2022). Hence societal knowledge means knowledge connected with society. In the present paper 'societal knowledge' term is used to represent the collective knowledge of human society, especially in the context of the information society.

1.5.2. Sustainability

According to the Oxford Learner's Dictionary, sustainability means 1. "The use of natural products and energy in a way that does not harm the environment" and 2. "The ability to continue or be continued for a long time" (Oxford Learner's Dictionary, 2022). Combining both meanings, we can say that sustainability means utilizing natural resources in a way that may continue forever. In the present research paper, sustainability means the sustainability of the human species with its technical evolutions.

2. RESEARCH METHODOLOGY

This study is based on review of the literature and co-relating the major results of these studies appropriately. Reasons for sustainability threats and its possible remedies have been discussed.

3. REASONS TO EMERGE GAPS BETWEEN SUSTAINABILITY KNOWLEDGE AND PRACTICE

Several prominent reasons may create a gap between the knowledge and practice of sustainability. Following are discussed some of those:

3.1. Wisdom Constraints in Collective Decisions

Wisdom is ability to apply knowledge in decisions. The studies clarify that societies may have collective knowledge, but no collective wisdom. Therefore, knowledge may fail to get appropriate results in practical applications. Baskarada & Koronios (2013) illustrates that massive knowledge within a society does not create great wisdom within it - since wisdom is limited to an individual (hence personal), not societal. Societies depend upon the wisdom of an individual (not of the entire society), making decisions on account of that particular society. Thus, personal wisdom is applied to utilize societal knowledge, which constrains the knowledge applications. Winterfeldt (2013) describes that decision-makers generally do not have scientific knowledge, and knowledgeable generally do not make decisions for society. Such gaps give inappropriate results to society. Values (from the side of users) and beliefs (from the side of experts) need being associated with scientific knowledge in the decision-making process. Fischer et al. (2012) clarify that the sustainability deficit arising in the present era - is not the result of knowledge deficits, but a lack of wisdom in the market-driven society and a lack of willpower in civil society and political structures. Information society could not prove itself an enlightened society. The evolution of collective consciousness in society is essential (Fischer et al., 2012) for future perspectives and solutions to major common problems.

3.2. Weak Role of the Knowledge Institutions in Sustainability

Some studies stressed that ways of producing knowledge by academic institutions are insufficient for sustainability. Epistemological pluralism and reflexivity are considered due necessities in this process (Miller et al., 2011). Maxwell (2012) finds that academic institutions need revolutions to bring revolutions in society since such institutions help in the mind-making of people who run democracy in reality. Environmental sustainability was found positively related to institutional quality in a study by Ahmad et al. (2021). Integrating science with society is essential through trans-disciplinary research with the co-production of knowledge through researchers, decision-makers, and stakeholders (Mauser et al., 2013; Miller et al., 2011). Dovers (2001) stressed the need for institutions for sustainable growth in Australia and recommended establishing an Australian Institute for 'ecologically sustainable development' along with the recruitment of a commissioner and legislative review to recommend necessary changes in law.

3.3. Role of Sustainable Knowledge in Societal Decisions

Palmer et al. (2005) stated that ecological knowledge has marginal roles in the decision-making process at present. They further envisaged that ecologists should come forward as purveyors too, rather than stellar researchers only since the individuals without knowledge about sustainability - are not supposed to discover the collective knowledge on sustainability. Ecological sciences should create a difference in the world's future by moving towards actions from focusing on pristine systems only (Palmer et al., 2005). Cash et al. (2003) suggest to manage boundaries between science and action so that salience, credibility and legitimacy of information produced by science could be enhanced. Learning how to harness boundary-spanning potential of multiple individuals and organizations in this procedure is essential. Arnott & Lemos (2021) observed that the relationship between scientific knowledge produced on sustainability and its use is hard to assess. However, they further clarify that complexities arising from the intersection of knowledge production and use may be more manageable if embraced or leveraged rather than reduced or ignored.

3.4. Exclusive Growth: Spangenberg

(2005) stressed for developing a knowledge society at the place of an information society for inclusive growth since information alone is meaningless without contexts of being used. He added that a knowledge society can't have only technological aspects, but socio-economic aspects as well. Fuchs (2006) assesses that economic, technological, ecological, political, and cultural sustainability is essential for prosperity to all and the usability of technologies. Egorov (2017) depicts that human have a low level of altruism in their genotype and believes in individual growth which obstructs the collective growth of the society.

4. REMEDIAL ACTIONS REQUIRED

Some urgent steps are essential to preserve the society and its growth. Many of the research studies are based on this issue, some of which have been elaborated in following sub-sections.

4.1. Environmental Education

Direct education system shapes the minds of young generations and provides a direction to them and the society. An effective environment education system may enhance public awareness and consciousness of the environmental issues. Some studies denote education as a significant factor in public consciousness; but others illustrate its limitations, that values are more important than the education. Holfelder (2019) stressed the need for environmental education for a better future because education has capacity to make changes in society and discover new paths to move on. Boca & Saracli (2019) performed a study for knowing the students' environment perception and how attitude towards environment education influenced their environment behavior. They found that higher education and universities have great potential to deal with environmental issues. Correlation analysis showed a positive relationship between perception, attitude, and behavior variables. Ajaps & McLellan (2015) analyzed that environmental education is indispensable, rather than only having environment knowledge within the professionals. An environment educated person may take appropriate actions for a healthier environment and judge the changes made after an activity. After a survey study, Hopper & Nielsen (1991) reported that recycling by households increased after providing proper recycling information to them and prompt reminders to them in the United States. The environmental attitudes and feedback were influential in energy use in the course of a study of nine-month duration among 120 households in the United Kingdom (Brandon & Lewis, 1999). Beyond the awareness to the public, Bazzaz et al. (1998) stressed that the ecologists should inform the people about the relevance of each research publication by them in a letter to the editors of a reputed journal.

However, Whitmarsh (2011) explored that the individuals' environmental values are more important than their education or knowledge. Age, gender, location, and lifestyle indirectly affect the individual interaction with nature. Heeren et al. (2016) also found that environmental education does not confirm creating consciousness among people (among students in their study); rather pro-behavioral control and social norms play more significant roles. Sustainability knowledge was found co-related, but least strongly co-related with pro-environment behavior.

It clears that environmental education plays a role in environmental awareness and pro-environmental behavior; however, values, social norms, and pro-behavioral control are more significant.

4.2. Weak or/and Strong Sustainability Approaches of Solution

There are two approaches to the sustainability – Weak sustainability and Strong sustainability. Weak sustainability estimates that there are no basic differences between well being generated by natural capital and manufactured capital (Ekins et al., 2003; Neumayer, 2012). Whereas strong sustainability considers that natural capital are product complexities between biotic and abiotic components of nature (Ekins et al., 2003) and manufactured capital are not alternative to natural capitals in each cases. Randall (2022) assesses by his mathematical formulations that weak sustainability is possible, only if business-to-usual economy, correcting market failures, and addressing impending resource crises is implemented. The strong sustainability seems hard to the author; however he is optimistic on solution of the present environmental crises by applying concept of 'Weak Sustainability plus' by applying strong sustainability patches to protect against impending resource crises. Ekins et al. (2003) estimates that strong sustainability needs to be applied in a number of environmental areas, which are based on maintenance of critical natural capital (CNC) with restoration of it to higher levels. Shang et al. (2019) explores relationship between energy consumptions, socioeconomic growth, and water and food footprint during 1987 to 2015 available in Inner Mongolia and finds that the region needs to balance better between socioeconomic development and environmental protection which needs strong sustainability-oriented development.

'Weak sustainability plus' concept, i.e., strong sustainability approach in selected areas with weak sustainability in general is recommendable.

4.3. Need of Value-Based Society

There is a need to change from a growth-centered society to an environment-conservationist society acknowledging biophysical limits on the Earth (Martin et al., 2016). The earlier African Ubuntu and other philosophies encouraged collective thinking to establish core values in those societies to promote human dignity and love to the nature (Preiser, 2021). Buddhism, Christianity, and other religions originated around two thousand years ago had roots in similar principles of love for all (Gassner, 2015). However, the individuals and governments are not sensitive to collective responsibilities yet. Knowledge is a social property; however, availability is also exclusive on the organizational, social, economic and national levels in any part of the world. How such dividing entities may deliberate equality within a community is a great question (Gill, 1996). Convergence of knowledge and technology for the benefit of society (CKTS) is critically required for sustainability in the present century (Roco& Bainbridge, 2013). Arogyaswamy (2020) suggests firms to create a culture of morality based on ethical standards with involvement of users and shareholders to decide future direction of the technology. Pattanayak (2021) assessed in an Academia Letters publication that human may get extinct due to its greed and aggressive behavior of the countries.

4.4. Sustainability not Enough, Regeneration Required

Habib (2015) discussed that working for sustainability is not enough; but regeneration practices to achieve stability in the biosphere are required. He further observes that the present world has one foot in fossil fuel-based capitalism but the other towards sustainability and where messages of sustainability for future are not being transferred from teachers to students (Habib, 2015). Morandín-Ahuerma et al. (2019) observes that the present socio-ecosystem crisis has been generated by high consumerism and great economic goals and ambitions, and cannot be returned back to the prior position by the same logic; rather ethics and strengthening relations with the nature needs to be more prominently followed. Human society should be treated as a sub-system of the natural system of life, and each components of the biosphere is essential for the planet. The human civilization is not an isolated system and needs a healthy bio-system everywhere for the present and upcoming future.

5. SUGGESTIONS

Decisions in a society depends on personal wisdom of decision makers hence the societal wisdom is abandoned despite existence of societal knowledge. The present information society is characterised by exclusive growth. The aim of sustainable development can be achieved only by collective efforts of all - from individuals to the policy makers in the society. Following are some suggestions extracted from the study:

1. Ethics is important for an information society besides knowledge. Market-driven consumerism should provide space for sustainability and inclusiveness.
2. Policy makers should frame policies consistent with the environment and nature. Deterioration of the environment needs to be assessed every year, and remedial actions should be taken.
3. Green technologies needs to be implemented by the states. Polluter pay policy should be implemented strictly by the governments.
4. Environmental education to masses is essential, rather than only having environmental knowledge within a limited community of learned persons (Ajaps & McLellan, 2015).

6. CONCLUSION

The technology in society is being deployed to fulfill the short-term needs of society and deteriorating life-supporting systems of the earth, developed in billions of years, and hence not scientific and therefore inappropriate. Ethics-driven culture needs to be adapted in place of market-driven consumer culture in society. Knowledge may be perilous to a society if environmental and ethical concerns are absent in decisions made on account of a society. Green technology products and a polluter pay policy are needed for sustainability. Strengthening inclusive growths may enhance the sustainable development of an information society where public education and library systems may play crucial roles in spreading consciousness on the environment and sustainability issues.

REFERENCES

Ahmad, M., Ahmed, Z., Majeed, A., & Huang, B. (2021). An environmental impact assessment of economic complexity and energy consumption: Does institutional quality make a difference? *Environmental Impact Assessment Review*, *89*, 9. doi:10.1016/j.eiar.2021.106603

Ajaps & McLellan. (2015). We don't know enough": Environmental education and pro-environmental behaviour perceptions. *Cogent Education, 2*(1), 1-17.

Arnott, J. C., & Lemos, M. C. (2021). Understanding knowledge use for sustainability. *Environmental Science & Policy*, *120*, 222–230. doi:10.1016/j.envsci.2021.02.016

Arogyaswamy, B. (2020). Big tech and societal sustainability: An ethical framework. *AI & Society*, *35*(4), 829–840. doi:10.100700146-020-00956-6 PMID:32218647

Attenborough, D. (2021). Foreword. In P. Dasgupta (Ed.), *Economics of biodiversity: Dasgupta review* (pp. 1–2). HM Treasury.

Baskarada, S., & Koronios, A. (2013). Data, Information, Knowledge, Wisdom (DIKW): A Semiotic Theoretical and Empirical Exploration of the Hierarchy and its Quality Dimension. *AJIS. Australasian Journal of Information Systems*, *18*(1), 5–24. doi:10.3127/ajis.v18i1.748

Bazzaz, F., Ceballos, C., Davis, M., Dirzo, R., Ehrlich, P. R., Eisner, T., Levin, S., Lawton, J. H., Lubchenco, J., Matson, P. A., Mooney, H. A., Raven, P. H., Roughgarden, J. E., Sarukhan, J., Tilman, C. D., Vitousek, P., Walker, B., Wall, D. H., Wilson, E., ... Woodwell, G. M. (1998). Ecological science and the human predicament. *Science*, *282*(5390), 879. doi:10.1126cience.282.5390.879c

Boca, G. D., & Sinan, S. (2019). Environmental education and student's perception for sustainability. *Sustainability (Basel)*, *11*(6), 1553. doi:10.3390u11061553

Brandon, G., & Lewis, A. (1999). Reducing household energy consumption: A qualitative and quantitative field study. *Journal of Environmental Psychology*, *19*(1), 75–85. doi:10.1006/jevp.1998.0105

Cash, D. W., Clark, W. C., Alcock, F., Dickson, N. M., Selin, N., Guston, D., Jager, J., & Mitchell, R. (2003). Knowledge systems for sustainable development. *Proceedings of the National Academy of Sciences of the United States of America*, *100*(14), 8086–8091. doi:10.1073/pnas.1231332100 PMID:12777623

Dasgupta, P. (2021). *Economics of biodiversity: Dasgupta review*. H M Treasury.

Dovers, S. (2001). *Institutions for sustainability*. Australian Conservation Foundation.

Egorov, V. V. (2017). Dozy-chaos end of the human civilization. [Section B]. *Journal of Ultra Scientist of Physical Sciences*, *29*(4), 87–96. doi:10.22147/jusps-B/290402

Ekins, P., Simon, S., Lisa, D., Folke, C., & De Groot, R. (2003). A framework for the practical application of the concepts of critical natural capital and strong sustainability. *Ecological Economics*, *44*(2-3), 165–185. doi:10.1016/S0921-8009(02)00272-0

Ewing, B., Moore, D., Goldfinger, S., Oursler, A., Reed, A., & Wackenagel, M. (2010). *Ecological footprint atlas 2010*. Oakland.

Fischer, J., Dyball, R., Faezy, I., Gross, C., Dovers, S., Ehrlich, P. R., Bulle, R. J., Christensen, C., & Borden, R. (2012). Human behaviour and sustainability. *Frontiers in Ecology and the Environment*, *10*(3), 153–160. doi:10.1890/110079

Fuchs, C. (2006). Social Informatics: An Information Society for All? In Remembrance of Rob Kling. In B. T., M. I. Numinen, & T. Impagliazzo (Eds), IFIP International Federation for Information Processing (pp. 219-230). Springer.

Gassner, F. (2015). *Comparative study between Buddhism and Christanity on peace and war [Masters degree dissertation]*. Portuguese Catholic University.

Gibbard, P. L., & Walker, M. J. C. (2013). Term 'Anthropocene' in the context of formal geological classification. *Special Publication - Geological Society of London*, *395*(1), 29–37. doi:10.1144/SP395.1

Gill, K. S. (1996). Knowledge and the Post-industrial Society. In Information Society. Human-centred Systems (pp. 3-29). Springer. doi:10.1007/978-1-4471-3249-3_1

Habib, B. (2015). Sustainability is not enough: A call for regeneration. *Ethos (Berkeley, Calif.)*, *23*(2 (Term 2)), 7–10.

Heeren, A. J., Singh, A. S., Zwickle, A., Koontz, T. M., Slagle, K. M., & McCreery, A. C. (2016). Is sustainability knowledge half the battle?: An examination of sustainability knowledge, attitudes, norms, and efficacy to understand sustainable behaviours. *International Journal of Sustainability in Higher Education*, *17*(5), 613–632. doi:10.1108/IJSHE-02-2015-0014

Holfelder. (2019). Towards a sustainable future with education? *Sustainability Science, 14*, 943-952. https://doi.org/https://doi.org/10.1007/s11625-019-00682-z

Hopper, J. R., & Nielsen, J. M. (1991). Recycling as altruistic behavior: Normative and behavioral strategies to expand participation in a community recycling program. *Environment and Behavior*, *23*(2), 195–220. doi:10.1177/0013916591232004

Hsu, K. F. (2020). Reframing public spaces: From recreational amenity to civic good in Resilient Cities. COVID-19 offers an opportunity to re-evaluate neighbourhood spaces to ensure they are available to all. *Journal of Public Space*, *5*(3), 67–82. doi:10.32891/jps.v5i3.1416

Lewis, S. L., & Maslin, M. A. (2015). Defining the Anthropocene. *Nature, 519*(7542), 171–180. doi:10.1038/nature14258 PMID:25762280

Martin, J. L., Maris, V., & Simberloff, D. S. (2016). Need to respect nature and its limits challenges society and conservarion science. *Proceedings of the National Academy of Sciences of the United States of America, 113*(22), 6105–6112. doi:10.1073/pnas.1525003113 PMID:27185943

Mathew, K. S. (2011). Professional associations and academic librarianship in Kerala. In Library 2.0 and Information Management: essays in honour of Dr. Jalaja (pp. 133-140). Atlantic Publishers.

Mauser, W., Klepper, G., Rice, M., Schmalzbauer, B. S., Hackmann, H., Leemans, R., & Moore, H. (2013). Transdisciplinary global change research: Co-creation of knowledge for sustainability. *Current Opinion in Environmental Sustainability, 5*(3-4), 420–431. doi:10.1016/j.cosust.2013.07.001

Maxwell, N. (2012). How universities can help humanity learn How to resolve crises of our times - From Knowledge to wisdom: University College London experience. In G. Heam, T. Katlelle, & D. Rooney (Eds.), *Handbook on the knowledge economy* (pp. 158–179). Edward Elgar. doi:10.4337/9781781005132.00019

Miller, T. R., Munoz-Erickson, T., & Redman, C. L. (2011). Transforming knowledge for sustainability: Towards adaptive academic institutions. *International Journal of Sustainability in Higher Education, 12*(2), 177–192. doi:10.1108/14676371111118228

Morandín-Ahuerma, I., Contreras-Hernández, A., Ayala-Ortiz, D. A., & Pérez-Maqueo, O. (2019). Socio–ecosystemic sustainability. *Sustainability*, 11.

Neumayer, E. (2012). Weak versus strong sustainability: Exploring the limits of two opposing paradigms. *Journal of Human Development and Capabilities, 13*(4), 561–579. doi:10.1080/19452829.2012.693067

Oxford Learner's Dictionary. (2022). *Societal*. Oxford Learner's Dictionary. Retrieved august 23, 2022, from https://www.oxfordlearnersdictionaries.com/definition/english/societal?q=societal

Oxford Learner's Dictionary. (2022). *Sustainability*. Oxford Learner's Dictionary. Retrieved august 23, 2022, from https://www.oxfordlearnersdictionaries.com/definition/english/sustainability?q=sustainability

Palmer, M. A., Bernhardt, E. S., Chornesky, E. A., Collins, S. L., Dobson, A. P., Duke, C. S., Gold, B. D., Jacobson, R. B., Kingsland, S. E., Kranz, R. H., Mappin, M. J., Martinez, M. L., Micheli, F., Morse, J. L., Pace, M. L., Pascual, M., Palumbi, S. S., Reichman, O. J., Townsend, A. R., & Turner, M. G. (2005). Ecological science and sustainability for the 21st century. *Frontiers in Ecology and the Environment, 3*(1), 4–11. doi:10.1890/1540-9295(2005)003[0004:ESASFT]2.0.CO;2

Pattanayak, S. (2021). *Homo sapiens: do the species making themselves extinct?* Academia Letters. doi:10.20935/AL1617

Preiser, R., Swilling, M., & Nnoli-Edozien, N. (2021). *Towards new narratives of hope for fostering transformative African futures*. Club of Rome. Retrieved November 05, 2021, from https://www.clubofrome.org/publication/towards-new-narratives-of-hope-for-fostering-transformative-african-futures/

Randall, A. (2022). Driving with eyes on the rear-view mirror—Why weak sustainability is not enough. *Sustainability (Basel), 14*(16), 13. doi:10.3390u141610203

Roco, M. C., & Bainbridge, W. S. (2013). The new world of discovery, invention, and innovation: Convergence of knowledge, technology, and society. *Journal of Nanoparticle Research, 15*(9), 15. doi:10.100711051-013-1946-1

Shang, C., Wu, T., Huang, G., & Ganlin, J. (2019). Weak sustainability is not sustainable: Socioeconomic and environmental assessment of Inner Mongolia for the past three decades. *Resources, Conservation and Recycling, 141*, 243–252. doi:10.1016/j.resconrec.2018.10.032

Spangenberg, J. H. (2005). Will the information society be sustainable? Towards criteria and indicators for a sustainable knowledge society. *International Journal of Innovation and Sustainable Development, 1*(1/2), 85–102. doi:10.1504/IJISD.2005.008082

. Stephen, W., Richardson, K., Rockström, J., Cornell, S. E., Fetzer, I., Bennett, E. M., Biggs, R., Carpenter, S. R., Vries, W. d., Wit, C. Ad., Folke, C., Gerten, D., Heinke, J., Mace, G. M., Persson, L. M., Ramanathan, V., Reyers, B., & Sörlin, S. (2015). Planetary boundaries: Guiding human development on a changing planet. *Science, 347*(6223).

United Nations. (2019, March 28). *Meetings coverage: Seventy third session, High level meeting on climate and sustainable development (am and pm).* United Nations. Retrieved August 15, 2021, from https://www.un.org/press/en/2019/ga12131.doc.htm

United Nations. (2023). *Sustainable development goals.* https://www.un.org/sustainabledevelopment/

Wagner, D. L., Grames, E. M., Forister, M. L., Berenbaum, M. R., & Stopak, D. (2021). Insect decline in the Anthropocene: Death by a thousand cuts. *Proceedings of the National Academy of Sciences of the United States of America, 118*(2), 1–10. doi:10.1073/pnas.2023989118 PMID:33431573

Whitmarsh, L. (2011). Scepticism and uncertainty about climate change: Dimensions, determinants and change over time. *Global Environmental Change, 21*(2), 690–700. doi:10.1016/j.gloenvcha.2011.01.016

Williams, B. (2021). *Saving us from ourselves: Can we repair 50 years of ecological overshoot.* https://poemsforparliament.uk

Winterfeldt, D. (2013). Bridging the gap between science and decision making. *Proceedings of the National Academy of Sciences of the United States of America, 110*(Suppl. 3), 14055–14061. doi:10.1073/pnas.1213532110 PMID:23940310

World Health Organization. (2022). *9 out of 10 people worldwide breathe polluted air, but more countries are taking action.* Retrieved October 14, 2022, from https://www.who.int/news/item/02-05-2018-9-out-of-10-people-worldwide-breathe-polluted-air-but-more-countries-are-taking-action

ADDITIONAL READING

Eggleton, P. (2020). State of world's insects. *Annual Review of Environment and Resources, 45*(1), 61–82. doi:10.1146/annurev-environ-012420-050035

Elhacham, E., Beb-Uri, L., Grozovski, J., Bar-On, Y. M., & Milo, R. (2020). Global human-made mass exceeds all living biomass. *Nature, 588*(7838), 442–444. doi:10.103841586-020-3010-5 PMID:33299177

Haberl, H., Marina, F. K., Krausmann, F., Martinez-Alier, J., & Winiwarter, V. (2011). A socio-metabolic transition towards sustainability? Challenges for another Great Transformation. *Sustainable Development (Bradford), 19*(1), 1–14. doi:10.1002d.410

Intezari, A. (2015). Integrating Wisdom and Sustainability: Dealing with Instability. *Business Strategy and the Environment, 24*(7), 617–627. doi:10.1002/bse.1892

Zalasiewicz, J. (2008). *Earth after s: What legacy will humans Leave in the rocks?* Oxford University Press. doi:10.1093/oso/9780199214976.001.0001

Chapter 4
Innovation, Sustainability, and Business Models

Namrata Pancholi
Amity University, Noida, India

Mona Chaudhary
Amity University, Noida, India

Seema Garg
https://orcid.org/0000-0002-4750-026X
Amity University, Noida, India

ABSTRACT

Technological innovation is changing the world, and these changes are bringing in new challenges. The most important among them is the customer perspective. This changing customer perspective is in demand for innovative business models. One of the changing trend is expecting sustainable models for business value proposition. Therefore, how far organizations are able to adapt to new business models leads to sustainable competitive advantage. However, research shows many business model innovations fail. The exact reasons for failure remain unexplored. Definitely, there is a lack of conceptual framework to guide the adoption and implementation of new sustainable business models. This has resulted into a need for designing new conceptual framework which can be a link between innovation and sustainability resulting in creating value for its customers. The chapter is written with the aim of providing a sustainable business model innovation which creates value for its stakeholders.

1. INTRODUCTION

In current competitive landscape business heavily relies on innovation. All industries have to constantly search creative and innovative methods as old methods are incapable of offering value to the stakeholders. Many business organizations have ambition to be innovative and creative as it offers economic benefits and competitive advantage. In current landscape sustainability is another crucial factor for organisational

DOI: 10.4018/978-1-6684-9979-5.ch004

success. The companies have increasingly aimed to maximise stakeholder value along with the financial performance. Today the investors and customer's perspective towards businesses has shifted and they increasingly take note of the sustainability efforts of companies (Freeman, 2020). Employees are willing to work with companies which adhere to sustainability development goals, customers are willing to pay more for it. Today more and more businesses are committed to net zero. For example, Apple is committed to be carbon neutral in its supply chain and products by 2030, Similarly Nissan motors is working for zero carbon emission from its manufacturing sites globally by 2050. Unilever is working with its partners to be carbon neutral by 2039. As there is no universal measure or method it is very challenging for businesses to identify and develop that balances the interest of owners, shareholders, employees, partners, consumers, suppliers, environment and coming generation along with financial performance.

Besides technological innovations stakeholders consider green innovation and environment friendly products and services to give competitive advantage. Quest for competitive advantage is considerably linked to business model. There is a strong relationship between the business models and the competitive advantage (Ranjith, 2016). An effective business model is a combination of deliberate alignment of resources and capabilities to strike competitive advantage. Firms in the emerging markets are striving to gain sustainable competitive advantage (Barney 1991). A business model describes how an organisation builds its systems, processes, culture, and leadership interacts with the ecological and business environment, does innovation, and manages all its stakeholders, organisational growth, profitability, and social impact. (Massa et al, 2017). Business Model Innovation is a critical activity undertaken by organisations to combat threat due to new entrants. (Lantano et al. 2022). One of the changing trends is expecting sustainable models for business value proposition. However, research shows many business model innovations fail (Geissdoerfer et al, 2018). The exact reasons for failure remain unexplored. There is a lack of conceptual framework to guide the adoption and implementation of new sustainable business model (Buchholz,2020). This has resulted into a need for designing new conceptual framework which can be a link between Innovation and Sustainability resulting in creating value for its customers.The book chapter is written with the aim of providing a sustainable business model that creates value for its stakeholders. (Evanset al, 2017).

2. INNOVATION

In today's competitive world innovation is a key to achieve growth for business organizations. Many business organizations heavily rely on innovation as it offers economic benefits and competitive advantage. Innovation is a product, business model, or a strategy that is both creative and useful. Innovation does not always require being a breakthrough in technology or new business models; it can be a small upgrade in a customer service provided by a company or small feature added to an existing product. But an idea that is creative idea and does not create a value or solve problem may not lead to be innovation. *Value innovations* the theory and framework developed as Blue Ocean Strategy by W. Chan Kim and Renee Mauborgne .As per the conventional framework of creating competitive advantage the business organization can choose between higher value in product and low cost. But with value innovation business successfully combines the low cost and higher value to the customer. This enables the firms to create uncontested market space and make competition irrelevant. Example in modern history Marvel movies has successfully transformed it from Red Ocean to Blue Ocean. The model of movie production was transformed by the company with an approach to pursue differentiation and low cost simultaneously.

2.1 Categories of Innovation

The business environment and the needs of customers are constantly changing. It is essential for businesses to improve different areas of business to solve emerging problems and keep creating new value for your customers. The approach for innovation varies from organization to organization. There are two broader dimensions of innovation in businesses – the technology used by them and the market they operate. Most innovations are smaller, gradual improvements on existing products, processes, and services while some innovations can be those ground-breaking technological inventions or business models that transform industries. Incremental innovation is gradually having lesser impact on the market but with low technology newness the organization keeps improving the existing products. By offering value to the customers, it keeps creating competitive advantage for its existing products or business operations. This enables the organizations to grow in the market and avoid stagnation. For example Gillette, a personal care brand, has uses incremental innovation strategy to stay on top of the industry. Its safety razors have continuously evolved with various features added to avoid razor burns. These improvements in the product have helped this brand stay relevant and maintain its market share.

Value innovation is a business strategy that focuses on generating new value for customers while at the same time lowering costs. It is a key concept in corporate strategy known as the Blue Ocean Strategy. Value innovation entails standing out from the competition by providing a special good or service that sets a business apart from rivals and draws in a new customer by:

Differentiation: Value innovation aims to create products or services that stand out in the market. This differentiation can come from unique features, functions, design, or other aspects that are not offered by existing competitors.

Cost Reduction: Value innovation works to cut expenses whenever possible at the same time. Companies can deliver their distinctive value proposition at a lesser cost, attracting customers, by reducing unnecessary features or processes.

Blue Ocean Strategy: A corporation should seek uncontested market space (a "blue ocean") rather than competing in crowded, fiercely competitive markets (a "red ocean"), according to the theory of "blue ocean" strategy. A business can do this to generate new demand and ease competitive pressures.

Value innovation has a strong focus on the client. It entails first knowing the wants and preferences of the customer before developing goods or services that creatively meet those needs. Disruptive inventions that redefine industries are frequently the outcome of this.

One classic example of value innovation is Cirque du Soleil. Instead of following the traditional model of circuses, Cirque du Soleil combined elements of theater, circus, and music to create a completely new form of entertainment that appealed to a different audience. By doing so, they eliminated many of the costly features of traditional circuses (like animal acts) while offering a unique and premium experience (Berry, et al, 2006).

In summary, value innovation is about redefining markets and industries by offering innovative products or services that break away from existing competition, create new customer demand, and simultaneously reduce costs. It's a powerful strategic concept for businesses looking to achieve sustainable growth and profitability.

Incremental innovations are small, incremental modifications or improvements to current goods, procedures, or services. Instead of delivering brand-new features or functionalities, these innovations are frequently evolutionary in character and work to improve already-existing ones. Incremental innovations are a crucial component of a company's continuous improvement initiatives and can boost productivity, cut costs, and improve customer satisfaction.

The following are some essential traits of incremental innovations:

Small Steps: They entail making minor, gradual adjustments to current goods, services, or procedures.

Low Risk: Compared to radical or disruptive breakthroughs, incremental improvements often carry lower levels of risk. The likelihood of failure is decreased because they build on already-existing infrastructure and knowledge.

Continuous improvement: Organizations frequently deploy small innovations as part of their continual efforts to improve.

Customer -Centric: Market demand and client input are frequently the driving forces behind these advancements. They are designed to cater to certain customer needs or preferences.

Cost-Effective: Because incremental innovations make use of already-existing resources and capabilities, they are frequently more affordable than major investments.

Some examples of incremental innovations are as follows:

Software updates are a typical example of incremental innovation because they regularly correct bugs, enhance performance, or include new features.

Product enhancements include the addition of new features or changes to the appearance of an existing product, such as the addition of better cameras or longer battery life to smartphones.

Process optimization: Simplifying and enhancing current manufacturing procedures to boost output and cut waste.

Enhancing customer service through greater employee training, quicker response times, or the addition of new support.

Incremental innovations are important for maintaining and growing a company's market share, as they allow organizations to keep up with changing customer preferences and stay competitive in their industries.

Radical Innovation makes use of technological breakthroughs that lead to transformation in the industry and creates new markets. In this type of the organization changes its processes, systems, and the way it interacts with its customers. The success of such a strategy is related to the firm organizational behavior and capabilities that create the ideal conditions transform the technology and new ideas to be successfully commercialized in the industry.

Disruptive innovation is a term coined by Harvard Business School Professor Clayton Christensen, and one that he believed to be widely misunderstood. Disruptive innovation rather is a process by which a smaller company, usually with fewer resources, that disrupts the existing market and challenges larger and well established businesses. The most common organizations exhibiting the characteristics of disruptive innovation are new entrants like startups and entrepreneurs who target the overlooked segment of the market to deliver more affordable or convenient offering than that of the established players. Examples of disruptive innovations taking place over the years are shown below:

Personal computers: By making computing power accessible to consumers and small organizations, personal computers like the Apple II and IBM PC caused a disruption in the mainframe computer sector.

Smartphones: A prime example of a disruptive breakthrough is the iPhone. It eventually caused disruption in the photography, GPS, and even the gaming industries by fusing a mobile phone, music player, and internet gadget into one.

Netflix: Companies like Netflix disrupted traditional cable TV and movie rental businesses by offering on-demand, affordable streaming services.

The literature on use of disruptive technology by Netflix suggests how emerging media technologies are reshaping on how individuals interact with television (Ellingsen, 2014; Tefertiller, 2018). There are various media companies that offer high-quality streaming services that can be accessed through various digital devices (Groshek and Krongard, 2016). Despite a growing number of people favoring this technology over traditional media (PwC, 2019), it represents a fiercely competitive sector within the industry. Numerous brands such as Amazon Prime Video, YouTube, and Netflix continually innovate their technology to provide consumers with a vast selection of media content (Tefertiller, 2018). Moreover, over time, these organizations have expanded their offerings by introducing mobile applications equipped with personalized recommendation systems to enhance user experiences, retain current customers, and attract new ones (Camilleri and Falzon, 2020).

Uber: The ride-sharing platforms brought about an entire disruption in car rental and taxi industries by providing a platform for bringing together the drivers and riders giving them the convenient and cost-effective way to get around. They made use of smartphone technology and the sharing economy.

Similarly, Amazon, Airbnb, Blockchain and Cryptocurrencies, Plant-Based Meat are examples that demonstrate on how disruptive innovations can reshape industries, create new opportunities, and change the way we live and do business.

Innovation requires massive investment and manager's effort and time. There are several challenges and risks associated to it. It is not an easy task to build and sustain innovation Not all innovations are successful. Companies can't claim that if they are disruptive in nature they are successful. This is because not necessarily that every disruptive path should lead to a triumph, and not every successful newcomer follows a disruptive path.

Both Uber and Apple's iPhone owe their success to a platform-based model: Uber digitally connects riders with drivers; the iPhone connects app developers with phone users. But Uber, true to its nature as a sustaining innovation, has focused on expanding its network and functionality in ways that make it better than traditional taxis. Apple, on the other hand, has followed a disruptive path by building its ecosystem of app developers so as to make the iPhone more like a personal computer.

Uber connects users with drivers online, while the iPhone does the same for users and app developers. Both Uber and Apple's iPhone credit their success to platform-based business models. But Uber has concentrated on growing its network and functionality in ways that make it superior to conventional taxis, staying true to its essence as a sustaining invention. On the other hand, Apple has taken a revolutionary approach by creating an ecosystem of app creators in order to make the iPhone more similar to a personal computer.

These instances show how disruptive technologies have the power to transform entire sectors, open up new opportunities, and alter how we live and conduct business. . Innovation and technology need to be integrated with business strategy. A clearly defined innovation strategy prioritizes goals of the organization integrating innovation and technological deliberations with functional strategies of the organization. As the strategy is formulated based on environmental factors and organizational capabilities, a company can create sustainable value through innovation and technology. Organizations need to build a supportive infrastructure for innovation like systems, tools, human resource, culture, and process and create competitive advantage in ever changing competitive marketplace.

2.2 Sustainability

A sustainable business strategy is a business strategy that incorporates economic, environmental, and social factors into an organization's policies, practices, and processes to create long-term benefits for the organization and its employees while being mindful of conserving and protecting resources. It encourages businesses to mitigate future risks to the environment and their individual negative impact on it. Therefore, integrate economic activities and planning, creating long-term value for the organization, its stakeholders, and society as a whole. A sustainable business strategy's goal is to meet the needs of the organization, its clients, employees, and stakeholders today, while protecting resources and creating policies and practices that are sustainable well into the future. It integrates economic activities and planning, creating long-term value for the organization, its stakeholders, and society as a whole. Business have been focusing on environmental, social, and governance (ESG) proposition in creating value. The environmental criteria, includes the energy a business takes in and the waste it discharges, the resources it needs, and the consequences for living beings as a result. Social criteria, addresses the relationships of business and the reputation it fosters with people and institutions in the communities where you do business. It includes labor relations and diversity and inclusion in workforce. Every company operates within a broader, diverse society. The governance is the internal system of practices, controls, and procedures a company adopts in order to govern itself, make effective decisions, comply with the law, and meet the needs of external stakeholders with fairness and transparency.

2.3 Traditional Profit Driven Models Vs Sustainable Business Model (SBM)

Traditional profit-driven business models can be problematic in terms of their social impact and unsustainability because they often gut resources, harm the environment, and leave employees burned out and disengaged. Sustainable business practices help to minimize these negative impacts, in the hopes of creating a better world and giving people purposeful, positive work. When businesses focus on profit alone, they can ignore or justify how their practices affect the environment and the people who work for them. Businesses that utilize sustainable business strategies and focus more on environmental impact demonstrate more accountability and transparency. Organizations that do this well have sustainability ingrained in every part of their culture, beginning with a mission or values statement with sustainability as its focus.

2.3.1 Benefits of Sustainable Value

1. A strong ESG proposition helps companies tap new markets and expand into existing ones. ESG can also drive consumer preference. McKinsey research has shown that customers say they are willing to pay to "go green."
2. ESG can also reduce costs substantially. Among other advantages, executing ESG effectively can help combat rising operating expenses (such as raw-material costs and the true cost of water or carbon), which McKinsey research has found can affect operating profits by as much as 60 percent.
3. ESG value an enable companies to achieve greater strategic freedom, easing regulatory pressure. In fact, in case after case across sectors and geographies, we've seen that strength in ESG helps reduce companies' risk of adverse government action. It can also engender government support.

4. A strong ESG proposition can help companies attract and retain quality employees, enhance employee motivation
5. A strong ESG proposition can enhance investment returns by allocating capital to more promising and more sustainable opportunities (for example, renewables, waste reduction, and scrubbers)

2.4 Sustainable Transformation

Although there is a broad consensus about the importance of sustainability for firms but transformation towards sustainability is a great challenge (Matos and Clegg, 2013). The academic studies on how to transform the firm into a sustainable organization has remained unclear. Nevertheless, the dynamic vision on how the business model changes to incorporate triple sustainability has received insufficient attention. Few companies are born with a broad-based commitment to sustainability.

2.5 Role of Leadership in Innovation

Leadership plays a very important role in making product innovation decisions leading to green environment (Khan etal,2022). By engaging employees, leaders can ensure green innovation and creativity. 80% of successful innovative companies have top leaders who reinforce the value and importance of innovation. Some classic examples of this are Steve Jobs, Bill Gates, and Thomas Edison. The role of a leader is to create an environment where employees can take risk in creating new products and processes which result in environmentally safe products. A good leader can communicate well with employees giving them a sense of ownership, responsibility and motivation for high performance. Thus, green leadership is related to improved firm performance and competitiveness.

3. COMPETITIVE ADVANTAGE

According to Michael Porter (1985) competitive advantage can be gained through differentiation and cost leadership. Firms can remain competitive in the marketplace by Integrating innovation and environmental sustainability into their business model (Porter & Van der linde, 1995). There is a strong relationship between the business models and the competitive advantage. Business innovation and sustainability should be integrated within the business model. It is only then it can give competitive advantage. For example innovations creating (Chang, 2011; Ismail et al., 2012; Joao et al., 2012; Ortega, 2010) product -based advantages comes through its superior quality, packaging, designs etc. Value based advantage comes through its cost and value it is offering for the price paid. (Morgan et al, 2004; Al-Bourini et al., 2013). E.g., Amazon has developed specialized distribution channels that allow them to reach customers in new and innovative ways.

4. OBJECTIVES OF THE STUDY

The chapter aims to achieve the following objectives:

1. To conduct a detailed review of existing literature for the purpose of understanding the link between Sustainability, Innovation and Business Model.
2. To design a Conceptual Framework of Sustainable Business Model Innovation.

5. LITERATURE REVIEW

5.1 Need for Innovation and Business Model Innovation

How can companies build competitive advantage today? Fierce competition and rivalry between companies is forcing companies to adopt newer ways of serving the customers. Thus, innovation is becoming the central concept for sustained economic gains and competitive advantage. (Schumpeter, 1934, Tushman et al., 1997). Innovation brings in change within the organizations (Drucker, 1985). Innovation is the major area to contribute to practice of management.

Academic literature and companies have shown interest in Business Model Innovation, as one it is a newer way of doing business (Markides,2006), second Industry 4.0 with its new technologies are considered essential for sustaining competitive advantage (Tohanean et al, 2020). Business Model Innovation is difficult to achieve (Tonder et al, 2020). Implementation of Business Model Innovation is a challenge for most organizations (Minatogawa et al, 2020). There is definitely a need for a conceptual framework to guide the implementation of BMI. Literature suggests that implementing new Business Model Innovation has a positive impact (Smajlovic, 2019) on firms competitive advantage but some researchers are of the view that in implementing new technologies can bring its own risks (Tohanean et al, 2020).

5.2 Sustainable Business Models (Evans et al., 2017)

A sustainable business model is that where sustainability drives competitive advantage, organization performance and decision making. There is a complete transformation of the product rather than just supplementing it with some variables of sustainability. Business model innovations for sustainability are defined as: Innovations that create significant positive and or significantly reduced negative impacts for the environment and or society, through changes in the way the organization and its value-network create, deliver value and capture value (i.e., create economic value) or change their value propositions" (Bocken et al, 2014).

Sustainable business models can be defined as giving five value propositions:

1. Sustainable economic, social, and environmental benefits
2. Sustainable value flows among multiple stakeholders
3. A value network with a new purpose, design, and governance.
4. Considering stakeholder's interests and responsibilities for mutual value creation.
5. Internalizing externalities through product-service systems enables innovation towards sustainable business models.

Sustainable business model innovation (SBMI) is a change in the way a firm operates in order to create positive impacts or to reduce negative consequences for the environment and the society. For example HealthMedia, a player in healthcare industry with a revenue of US$6 million, was drowning in a red ocean of intense competition. At the time, the industry had two strategic groups, one offering telephonic counseling predominantly for severe medical conditions, the other offering generic, digitalized content like you find on WebMD. It created digital health coaching through interactive online questionnaire. Within just two years, Health Media created a value proposition so compelling that Johnson and Johnson swooped in and purchased the company for US$185 million.

5.3 Sustainable Business Model Innovation

Sustainable Business Model Innovation should help companies reduce the adverse effects of external environment and rather create positive effects of external environment. SBMI research has drawn attention in the field of technology innovation, sustainable innovation and business models (Cillo, Messeni Petruzzelli, Ardito, & Del Giudice, 2019). Thus, the research helps provide organizations with guidelines to integrate sustainability into their existing or modified business models.

Firms built competitive advantage through green innovation integration in their systems and processes (Presenza et al., 2019). Through Sustainable Business Model Innovation the core activity shall transform value proposition towards sustainable value proposition (Baldassarre, Calabretta, Bocken, & Jaskiewicz, 2017). This should allow mangers to make decisions which are non-financial and promote a mindset of sustainability for the whole organization. SBMI should create sustainability considering the interests of all stakeholders of business.

Sustainable business model innovations enable organisations to generate profit while reducing their adverse effects on the environment, society, and economy. These developments are crucial for solving urgent global issues like social injustice, resource depletion, and climate change. Here are a few examples of innovative sustainable business models:

Product-as-a-Service: Businesses provide products as a service rather than selling them, encouraging recycling and reuse. For appliances and electronics, for instance, subscription-based business models or leasing.

Take-Back and Recycling Programmes: Businesses are in charge of gathering and recycling their products at the end of their useful lives, thereby minimising waste and fostering resource effciency.

Platforms for the sharing economy: Platforms like Airbnb and Uber allow people to share their resources (like lodging or rides) with other people.

Consumers who exchange goods and services within a community cut down on the demand for ownership and the creation of brand-new things.

Distributed energy generation: By producing their own renewable energy, businesses cut back on their reliance on fossil fuels and greenhouse gas emissions.

The Business Model Innovation is essential for survival in today's turbulent environment (Velu, 2015), and a source of a firm's competitive advantage. (Pieroni, et al, 2019). It is also a significant contributor to improved firm performance. (Hamelink et al, 2019), and an efficient method for generating customer and social value by fusing environmental, social, and business activities (Schaltegger, et al, 2012). All the above knowledge helped the researchers to develop a conceptual framework of Sustainable Business Model Innovation.

6. CONCEPTUAL FRAMEWORK

The above framework reveals the key elements of Sustainable Business Model Innovation which are sustainability (environment, social, economic aspects), Technological innovation, ethics and organizational values, culture, leadership, key activities of marketing, sales all leading to revenue generation, stakeholder involvement and expectations leading to changes in business model. The framework is a guide to understand about the value definition, value creation, delivery and value capture in an organization.

A conceptual framework should be able to define how a company creates, delivers and extracts value through its whole set of integrated components, all of which can be looked on as opportunities for innovation and competitive advantage (Skarzynski et al, 2013).

In the value definition (Richardson, 2008) section the authors have described about integration of business model with the concept of sustainability. Innovation is the key to corporate success. Innovation is aligned with the vision, mission statement of the company.

Strategic Intent can be understood as the philosophical base of the strategic management process. It implies the purpose, which an organization endeavours to achieve. It is a statement that provides a perspective of the means, which will lead the organization to reach the vision in the long run. The concept 'strategic intent' was first popularized in a 1989 Harvard Business Review article by Gary Hamel and Prahalad (1989). Hamel and Prahalad argue that Western companies focus on trimming their ambitions

Figure 1. Key elements of sustainable business model innovation

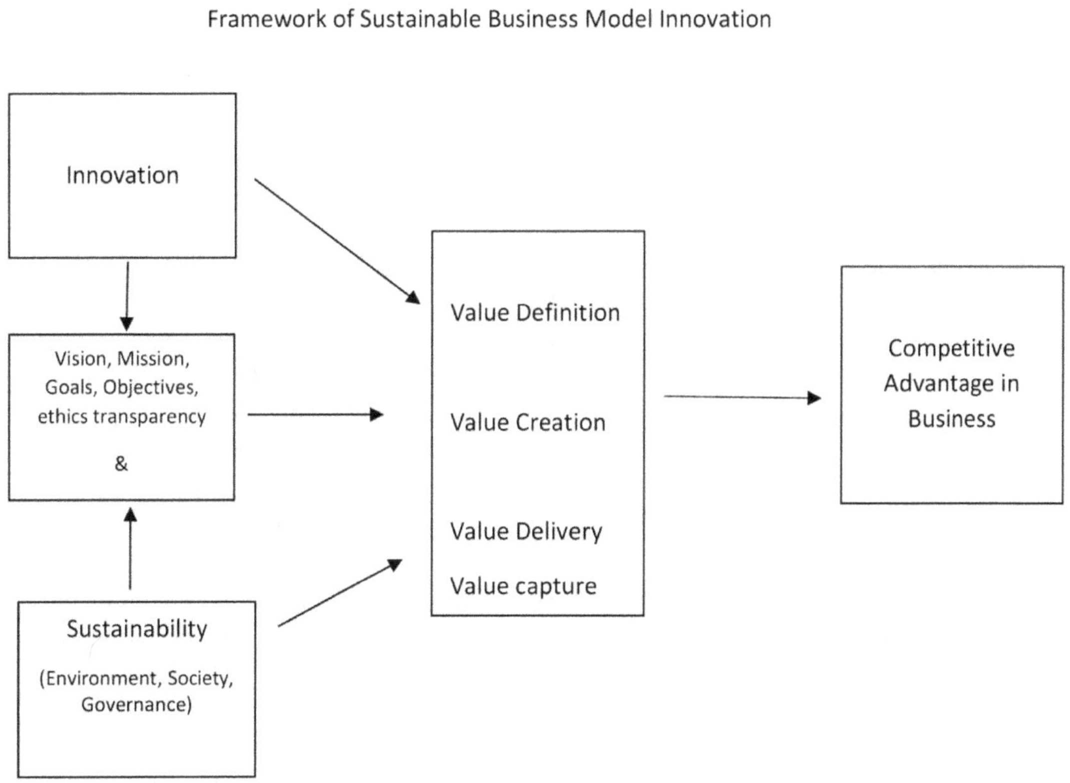

to match resources and, as a result, search only for advantages they can sustain. By contrast, Japanese corporations leverage resources by accelerating the pace of organizational learning and try to attain seemingly impossible goals. Strategic intent gives an idea of what the organization desires to attain in future. It answers the question what the organization strives or stands for? It indicates the long-term market position, which the organization desires to create or occupy and the opportunity for exploring new possibilities. Hierarchy of strategic intent is vision, mission, business definition, business model, goals and objectives. This strategic intent usually incorporates stretch targets, which force companies to compete in innovative ways. In this McKinsey Award–winning article, Hamel and Prahalad describe four techniques that Japanese companies use: building layers of advantage, searching for "loose bricks," changing the terms of engagement, and competing through collaboration.

Transparency is important as it defines the amount of information organisations can disclose to the stakeholders. This is not possible without ethical practices (Depken & Zeman, 2018). Leadership and governance are important in building an appropriate organisation culture and talent retention. Value creation and delivery are important from all stakeholders view point. Innovation should create products that are ecologically sustainable supporting health and welless. This will add value accepted by the changing consumer attitude today. The marketing efforts can be directed in the required direction. Sales team can play an important role in promoting firms value definition. This also enables employee retention.

Sustainability and Innovation have meaning for a firm if they are able to reduce cost and deliver value for customers. This can be measured with the help of various performance measurement tools and metrics. Example can be last year's turnover and net income, the number of new offices, the hiring of new workers, the new investments attracted, the environmental indices and the customer feedback.

The above framework defines how a business model for a company can create, deliver and extract value. It includes organisation mission, vision, ethics and governance- the business model integrated with the sustainability and innovation components resulting into competitive advantage for the organisations.

7. DISCUSSION AND CONCLUSION

To maintain competitive advantage it is significant for companies to transform their business models to value creation, delivery and capture for all stakeholders who are connected to business directly or indirectly. The innovation and technology have significant role in implementing the value created by the company. Companies like Indigo Ag, Agriculture industry start up, Ajinomoto a global food and Biotech Company have gone one step ahead to help its customers giving sustainable solutions, creating value for the organisation and customers both. Indigo, an agriculture technology-based company launched a service to help farmers regenerate agriculture practices. Similarly a Japanese food and biotech company, Ajinomoto helps the customers with digitally enabled diagnostics, and product recommendations. Technological disruptions have played a significant role in driving sustainability initiatives within all industries. Hotels are examples where they are increasingly adopting innovative technologies to minimize their environmental impact, reduce energy consumption, promote waste reduction, and implement ecofriendly practices.

REFERENCES

Baldassarre, B., Calabretta, G., Bocken, N. M. P., & Jaskiewicz, T. (2017). Bridging sustainable business model innovation and user-driven innovation: A process for sustainable value proposition design. *Journal of Cleaner Production*, *147*, 175–186. doi:10.1016/j.jclepro.2017.01.081

Barney, J. (1991). Firm resources and sustained competitive advantage. *Journal of Management*, *17*(1), 99–120. doi:10.1177/014920639101700108

Berry, L. L., Shankar, V., Parish, J. T., Cadwallader, S., & Dotzel, T. (2006). Creating new marketsthrough service innovation. *Sloan Management Review*, *47*(2), 56–62.

Bocken, Short, Rana, & Evans. (2014). A literature and practice review to develop sustainable business model archetypes. *Clean. Prod.*, *65*, 42-56. U6 doi:10.1016/j.jclepro.2013.11.039

Buchholz, H., Eberle, T., Klevesath, M., Jürgens, A., Beal, D., Baic, A., & Radeke, J. (2020). Forward Thinking for Sustainable Business Value: A New Method for Impact Valuation. *Sustainability (Basel)*, *12*(20), 8420. doi:10.3390u12208420

Camilleri, M.A. & Falzon, L. (2020). Understanding motivations to use online streaming services: Integrating the technology acceptance model(TAM) and the uses and gratifications theory (UGT). *Spanish Journal of Marketing*. doi:10.1108/SJME-04-2020-0074

Cillo, V., Messeni Petruzzelli, A., Ardito, L., & Del Giudice, M. (2019). Understanding sustainable innovation: A systematic literature review. *Corporate Social Responsibility and Environmental Management*, *26*(5), 1012–10. doi:10.1002/csr.1783

Depken, D., & Zeman, C. (2018). Small business challenges and the triple bottom line, TBL: Needs assessment in a Midwest State, USA. *Technological Forecasting and Social Change*, *135*, 44–50. doi:10.1016/j.techfore.2017.05.032

Ellingsen, S. (2014). 'Seismic Shifts: Platforms, Content Creators andSpreadable Media'. *Media International Australia, Incorporating Culture & Policy*, *150*(1), 106–113. doi:10.1177/1329878X1415000121

Evans, S., Vladimirova, D., Holgado, M., Van Fossen, K., Yang, M., Silva, E., & Barlow, C. (2017). *Business model innovation for sustainability: towards a unified perspective for creation of sustainable business models*. Bus. Strat. Environ. doi:10.1002/bse.1939

Freeman, R. E., Martin, K. E., & Parmar, B. L. (2020). *The Power of and: Responsible Business without Trade-Offs*. Columbia University Press.

Groshek, J., & Krongard, S. (2016). Netflix and engage? Implications fo rstreaming television on political participation during the 2016 US presidential campaign. *Social Sciences (Basel, Switzerland)*, *5*(4), 65–83. doi:10.3390ocsci5040065

Hamel, G., & Prahalad, C. K. (1989). Strategic intent. *Harvard Business Review*, (May–June), 63–76. PMID:10303477

Hamelink, M., & Opdenakker, R. (2019). How business model innovation affects firm performance in the energy storage market. *Renewable Energy*, *131*, 120–127. doi:10.1016/j.renene.2018.07.051

Lantano, F., Petruzzelli, A. M., & Panniello, U. (2022). Business model innovation in video-game consoles to face the threats of mobile gaming: Evidence from the case of Sony PlayStation. *Technological Forecasting and Social Change*, *174*, 121210. doi:10.1016/j.techfore.2021.121210

Markides, C. (2006). Disruptive Innovation: In Need of Better Theory*. *Journal of Product Innovation Management*, *23*(1), 19–25. doi:10.1111/j.1540-5885.2005.00177.x

Martin & Evans. (2018). Sustainable business model innovation: A review. *Journal of Cleaner Production*, *198*, 401-416. https://www.sciencedirect.com/science/article/pii/S0959652618318961 doi:10.1016/j.jclepro.2018.06.240

Massa, L., Tucci, C., & Afuah, A. (2017). A critical assessment of business model research Acad. *The Academy of Management Annals*, *11*(1), 73–104. doi:10.5465/annals.2014.0072

Minatogawa, V. L. F., Franco, M. M. V., Rampasso, I. S., Anholon, R., Quadros, R., Durán, O., & Batocchio, A. (2020). Operationalizing Business Model Innovation through Big Data Analytics for Sustainable Organizations. *Sustainability (Basel)*, *12*(1), 277. doi:10.3390u12010277

Pieroni, M. P., McAloone, T. C., & Pigosso, D. C. A. (2019). D.C. Business model innovation for circular economy and sustainability: A review of approaches. *Journal of Cleaner Production*, *215*, 198–216. doi:10.1016/j.jclepro.2019.01.036

Presenza, A., Messeni Petruzzelli, A., & Natalicchio, A. (2019). Business model innovation for sustainability. Highlights from the tourism and hospitality industry. *Sustainability (Basel)*, *11*(1), 2–5. doi:10.3390u11010212

PwC. (2019). *A period of video streaming peace: Consumers have found their video consumption groove: PwC*. PwC. Available at: https://www.pwc.com/us/en/services/consulting/library/consumer-intelligence-series/consumer-video-streaming-behavior

Ranjith, V. K. (2016). Business Models and Competitive Advantage. *Procedia Economics and Finance*, *37*, 203-207.

Richardson, J. (2008). The business model: An integrative framework for strategy execution. *Strategic Change*, *17*(5-6), 133–144. doi:10.1002/jsc.821

Schaltegger, S., Freund, F. L., & Hansen, E. G. (2012). Business cases for sustainability: The role of business model innovation for corporate sustainability. *International Journal of Innovation and Sustainable Development*, *6*(2), 95–119. doi:10.1504/IJISD.2012.046944

Skarzynski, P., & Gibson, R. (2013). *Innovation to the Core: a Blueprint for Transforming the Way Your Company Innovates*. Harvard Business Press.

Smajlović, S., Umihanić, B., & Turulja, L. (2019). The Interplay of Technological Innovation and Business Model Innovation toward Company Performance. *Management*, *24*(2), 63–79. doi:10.30924/mjcmi.24.2.5

Tefertiller, A. (2018). Media substitution in cable cord-cutting: The adoption ofweb-streaming television. *Journal of Broadcasting & Electronic Media*, *62*(3), 390–407. doi:10.1080/08838151.2018.1451868

Tohãnean, D., Buzatu, A. I., Baba, C.-A., & Georgescu, B. (2020). Business model innovation through the use of digital technologies: Managing risks and creating sustainability. *Amfiteatru Economic*, *22*(55), 758–774. doi:10.24818/EA/2020/55/758

Van Tonder, C., Schachtebeck, C., Nieuwenhuizen, C., & Bossink, B. (2020). A framework for digital transformation and business model innovation. *Management*, *25*(2), 111–132. doi:10.30924/mjcmi.25.2.6

Velu, C. (2015). Business model innovation and third-party alliance on the survival of new firms. *Technovation*, *35*, 1–11. doi:10.1016/j.technovation.2014.09.007

Chapter 5
Green Technology Innovations

Ingrid Yadibel Cuevas Zuñiga
Instituto Politécnico Nacional, ESCA, Santo Tomás, Mexico

Jessica Aimeé Gómez Quiroz
https://orcid.org/0000-0003-0061-6852
Instituto Politécnico Nacional, ESCA, Santo Tomás, Mexico

María del Rocío Soto Flores
https://orcid.org/0000-0002-6033-1472
Instituto Politécnico Nacional, ESCA, Santo Tomás, Mexico

Samantha Cuevas Antunez
Instituto Politécnico Nacional, ESCA, Santo Tomás, Mexico

ABSTRACT

Derived from the environmental crisis and global warming, the demands of the markets have increased by establishing as a requirement the use of more rational and efficient processes that ensure the quality of the products and mitigate the environmental impact of the daily operations of the organizations. This has led to the emergence of technologies related to process optimization. Therefore, the objective of this chapter is to identify green technological innovations as a tool for organizations to assume their commitment to sustainable development, obtaining economic, social, and environmental benefits by incorporating them into their production processes. The results show that green technological innovations are mainly used to minimize energy and water consumption, optimization of raw materials, reduction of waste and emission of polluting gases, soil maintenance, and the proper use of bioinformatics, nanotechnology, and biotechnology.

INTRODUCTION

Derived from the environmental crisis and global warming, the demands of the markets have increased by establishing as a requirement the use of more rational and efficient processes that ensure the quality of the products and mitigate the environmental impact of the daily operations of organizations, which

DOI: 10.4018/978-1-6684-9979-5.ch005

improves the life quality of society in general. This has caused technologies related to the optimization of processes to emerge that seek not only efficiency and effectiveness but also effectiveness in business, the so-called green technologies.

Therefore, the objective of this chapter is to identify green technological innovations that have emerged as a tool for organizations to assume their commitment to sustainable development, obtaining economic, social and environmental benefits by incorporating them into their production processes.

The research is documentary type, since an-in-depth analysis of the literature found in the main databases and with the highest number of citations between 2015-2022 was carried out. The results show that green technological innovations are mainly used to minimize energy and water consumption, optimization of raw materials, reduction of waste and polluting gas emissions, soil maintenance, and the appropriate use of bioinformatics, nanotechnology and biotechnology. This research provides evidence on the need for companies to invest in technological innovations that optimize processes and products in a sustainable manner; furthermore, it is considered that these contribute to society having a better quality of life, giving them greater prestige and brand recognition.

BACKGROUND

In this radiant era of the fourth Industrial Revolution, a large amount of technology is observed as part of the daily life of human beings, its development and the advances that are generated create expectations of what life will be like in some years. The integration of green technological innovations is important in organizations because climate change occurs with greater intensity, since they face greater barriers to obtaining their raw materials, in addition to the transformation process to have the final product requires using the least amount of inputs and reducing its environmental impact, the distribution of goods and/or services must consider the amount of polluting gas emissions, and the waste generated once the product is used must also be considered. In this sense, green technological innovations are essential to face the mentioned before adversities, as a proactive measure for organizations to assume their commitment to sustainable development, but at the same time, have added value and are more competitive in these increasingly globalized markets.

THE FOURTH INDUSTRIAL REVOLUTION AND THE ADVANCE OF TECHNOLOGICAL INNOVATION

Before talking about the fourth industrial revolution, it is pertinent to define the word revolution which comes from the Latin revolution, which is the action and effect of causing a radical change (Real Academia Española, 2023). In this sense, the term Industrial Revolution emerged at the beginning of the 19th century; Bezancon claims to have found it in writings of the Elboeuf Chamber of Commerce in 1806 (Otero & de Grossi, 1998).

Southcliffe in his book "The Industrial Revolution 1760-1830" mentions that the term Industrial Revolution refers to changes not only industrial, but also social and intellectual (Otero & de Grossi, 1998). Different authors agree that the term industrial omits economic and social changes, and even consider that instead of a revolution it was an accelerated evolution (Blanco, Castro, Gayoso, & Santana, 2020; Useche, Juarez, & Ramirez, 2022) Landes (1979), mentions that the term Industrial Revolution usually refers to the complex of technological innovations that, by replacing human skill with machinery and human and animal strength with mechanical energy, causes the transition from artisanal to factory production thus giving rise to the birth of the modern economy.

In this context, the Industrial Revolution refers to the abrupt and historical change or transformation that mainly impacts three major fields: the economy, technology and society. Therefore, technology is a basic aspect that promotes an industrial revolution, since innovation refers to a process that involves transforming a creative idea into a solution for a problem or need. This process is achieved not only by creating something completely new, but also through improvements to something already existing. (OECD, 2005).

Innovation is a complex process that involves different interconnected stages, such as the generation of ideas, the creation of new or improvements in products, processes, forms of organization, and the exploration of novel markets. Although one organization may be the leader of this innovation process, it often benefits from external contributions from different sources. This means that innovation can arise from both internal efforts and external influences (Galicia, 2015).

Now, a technological innovation occurs when technology is used to implement changes, and this process is considered continuous, and its success depends on the relationship between science and technology (RIDAA, 1996). Technological innovation begins with the identification of technology needs in industrial organizations, until the introduction into the market of products, processes or equipment developed through research and development and other initiatives, to satisfy the demands of the productive sector through technical changes that, after entering the market, they generate an impact on economic and social issues. According to the OECD (2005), technological innovation should be defined as the first application of science and technology in a new direction, followed by commercial success.

Therefore, it is important to highlight that the technological innovation process is conditioned by historical factors and plays a central role in the evolutionary approach. This means that the evolution of a technology depends largely on the path it has followed in the past. Furthermore, the technological innovation process is partially irreversible, and the development of technology within a certain technological trajectory generates new knowledge through various feedback mechanisms, which improves its performance. These mechanisms reinforce the dominant position of a technology over others with which it competes in the market.

Based on the above, and considering that technology is a precursor element of any Industrial Revolution, those that have existed to date are briefly described below:

- The First Industrial Revolution has its origins in the second half of the 17th century in Great Britain, spreading throughout Europe (Martín, 2018). This was characterized by the change from manual operation to machinery through technological development, the main energy source was steam. With industrialization, employment grew, and notable technologies include the steam engine, the railroad, the telegraph and the power loom (Useche, Juarez, & Ramirez, 2022).
- The Second Industrial Revolution covered the years from 1870 to 1914. In this, the main energy source was based on oil and electricity, with emphasis on mass production. During this period, better conditions were developed for workers, automobiles, radio, and the rise of household appliances arose (Blanco, Castro, Gayoso, & Santana, 2020).
- The Third Industrial Revolution covered the years from 1945 to 1970 and was characterized by electronic implementations, the development of the first computers applied to the mechanization of business processes, focused on administration. Likewise, there was an increase in consump-

tion and, therefore, in the demand for products. Among the notable inventions are television, the Internet and microprocessors. With the emergence of the internet, there is no doubt that technological development grew exponentially (Concha & Monsalvo, 2020).

Thus, it can be seen that through the different Industrial Revolutions automation has been developed through different energy sources and that these, in turn, lead to substantial changes in society, particularly in the way of relating, access rapid access to information, improvement in decision making and even the development of new skills to adapt to these technological changes (García & Oliveira, 2023).

Currently, we are experiencing the fourth Industrial Revolution Schwab (2016), founder and director of the World Economic Forum, mentions that this revolution has its origins in the early 21st century and maintains that it is based on the digital age. Among its most notable features are the improvement of the internet, artificial intelligence, and sensor innovations to make them smaller, cheaper and more powerful.

The fourth Industrial Revolution is characterized by three elements that make it unique:

1. Speed. Innovation, adaptation and dissemination in terms of technology went from having an exponential linear advance. The above is observed with the Internet, which spread globally in less than a decade. However, this characteristic also comes into debate because the advancement of technology may remain in a small group of societies, which will increase the already existing social inequality (Maximo, 2017).
2. The scale. This is related to the scope and dissemination given to technological advances that are directly related to globalization.

The impact. It refers to the transformation of complex systems between countries, companies, industries and society.

The term of the fourth Industrial Revolution is associated with Industry 4.0, where most authors consider both terms as synonyms. However, there are those who maintain that this Industry 4.0 is a phase of said revolution. Since the term Industry 4.0 emerged in Germany, at the Hannover Fair in 2011 and was introduced by a group of specialists to generate a program that would improve the productivity of the manufacturing industry (García & Oliveira, 2023), this with the objective of position itself in the market that has become increasingly demanding and changing.

The United Nations Industrial Development Organization defines industry 4.0 as the increase in convergence between different emerging technological fields, such as: digital production technologies, nanotechnologies, biotechnologies and new materials and their complementarity in production (Dutrénit, 2020). In such a way, this term refers to when the supply chain is intelligent in all its stages, in other words; It is a systematic process of innovation based on the skills, performance and knowledge of the workforce of a nation or organization (Blanco, Fontrodona, & Poveda, 2017).

The main objectives of Industry 4.0 are:

- Digital transformation of industrial production and communication.
- Chaining, optimization and flexibility of the value chain.
- Decentralized (collaborative) production, effective, flexible, sustainable and with greater customization.
- Higher product quality, shorter product launch time, new services and business models (Sánchez, 2019).

According to Colotla and his collaborators (2016), among the benefits offered by Industry 4.0 are:

- Higher quality of products.
- The expansion of productivity.
- Greater production flexibility.
- Fastest products on the market.
- Environmental Protection.

In this context, the new technologies present in the Fourth Industrial Revolution and Industry 4.0 have the key characteristic of taking advantage of digitization, which consists of converting something (such as a text or image) in such a way that it can be processed by a computer. Both Schwab (2016) and Guoping and his collaborators (2017), classified the precursor technologies of the fourth Industrial Revolution into three types:

1. Digital technologies. They are non-tangible technologies, such as the Internet of Things (IoT), sensors and the blockchain, among others. This present greater sophistication compared to those developed in the third Industrial Revolution.
2. Physical technologies. They are tangible ones such as autonomous vehicles, 3D printing, advanced robotics, new materials and neuro technologies.
    a. Autonomous vehicles.
    b. 3D print.
    c. Advanced robotics.
    d. New materials.
3. Biological technologies. These technologies have their origin in the biological field, such as genomics and synthetic biology.

However, there are multiple lists of the technologies of this era, for example, Ruessmann et. al. (2015) considers that there are nine fundamental technologies that companies must use and that modify the way in which they operate and relate to their suppliers, producers and customers, as well as the new relationship that exists between people and machines. These technologies are the following:

1. Big data and analytics: Involves the analysis of massive data sets to support real-time decision making.
2. Autonomous robots: Robots are becoming increasingly autonomous, flexible and cooperative, interacting with each other and working alongside humans safely. Prices for robots and software are expected to decrease, allowing for greater adoption in various tasks.
3. Simulation: 3D simulations are used not only in the engineering phase, but also in production operations. This allows machines to be tested and optimized in a virtual environment before deploying them in the real world.
4. Horizontal and vertical integration of systems: Manufacturers, suppliers and customers will be closely linked through computer systems, facilitating fully automated value chains. The same will happen between departments within a company.

5. Industrial Internet of Things (IoT): More and more devices will be connected through standard technologies, allowing communication and interaction between devices and central controllers, decentralizing decision making and allowing real-time responses.
6. Cybersecurity: With the increasing connectivity of Industry 4.0, it is necessary to protect critical industrial systems and production lines against cyber threats, as well as safeguard intellectual property and privacy.
7. The cloud: Production-related tasks will increasingly require data exchange, and cloud technologies will improve their responsiveness. This will enable the transfer of computing jobs to the cloud and make it easier for more IT services to go into production.
8. Additive manufacturing: 3D printing will extend beyond the creation of individual components and prototypes, enabling the production of small batches of customized products. This will reduce the use of raw materials, inventory levels and the logistics to be used.
9. Augmented reality: The use of augmented reality glasses will provide workers with real-time information to improve decision making and work procedures. Greater use of augmented reality is expected in areas such as repairs and human capital formation.

Table 1 shows how countries have adapted to the fourth Industrial Revolution where the United States of America, Japan and Germany are seen as leaders through the creation of new technologies, other countries such as Israel, Australia and Mexico are those that are in the process of creating technologies and other countries that do not have the capacity to create them only use them.

Technological innovation has generated numerous benefits and advances in various areas. However, it is also important to consider the possible negative consequences associated with these transformations. One of the most notable negative effects is unemployment and job displacement. Automation and

Table 1. Elements of the fourth industrial revolution

Elements	Description	Criteria
1. Leading economies (10): United States of America, Japan, Germany, China, Taiwan, France, Switzerland, the United Kingdom, the Republic of Korea and the Netherlands.	There are 10 major economies actively participating in the creation and use of new technologies.	They have more than 100 patent applications.
2. Second-tier economies in terms of production and use (40 countries): a) Innovative (example: Israel, Italy, Sweden). b) Exporters (example: Australia, Canada). c) Importers (example: Mexico, Thailand, Türkiye).	As innovators, they are actively in the process of creating patents for new technologies. As exporters, they are economies actively involved in the export of these technologies and sell large volumes to international markets. As importers, they are economies actively involved in the import of new technologies.	They have 20 or more patent family applications. Relatively specialized in export of related goods to these technologies. Relatively specialized in import, which buy large volumes in international markets.
3. Industrialization countries Late in terms of production and use (29 countries): d) Innovative. e) Exporters. f) Importers.	As innovators, they are economies with some patent creation activity related to these technologies. As exporters, they are economies with some export activity related to new technologies. As importers, they are economies with some import activity related to these technologies.	Economies with at least some patent applications. Relative specialization in the export of goods associated with these technologies and sold within international markets. Relative specialization regarding the purchase of new technologies.

Font: García & Oliveira (2023), based on ONUDI (2020)

robotization of processes can lead to the replacement of human workers with machines, resulting in structural unemployment and hardship for those whose skills become obsolete (Organizaci6n Mundial de la Propiedad Intelectual, 2022).

Furthermore, the adoption of new technologies can widen the economic and social gap for those people or regions with limited access to technology and the necessary training can be left behind, generating inequalities and increasing the division between those who benefit from them and those who do not. Privacy and security are also important concerns, technological advancement poses risks in terms of privacy violation and the possibility of cyber-attacks. The collection and use of personal data can compromise people's privacy, in addition to unauthorized access to sensitive information.

Another aspect to consider is the growing technological dependence. Although technology provides great advantages, it can also make societies more vulnerable to failures or interruptions in systems (Wood & Berge, 1997). This has negative consequences in areas such as infrastructure, public services, and national security.

The environmental impact is another relevant concern since some technologies and production processes associated with the fourth Industrial Revolution can contribute to pollution and the depletion of natural resources. The extraction of resources for the manufacture of electronic devices or the generation of waste of this type are examples of how technology can have a negative impact on the environment.

Therefore, there is a risk of dehumanization and loss of skills, as tasks become automated, some human skills and knowledge may become obsolete. Furthermore, the increasing interaction with digital technologies can lead to the loss of human connection and social interactions (Villa de Prado, 2019).

It is essential to address these negative aspects to ensure equitable and sustainable development in the context of the fourth Industrial Revolution and technological innovation. While these challenges are neither inevitable nor exclusive to these transformations, it is important to take steps to mitigate them and promote a responsible approach to the use of technology.

Therefore, it is noted that the fourth Industrial Revolution involves a digital transformation that not only modifies the economy of countries, but also the nature of labor markets and the workforce. There are many challenges that the world faces in facing this revolution, but one of the most important is understanding the new technological revolution that leads to the transformation of humanity, due to the convergence of digital, physical and biological systems who star in it. New technologies are changing the way people live, work and relate to each other, forcing companies to rethink how they can generate value through it.

SUSTAINABLE DEVELOPMENT AND THE EMERGENCE OF GREEN TECHNOLOGICAL INNOVATIONS

Since the first industrial revolution, it stands out for the beginning of mass production of goods, which at first were necessary, however, they ended up overproducing more than what was needed, which caused society to not only satisfy its needs also what is known as consumerism, this being an artificial act that generates problems that cannot later be solved by acquiring, consuming or accumulating goods and services that, in many cases, are not necessary for people.

In this sense, the expansion and generalization of capitalist relations forced the world to produce excessively, consequently the consumption of natural resources increased and at the same time the generation of waste was favored, in this sense industrial revolutions potentiated the use of new sources of energy and machinery that changed the economic and social structure; likewise, due to the expansive nature of capitalism, production relations expanded throughout the planet, which caused environmental problems to become global problems (Juré & Fittipaldi, 2017).

The emergence of interest in caring for the environment dates back to the moment when humanity began to realize its deterioration as a consequence of the irrational use of resources, which led to linking development goals with the control of negative impacts on the environment, for that reason people start to put value on the use of natural resources that still exist in a sustainable way and putting social responsibility, justice and equity first in the productive activities of human beings that affect the future of the planet (Peña, Ortega, & San Juan, 2020).

Derived from the last, sustainable development has the purpose of reconciling environmental protection and social justice with the economic activities of each nation, allowing "to satisfy the needs of present generations without compromising the satisfaction of the needs of future generations" in accordance with what was established by the Bruntland commission in 1987, when considering intergenerational equity, a fair distribution of environmental benefits and costs (ONU, 2023).

The comprehensive approach between equality and sustainability involves imposing structural changes in the economic system, which requires important social pacts. Therefore, in September 2015, world leaders adopted a set of global objectives to unify the international process towards sustainable development and the international development agenda to integrate and achieve sustainable development as a fundamental concept that goes beyond caring for the environment and with specific goals to be achieved in the subsequent 15 years (ONU, 2023).

The growing integration and interaction between countries, people and institutions have allowed the economic, political, social, technological and cultural processes to integrate the care of the environment; however, during this process the exploitation of resources has intensified due to the fiscal conditions of various countries and in the absence of environmental procedures. Therefore, environmental problems have come to occupy one of the main places in international priorities (del Valle Inclán, 2016).

In this sense, Ramírez (2020), denotes the interest and awareness in achieving sustainable development in the face of threats that have led countries, international organizations and decision makers to examine means that allow them to evaluate, monitor and control the evolution of environmental trends according to development processes through the use of sustainability indicators as a measure for monitoring and defining new policies, actions and implementation of sustainable strategies and the analysis of costs and benefits.

Globalization in business has made the panorama more complex; the increase in competitors, access to new markets and the development of new technologies have led businesses to focus efforts on building an identity that strengthens their competencies, based on the competitive advantage that, according to Porter, exists when there is an equivalence in the various distinctive and convincing advantages to settle in a place (Gómez, Cuevas, & Soto, 2022).

For the industry, this means decentralization and flexibility in its productive units to diversify the production of goods and shorten their life cycles. In this sense, a globalization strategy finds support in modern technologies as one of the ways to reduce utilization. of production processes or polluting energies, with viable alternatives capable of being integrated into development and thus reducing emissions and waste to the environment (Ramos, 2020).

Therefore, innovation is considered one of the key components of sustainable development to find new technologies and ways to solve environmental issues through effective solutions, for the United Nations (2019), innovation is "the process of using knowledge and technology to develop or improve the production or performance of products, services and processes that have value in commercial terms, social impact or benefit".

In this sense, technology aims to search for efficient resources that better satisfy existing consumer demand while addressing new needs or existing unsatisfied needs, likewise according to Forbes (2016), the importance of innovation and therefore of technology lies in the value that organizations give it since they seek to advance in a world that constantly changes and, as mentioned above, in building an identity that strengthens their competencies, however, some of these changes They can represent threats to the environment and society, which has a direct impact on the well-being of the organization since its integrity, values and transparency can be questioned.

Historically, technology causes great changes in the landscapes of humanity, through Research and Development processes to promote continuous improvement, technological changes are considered important since they are part of the industries and help the economy of the countries grow, for example, by renewing machinery, processes or in the implementation of new methods or models that facilitate production, these new implementations in turn contribute to saving time, improving efficiency and in some way optimizing quality by promoting that the economy advances more quickly.

In a sustainable context, when studying technological innovation, the analysis of the processes of generation, adoption, implementation, monitoring and evaluation of technologies considered green or clean must be included (Gómez, Cuevas, & Soto, 2022). From a development perspective, green technologies related to sustainable development are a fundamental element in the competitiveness of companies since their use will allow them to position themselves in the demands of changing markets, but at the same time it offers them the improvement of their efficiency in the different sectors of the country and with this the reduction of its environmental impact (Cuevas, 2016).

In this sense, green technology is defined as the study of techniques that allow the growth of life through the creation, design or construction of artifacts, devices or objects that allow life to be preserved and promote its growth through the use of scientific knowledge, reducing or mitigating the environmental impact through various technological advances or with the creation of new technologies (Mora, 2022).

Therefore, green technology is the set of industrial instruments and procedures of a certain sector or product that have beneficial properties for humans and that are not harmful to the environment. It is important to note that the term was introduced by the agenda 21 of the United Nations Conference in 1992, to designate those technologies that protect the environment, that are less polluting, those that use all resources in a sustainable way, recycle their waste and products and treat residual waste in a more sustainable way. acceptable compared to the technologies they replaced, not only referring to isolated technologies, but to those that encompass comprehensive systems established by the Paris Agreement in 2015 to reduce greenhouse gas emissions, through mitigation, adaptation and resilience of ecosystems to the effects of global warming (Sass & Carboni, 2020).

In this way, green technologies are defined as goods or services that have the purpose of measuring, preventing and limiting the generation of pollution with the aim of improving environmental conditions in water, soil, waste and noise pollution in an affordable way., adaptable and available in the different markets of developing countries for use, distribution and export, and focus on conserving and protecting the environment to reduce pollution through its use in different business activities that meet the needs of the human being.

In a broader sense, green technologies make use of various tools and strategies that help minimize waste and prevent pollution since according to Villagaray & Bautista (2011), these technologies involve processes that consume less materials or energy, which allows low waste generation compared to existing technologies.

The definition is expanded by Hottenrot, Rexhausser, & Veugelers (2016), who consider green technologies as those that reduce CO2 (carbon dioxide) emissions by replacing the use of fossil fuels or by making efficient use of them. of the same with the purpose of reducing Greenhouse Gas (GHG) emissions through sustainable innovation that involves materials that save and make the use of resources efficient through integrated processes, which favors green efficiency or eco-efficiency which is defined by Wossen, Muller, & Nichterlein (2019), as the maximum use and development of natural resources while minimizing environmental damage, since the creation of sustainable value prevails in the production of goods and services, in this context the use of these technologies in companies, favors organizational change by understanding the application of new business practices, changes in production and quality improvement (Sass & Carboni, 2020).

The main objective of the use of green technologies is to control climate change, protect the environment and reduce dependence on non-renewable resources through the development of innovative inventions that do not affect natural resources, for Qamar, Noor, Ali, & Obaid (2020), these technologies continuously develop products, systems and equipment that:

- They limit the consumption of non-renewable materials; therefore, they reduce the possibility of depleting resources in the understanding that the world has limited resources.
- They considerably prevent GHG emissions.
- They promote energy production using renewable resources and energies.

These objectives within organizations create competitive advantages since they represent new starting points through which new business models can be implemented whose purpose is to incorporate sustainable development throughout the value chain of different products and services. that are offered, for this it is necessary that companies identify the needs of their consumers in relation to sustainable development and thus modify or adapt their operations using new technologies that allow them to reduce their environmental impact. According to Vivanco (2020), these technologies have the property of renewing themselves despite their use, however, it is important to mention that their rate of exploitation can also exceed overuse, which is why regeneration occurs naturally. or artificially.

Under this context, technological change then represents a fundamental force to shape the transformation patterns of the economy and the restructuring of old industrial branches to create new ones that make it possible to integrate new processes and devalue traditional raw materials from non-renewable sources, in this sense green technological innovation is a process that, through creativity and implementation of knowledge based on organizational strategies, will allow us to achieve the objectives of sustainable development (Ramos, 2020).

According to Karimi, Sayyadi, & Shahabaldini (2021), research in green technological innovations is of great importance in terms of sustainability and economic profitability since it guides companies towards achieving sustainable competitive advantages, and increases the participation of companies in the market, attract different customer segments and provide a differentiating element to the goods or services offered.

For Castellacci & Mee (2017), green innovation, also known as eco-innovation, is defined as a process that contributes to the creation of new technologies and products with the aim of reducing environmental risks such as pollution and the negative consequences of resource exploitation. It is divided into green innovation. a product or service and process innovation, which have the objective of improving the functioning of products or services for consumers and clients based on cost efficiency and organizational flexibility, likewise, the main objectives of green innovation fall on:

- Contribute to mitigating environmental risks.
- Increase resource efficiency.
- Create new opportunities to exercise environmentally friendly practices.
- Reduce pollution rates.
- Increase recycling.
- Achieve competitive advantages in the company and sector.
- Improve the environmental performance of companies.
- Support strategic objectives through improving positive performance in organizations.
- Improve the quality-of-service delivery.
- Boost green economic performance.
- Design and produce environmentally friendly products or services.
- Generate a beneficial ecological reputation.
- Reduce the use of dangerous products and non-operational technologies, among others.

Therefore, green technological innovations are important tools that can help society, organizations and companies achieve environmental sustainability as they play an important role in achieving competitive advantage (Chu, Wang, & Lai, 2018), and improve economic performance, by facing the challenges of technological innovation and the environment, however, it is important to mention that organizational success in terms of sustainable innovation requires the acceptance of green technologies, organizational collaboration and the implementation of a sustainable culture in companies.

In productive terms when we consider green innovation in processes, technology has the possibility of becoming a tool to make a change towards sustainable development through four ways in which technological innovation can be integrated into the productive processes that involve:

I. Substitution of inputs; when materials are used that produce less pollution or waste or introduce a smaller quantity of the same polluting materials, reducing the magnitude of waste.
II. Pollutant control and prevention technologies; those that deal with emissions resulting from the production process.
III. Recycled in the middle of the process; This involves the recovery of materials used in the same production cycle.
IV. Radically new production processes: This requires radical innovation and uses alternatives such as green or clean energy or technologies, change in raw materials, and process reengineering can even be considered (Ramos, 2020).

The use of green technologies in companies also implies certain disadvantages, since their creation they tend to be more expensive than conventional technologies, which causes low demand for them, which in turn translates into little availability or low reach of these. For that, Del Río (2003), proposes a triangular model of the factors that influence companies to make the decision to adopt green technologies, which consist of applying a dynamic and systemic approach to technological change and the impacts they have on the following factors:

- Proactive environmental strategies. Through the setting of environmental objectives and the establishment of an organizational culture that clearly defines environmental protection responsibilities. In this context, the adoption of green technologies as part of the incorporation of sustainable development should be considered one of the priorities of those in charge of company administration, for this reason, company management represents a fundamental part for the consideration of business opportunities that involve the implementation of technological strategies.
- Technological capacity. The lack of technological skills is related to the main obstacle to the adoption of green technology innovations, this is related to the technical capacity of operational personnel and their ability to operate a new technology, such is the case of green technologies, which due to Their nature tends to be complex and sophisticated due to the substantial change in the production process; therefore, to choose a green technology, companies must have the necessary information about technological alternatives as well as receive relevant training in the use of the technologies. themselves, this must be accompanied by the correct identification of the technology to be used in the market.
- The competitive structure. It can affect the decision of companies to implement sustainable alternatives due to the lack of strategies and organization of the company, this considers its infrastructure and training of designated personnel in order to make the most of the opportunities offered by the new technologies (Del Río, 2003).
- Financial situation. Wood & Berge cited by Del Río (2003), indicate that new technologies tend to be more expensive than conventional technologies due to their low availability derived from the low demand they have since they generally entail extra expenses for their implementation, installation, training and use, because sometimes they can modify processes from start to finish, which is why it is important for companies to organize and manage their financial resources so that investment in green technological innovations is possible.

It is important to mention that the factors mentioned above depend on the current regulations regarding sustainable development, as well as the corporate strategies and objectives accompanied by the management vision, as this will allow identifying the most convenient technology according to the productive activities. of each company and according to its needs and search for new competitive strategies. On the other hand, among the external factors of companies that influence the decision to use green technologies, there are mainly environmental behavior and pressure from outside that directly influence the decisions of companies.

As mentioned above, the adoption and/or use of green technological innovations has different techno-economic characteristics that give rise to tangible and intangible benefits. However, despite what has been described, there are different barriers to the use of this type of technology. These mainly include:

- Uncertainty and risks associated with environmental, governmental and social regulations, as well as acceptance by consumers and different psychosocial factors that may not allow their acceptance
- Drastic changes in companies that staff are not willing to accept.
- Market uncertainty.
- Investment and transition costs.

- Adoption of a sustainable culture by companies, society, and governments.
- Expectations.

In addition to the advantages and disadvantages of adopting green technologies in companies, technological innovations are increasingly used in organizations with the purpose of contributing to sustainable development, on the one hand, the more entrepreneurs commit to the protection of the environment, the more innovative they will be in the implementation of green innovations (Suasana & Ekawati, 2018).

For green technological innovations to transcend with true social appropriation, a strategy and support of education, training, transfer and monitoring carried out in a transcendental and continuous manner is required. It is also important to mention that within the technological environment the role of green technology policy which will seek to achieve environmental, economic, and social objectives to confront globalization but at the same time address climate change (Gavito, y otros).

Derived from the above, Galleguillos & Santelices (2012), point out that the design of green technologies in the future is based on the following:

- Advanced sensors.
- Biotechnology.
- Green technologies applied in automobiles.
- Recycling of products.
- Intelligent water purification
- Intelligent waste treatment.
- Clean industrial processes and micro manufacturing.
- New renewable energy technologies.
- Photovoltaic cells.

The 2030 Agenda for Sustainable Development and the Sustainable Development Goals (SDGs) raise the urgency of moving towards new models of growth and development, with more sustainable and inclusive consumption and production patterns, with the purpose of encouraging the use of services and products that respond to the basic needs of the population and the quality of life and that at the same time allow reducing the use of natural resources and toxic materials, as well as the emissions of waste and polluting substances during the life cycle of the service or product, in this regard, the development of science, technology and innovation play a fundamental role in the search for sustainable production models (Comisión Económica para América Latina y el Caribe, 2021).

Consumption and quality of life expectations can only be met in a sustainable way if green innovation is used accompanied by technological change where companies must improve efficiency in the use of natural resources and reduce pollution through the incorporation of technologies. green in new or existing processes, products and organizational and commercial practices that are more respectful of the environment, as a consequence of these changes, they will have added value that will make them more competitive in the markets.

OBJECTIVE AND RESEARCH QUESTIONS

Therefore, the general objective of this chapter is to identify green technological innovations that have emerged as a tool for organizations to assume their commitment to sustainable development by obtaining economic, social and environmental benefits by incorporating them into their production processes.

The specific objectives of the chapter are:

- Describe the fourth industrial revolution and the progress that technological innovation has had with it.
- Point out the conceptualization of the concept of sustainable development and as a derivative of this, green technological innovations have emerged.

The general question of this chapter is: ¿What green technological innovations have emerged as a tool for organizations to assume their commitment to sustainable development by obtaining economic, social, and environmental benefits by incorporating them into their production processes?

The specific questions of the chapter are:

- What is the fourth industrial revolution and what has been the progress that technological innovation has had with it?
- What is sustainable development and how has it caused the emergence of green technological innovations?

METHOD

The research is documentary type, since an in-depth analysis of the literature found in the main databases and with the highest number of citations between 2015-2022 was carried out, it is also qualitative, since Marshall & Rossman (2016), they point out that this allows us to explore the complexity and richness of the phenomena, through in-depth data collection and detailed analysis.

Likewise, it is observational since Bogdan & Biklen (2007), mention that this type of research allows the collection of information about human and social behavior through the systematic and structured observation of events, behaviors, activities and situations in their context. natural; characterized by being flexible and adaptable to different situations. It is descriptive since it seeks to refer to and understand the characteristics of a particular phenomenon by identifying its main characteristics, which allows patterns and trends to be identified (Creswell & Poth, 2017).

The process was carried out as follows:

- Search and selection of existing literature in various sources of scientific information regarding the fourth industrial revolution, sustainable development and green technological innovation from 2015-2022.
- Critical analysis of the information, through which connections between themes were found and relationships between them were established, by structuring the information to discover patterns, trends and key concepts.

- Synthesis of information in which the analyzed information was integrated, and a deeper understanding of the topics was generated. The main findings were identified, and the different perspectives of the authors were compared.

THE APPLICATION OF GREEN TECHNOLOGICAL INNOVATIONS IN ORGANIZATIONS.

The environmental problems that affect globalized society have caused the various sectors of the economy to define actions and standards that minimize the pollution rate, caused mainly by the activities emanating from industries (Silva, 2011). Likewise, the growth of societies and the emergence of green technologies seek to be a solution for organizations to continue developing and at the same time minimize their environmental impact, making them more competitive (Paredes & Tigrera, 2006).

In this sense, for the application of green technological innovations in organizations, it is necessary to promote managerial management with an integrative vision, where the environmental dimension and technological development make it possible to achieve growth in terms of knowledge, organization, innovation, creativity, technical and economic efficiency. Considering the environmental dimension means for organizations a change in management and business culture that allows them to understand that quality can be improved and expanded.

In this way, the application of green technological innovations must be supported by effective processes of innovation and permanent knowledge transfer, which represents a new challenge that organizations must take on (UNCTAD, 2023) Therefore, there is a need and, consequently, an opportunity for the development and diffusion of these types of innovations in response to climate change, to the point that human survival depends on the improvement and diffusion of technology.

Recent green technology innovations include energy storage, light-emitting diodes, solar energy, and carbon capture and storage. The cost of climate change adaptation, deployment and mitigation will depend on how they are implemented and managed. One of the main challenges for this is determining whether private markets or patents or shared resources will be used to achieve sustainable development. This decision depends on national and international policies and the management adopted by the organization in search of new and creative approaches to take advantage of technological ecological opportunities (Lema, Fu, & Rabellotti, 2020).

In this sense, and considering that until 2030 the world population will increase, global consumption expectations will be impacted by green technological innovations that support the preservation of the environment, some that have been implemented are mentioned below:

- Elimination of dioxide carbon, by using injection methods that allow it not to reach the atmosphere, likewise, technological innovations hope to be able to separate it in an elemental way to mitigate the polluting effect it has.
- Recovery and bio-elimination technologies of contaminants in the environment, through the use of microorganisms and plants that allow contamination levels to be assimilated through bioremediation (microbes) and phyto-remediation (plants).
- Harnessing tidal and wave energy from the energy generated by the movement and collision of waves in the oceans.

- OTEC (Ocean Thermal Energy Conversion) technologies convert the thermal energy contained in the oceans into electricity using the temperature difference between the surface of the water (hot) and in the depths (cold).
- Harnessing the power of hydrogen, as a green alternative to traditional fossil fuels, it can be harnessed by generating energy from an electrochemical reaction between oxygen and hydrogen.
- Greater advances in the desalination process where salt and minerals are extracted from seawater to obtain drinking water, among some others that currently have applications and that focus mainly on the energy sector (Instituto Nacional de Contadores Públicos, 2022).

In this same sense, Wu & Strezov (2023) point out that the application of green technological innovations in organizations is classified as follows:

- Concept, theory, metric, analysis
    - Analysis of green and sustainable systems
    - Green metrics
    - Sustainable green environments
    - New green and sustainability concepts
    - Carbon neutrality
    - Energy efficiency and resource efficiency
    - Mathematical models for green and sustainable technologies
    - Analysis, evaluation, and life cycle approaches
    - Green sustainable engineering
    - Modeling and theory for green and sustainable management
    - Circular economy
    - Dynamic complex and sustainable environmental/economic/social systems
- Issues for sustainable development goals
- Architecture, designs, algorithms, approaches
    - Architecture, strategies, algorithms, protocols, programming and/or ecological and sustainable designs
    - Measurement and profiling of ecological problems
    - Low cost and energy efficient designs
    - Performance characterization, evaluation, optimization and design Tradeoffs
    - Use of cognitive principles to achieve ecological goals
    - Relevant experimental testbeds and results
    - Relevant standardization, policy and regulation
    - Pollution control technologies
    - Recycling and reuse
    - Strategic management for sustainability
- Approaches to achieve sustainable development goals.
    - Domain technologies.
        - Environmental engineering.
        - Air and water pollution and control, including sustainable wastewater treatment.
        - Generation, separation, processing and disposal of green and sustainable waste.

    - Sustainable waste management technologies that promote the reuse and recycling of waste.
    - Green and sustainable manufacturing.
    - Green and sustainable chemical engineering.
    - Sustainable transport and logistics.
    - Environmental management and decision making for sustainability.
    - Carbon neutral technologies.
    - Detection, evaluation, and control of noise pollution.
    - Ocean and coastal pollution and mitigation.
    - Green and sustainable supply chain management.
- Green electronics, devices, and systems.
- Environmental sciences.
    - Green chemistry.
    - Soil, land contamination and remediation.
    - Environmental modeling.
    - Life cycle impact assessment and sustainability assessment.
    - Green and sustainable water processing.
- Green energy technologies.
    - Renewable and sustainable energy, including solar, wind, hydroelectric, geothermal, biomass and ocean energy, and renewable energy and hydrogen systems.
    - Green and sustainable energy for transportation, including biofuels, electric vehicles and hydrogen-powered vehicle solution.
    - Generation and transmission of green, clean, and sustainable energy
    - Battery and energy storage technologies.
    - High performance sustainable energy systems.
    - Smart grids and networks of smart grids.
    - Advanced metering infrastructure and smart meter technologies.
    - Energy harvesting, recycling and wireless energy transfer.
- Green digital and information technologies.
    - Green and sustainable communications, computing, and systems.
    - Big data and artificial intelligence meet ecological challenges.
    - Environmental imaging assisted by artificial intelligence or data analysis and decision making.
    - Information and communications technologies for sustainable development goals.
    - Mitigation of electromagnetic pollution.
    - Sustainable security and privacy.
    - Security and privacy issues for green and sustainable technologies.
    - Green and sustainable intelligent transportation systems.
    - Green data centers and cloud computing.
- Green finance and economy.
- Green sustainable materials.
- Green and sustainable construction and civil engineering.
- Green sustainable cities.

It is important to note that green technological innovations can be different depending on the industry to which a company belongs. Furthermore, with the emerging importance of eco-innovation, researchers have approached eco-innovation from different perspectives. An analysis of the existing literature carried out by Bigliardi & Bertolini (2012) point out that the application of green technological innovation depends on:

- The factors driving green technological innovation and the performance outcomes that arise from this.
- The dimensions of green technological innovation.

Based on the aforementioned, the need to characterize the external environment such as government environmental protection laws and regulations for the application of green technological innovations becomes imperative since companies may be forced to carry out innovation for a external issue. However, to achieve short-term performance results, some companies can apply green technological innovations in a proactive way, allowing managers to perceive greater economic, social, and environmental benefits by considering that good management of these will be reflected. in consumer preferences, greater market acceptance, a source of competitive advantage and greater prestige and brand recognition.

As shown in Figure 1, the results of the reviewed literature suggest important implications for business strategy since decision makers need to apply green technological innovations to improve organizational performance. However, the application of these innovations carries an element of risk: with green technological innovation in products and processes it is necessary to consider both the cost of inputs and the costs of conversion and risks of consumer acceptance (Tang, Walsh, Lerner, Fitza, & Li, 2017). Therefore, companies face an increasingly changing market and a group of consumers that grows more concerned every day about caring for the environment, where this is already an element to consider when choosing a product that meets their needs (Nikolai, 2009). Added to the above is the fact that companies are finding more and more barriers to obtaining raw materials for the production of goods and/or services; so, green technological innovation leads to an increase in the efficient use of inputs and greater efficiency in the conversion process, therefore, it depends less on external factors of the company and, the environment is not an afterthought dissociated but a decision with strategic significance.

In this sense, for the application of green technological innovations in organizations, it is necessary to identify the sector to which it belongs, to know the innovations of this type that can be applied, establishing networks of experts in the different specialized technological areas or related to green technological innovation, optimizing the transfer of specialized knowledge, which results in faster progress in the learning curve and would guarantee a much faster introduction to the market of green technological innovations so that they can begin in the short term with the obtaining of economic, social and environmental benefits.

SOLUTIONS AND RECOMMENDATIONS

Green technological innovations are drivers of environmental and socio-ecological transformation and are created by the conjunction of scientific and technical capabilities. The scientific community points out that the application of this type of technology requires creative, innovative management with firm foundations in sustainable development that establishes methods, processes and tools to incorporate them into the organization, not only in production processes, but also in all areas of this, such as the administrative or managerial part, logistics, finance, marketing, among others. For the above, it is necessary

Figure 1. Published articles in the last 20 years
Source: Aponte (2021)

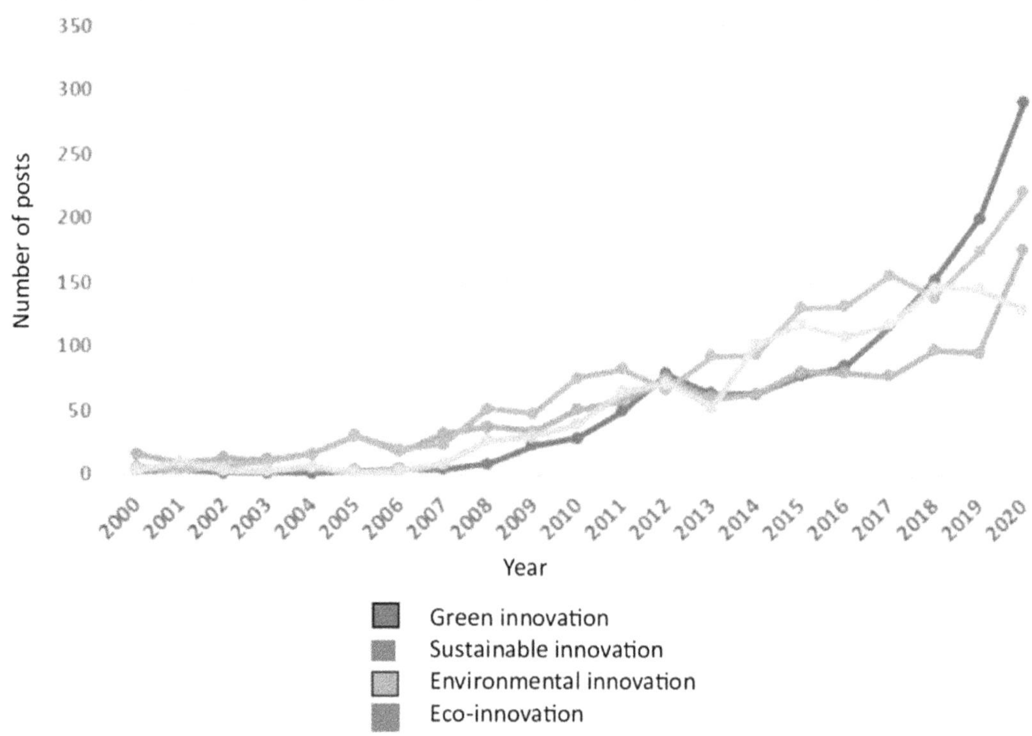

to be clear about the sector to which it belongs to identify which are the most appropriate technological innovations for the sector, know the cost-benefit relationship, the return on investment, and above all the social and environmental benefits that are derived from these so that decision makers see their application as viable and profitable.

FUTURE RESEARCH DIRECTIONS

Scientists worldwide should be more involved in the construction of methods and tools that allow improving the application of green technological innovations in the productive sector from a management point of view, not only at a technical or operational level, for this it is necessary to promote strategic alliances between company-company and university-company. In this same way, international standards that correspond to the contexts of groups of developed, emerging and undeveloped countries are required to facilitate the implementation of green technological innovations based on the development and growth of each country. Probably, the greatest challenge that scientists and researchers have is that to contribute to the construction of sustainability, they have to work together with politicians, producers, businessmen, teachers, rural and urban communities, among other actors and social sectors.

CONCLUSION

Green technological innovation is a term whose rise is closely associated with the obligation that countries have to comply with the Sustainable Development Goals. In that sense, these innovations are those goods or services that have the purpose of measuring, preventing and limiting the generation of pollution with the aim of improving environmental conditions in water, soil, waste and noise pollution in an affordable, adaptable way and available in the different markets of developing countries for use, distribution and export, and focus on conserving and protecting the environment to reduce pollution through its use in different business activities.

This term is not the only one used in the literature to distinguish this type of innovations; the terms environmental innovation, sustainable innovation and eco-innovation are also used interchangeably. Other terms used in the literature that are related to green innovation are: Green technology, environmental technologies, climate-related technologies and adaptation and mitigation/environmentally friendly technologies, energy saving, pollution prevention, waste recycling, green product design.

To the extent that environmental aspects have gained importance in the development model of countries, they have sought ways to monitor the progress of their achievements in this matter, as well as optimize the learning curve to obtain better results in less time. Therefore, green technological innovations are considered pressing for the development and growth of organizations, their advancement and implementation is increasingly reflected through environmental certifications, published patents, and the capacity to generate renewable energy. and foreign direct investments in matters related to technology or green innovation. Technological green innovations are a reality that is advancing more and more and to which a greater number of companies in the world are joining. Thus, it is more common to see projects that involve sustainable development in their business strategy through green technological innovations since the transition towards a green economy implies a true technological revolution that will have a profound impact on production structures and patterns of consumption.

REFERENCES

Aponte, F. G. (2021). Innovación verde: Indicadores y principales tendencias. *Gestión I+D*, 12-48.

Bigliardi, B., & Bertolini, M. (2012). Green innovation management: Theory and practice. *European Journal of Innovation Management, 15*(4). Advance online publication. doi:10.1108/ejim.2012.22015daa.001

Blanco, F., Castro, J. M., Gayoso, R. A., & Santana, W. (2020). *Las claves de la cuarta revolución industrial : ¿cómo afectará a los negocios y a las personas?* ECOE Ediciones.

Blanco, R., Fontrodona, J., & Poveda, C. (2017). *La Industria 4.0: El estado de la cuestión*. Academic Press.

Bogdan, R., & Biklen, S. K. (2007). Investigación Cualitativa en Educación: Una introducción a teoría y métodos (4ta. ed.). Alfaomega.

Castellacci, F., & Mee, C. (2017). A taxonomy of green innovators: Empirical evidence from South Korea. *Journal of Cleaner Production, 143*, 1036–1047. doi:10.1016/j.jclepro.2016.12.016

Chu, Z., Wang, L., & Lai, F. (2018). Customer pressure and green innovations at third party logistics providers in China: The moderation effect of organizational culture. *International Journal of Logistics Management, 30*(1), 57–75. doi:10.1108/IJLM-11-2017-0294

Colotla, I., Fæste, A., Heidemann, A., Winther, A., Andersen, P., Duvold, T., & Hansen, M. (2016). *Winning the Industry 4.0 race. How ready are Danish manufacturers?* The Boston Consulting Group.

Comisión Económica para América Latina y el Caribe. (2021). *Innovación para el Desarrollo.* Obtenido de Conferencia de Ciencia, Innovación y Tecnologías de la Información y las Comunicaciones de CEPAL: https://innovalac.cepal.org/3/sites/innovalac3/files/c2100805_web.pdf

Concha, J. M., & Monsalvo, M. G. (2020). *Cuarta Revolución Industrial y el Trabajo en México.* UNAM. Ciudad de México: Dirección General de Bibliotecas. Tesis Digitales. Obtenido de http://132.248.9.195/ptd2020/enero/0800339/Index.html

Creswell, J. W., & Poth, C. N. (2017). *Qualitative Inquiry and Research Design: Choosing Among Five Approaches.* Sage Publications.

Cuevas, I. (2016). *El uso de las Tecnologías Verdes y sus efectos en el desempeño del sector agroindustrial en México.* Obtenido de Repositorio Instituto Politécnico Nacional: https://tesis.ipn.mx/handle/123456789/22783

Del Río, P. (2003). La adopción y difusión de tecnologías limpias. *Economía Industrial*, 47-58. Obtenido de Economía Industrial: https://digital.csic.es/handle/10261/10419

del Valle Inclán, D. (2016). La globalización y su incidencia en el desarrollo sostenible. In A. Hérnandez, H. González, & G. Tamez (Eds.), *Desarrollo sustentable: de la teoría a la práctica (págs* (pp. 37–56). Universidad Auntónoma de Nuevo León.

Dutrénit, G. (2020). In R. Álavrez (Ed.), *Nuevas tecnologías, industria 4.0 y sociedad* (pp. 6–7). Ciencia.

Forbes, D. (2016). How Can We Define 'Innovation'? *Entrepreneur & Innovation Exchange, 1-4.* Advance online publication. doi:10.17919/X9DW2X

Galicia, R. (2015). Innovación Tecnológica. *Ecorfan Journal.* Obtenido de https://www.ecorfan.org/proceedings/CTI_I/3.pdf

Galleguillos, L., & Santelices, I. (2012). Cambio tecnológico y desarrollo sostenible. *Revista Ingeniería Industrial, 1*(1), 59-66. Obtenido de https://dialnet.unirioja.es/descarga/articulo/3995770.pdf

García, M., & Oliveira, A. (2023). Industria 4.0: ¿una nueva revolución tecnológica? (R. Álvarez, Ed.) *Ciencia. Nuevas tecnologías, industria 4.0 y sociedad, 74*(2). Obtenido de https://www.revistaciencia.amc.edu.mx/images/revista/74_2/PDF/Ciencia74-2.pdf

Gavito, M., van der Wal, H., Aldasoro, M., Ayala, B., Bullén, A., & Cach, M. (n.d.). *Ecología, tecnología e innovación para la sustentabilidad: retos y perspectivas en México.* Academic Press.

Gómez, J., Cuevas, I., & Soto, M. d. (2022). El cambio tecnológico y su incidencia en la incorporación de estrategias de sustentabilidad en la Industria de los Detergentes en México. *Transformación digital como propuesta de valor para la competitividad, 16*(16), 89-106. Obtenido de https://www.riico.net/index.php/riico/article/view/2102/1918

Guoping, L., Yun, H., & Aizhi, W. (2017). Fourth Industrial Revolution: Technological Drivers. *Impacts and Coping Methods., 27*(4), 626–637.

Hottenrot, H., Rexhausser, S., & Veugelers, R. (2016). Organisational change and the productivity effects of green technology adoption. *Resource and Energy Economics, 43*, 172–194. doi:10.1016/j.reseneeco.2016.01.004

Instituto Nacional de Contadores Públicos. (2022). *Las 10 tecnologías verdes que pueden ayudar a preservar el medio ambiente.* Obtenido de https://incp.org.co/las-10-tecnologias-verdes-que-pueden-ayudar-a-preservar-el-medio-ambiente/#:~:text=Estas%20tecnolog%C3%ADas%20incluyen%20bio%2Dremediaci%C3%B3n,si%20mismo%20con%20el%20tiempo

Juré, G., & Fittipaldi, I. (2017). Del origen y tratamiento de la cuestión ambiental a la propuesta del desarrollo sustentable. In Introducción al desarrollo sustentable (pp. 63-84). Universidad Nacional Arturo Jauretche.

Karimi, S., Sayyadi, H., & Shahabaldini, Z. (2021). Green innovation: A systematic literature review. *Journal of Cleaner Production, 279*, 122474. Advance online publication. doi:10.1016/j.jclepro.2020.122474

Landes, D. (1979). *Proceso Tecnológico y Revolución Industrial.* Tecnos.

Lema, R., Fu, X., & Rabellotti, R. (2020). Green windows of opportunity: Latecomer development in the age of transformation toward sustainability. *Industrial and Corporate Change*, 1193–1209.

Marshall, C., & Rossman, G. B. (2016). *Designing qualitative research.* Sage Publications.

Martín, M. (2018). ¿Por qué Gran Bretaña?: Revolución. *Universidad de Valladolid. Facultad de Ciencias Económicas y Empresariales. Grado en Marketing e Investigación de mercados.*

Maximo, B. (2017). *Impacto industrial y laboral.* CCOO de Industria.

Mora, H. (2022). *Diccionario Etimológico.* Obtenido de Tecnologías: http://etimologias.dechile.net/?tecnologi.a

Naciones Unidas. (2019). *Science, technology & innovation capacity development course.* Obtenido de United Nations Conference on Trade and Development: https://unctad.org/system/files/official-document/dtlstictinf2019d1_en.pdf

Nikolai, S. (2009). *Green Technologies: Opportunities for South-South Trade.* International Trade Forum.

OECD. (2005). *Manual de OSLO.* OECD.

ONU. (2023). *Objetivos del Desarrollo Sostenible.* Obtenido de https://www.un.org/sustainabledevelopment/es/objetivos-de-desarrollo-sostenible/

ONU. (2023). *Sostenibilidad*. Obtenido de https://www.un.org/es/impacto-acad%C3%A9mico/sostenibilidad#:~:text=En%201987%2C%20la%20Comisi%C3%B3n%20Brundtland,mundo%20que%20buscan%20formas%20de

ONUDI. (2020). *Informe sobre el Desarrollo Industrial 2020. La industrialización en la era digital.* Viena.: ONUDI.

Organización Mundial de la Propiedad Intelectual. (2022). *Índice Mundial de Innovación.* Obtenido de https://www.wipo.int/global_innovation_index/es/#:~:text=El%20%C3%8Dndice%20Mundial%20de%20Innovaci%C3%B3n%202022%20toma%20el%20pulso%20de,de%20suministro%20y%20la%20energ%C3%ADa

Otero, A. S., & de Grossi, M. (1998). *La llamada revolución industrial.* Universidad Catolica Andres. Obtenido de https://books.google.es/books?id=YmbEneoFEI0C&dq=revoluci%C3%B3n+industrial+definicion&lr=&hl=es&source=gbs_navlinks_s

Paredes, C. A., & Tigrera, C. R. (2006). Gerencia del proceso de innovación: Un aporte para el desarrollo de tecnologías verdes en el Parque Tecnológico Universitario. *Revista de Ciencias Sociales*, 6–23.

Peña, C., Ortega, L., & San Juan, L. (2020). Desarrollo sustentable: un detonante para el crecimiento. *Boletín Científico INVESTIGIUM de la Escuela Superior de Tizayuca, 6*(11), 15-23. Obtenido de https://repository.uaeh.edu.mx/revistas/index.php/investigium/issue/archive

Qamar, M., Noor, M., Ali, W., & Obaid, M. (2020). Green Technology and its implications Worlwide. *The inquisitive Meridian.* Obtenido de https://www.researchgate.net/publication/350443477_Green_Technology_and_its_Implications_Worldwide

Ramírez, J. (2020). *La globalización y el desarrollo sustentable.* Universidad del Valle de México.

Ramos, D. (2020). *Innovación en tecnología verde y las restricciones estructurales para su desarrollo en los países de periferia.* Universidad Nacional Autónoma de México. Obtenido de http://ru.iiec.unam.mx/5251/1/2-096-Ramos.pdf

Real Academia Española. (2023). *Real Academia Española.* Obtenido de https://dle.rae.es/revoluci%C3%B3n

RIDAA. (1996). La innovación tecnológica: definiciones y elementos de base. *Redes, 6*(3), 131-175. Obtenido de https://ridaa.unq.edu.ar/bitstream/handle/20.500.11807/500/07R1996v3n6.pdf?sequence=1&isAllowed=y

Ruessmann, M., Lorenz, M., Gerbert, P., Waldner, M., Engel, P., Harnisch, M., & Justus, J. (2015). *Industria 4.0: El futuro de la productividad y el crecimiento en las industrias manufactureras.* Obtenido de The Boston Consulting Group.: https://www.bcg.com/publications/2015/engineered_products_project_business_industry_4_future_productivity_growth_manufacturing_industries

Sánchez, I. (2019). La Cuarta Revolución Industrial en México. *OBSERVATORIO DEL DESARROLLO, 8*(24), 58-64. Obtenido de https://estudiosdeldesarrollo.mx/observatoriodeldesarrollo/wp-content/uploads/2021/02/OD245.pdf

Sass, L., & Carboni, G. (2020). Acesso a tecnologias verdes e propriedade intelectual. *JOTA Desenvolvimiento Sustentável*, 1-10. Obtenido de https://www.jota.info/opiniao-e-analise/artigos/acesso-a-tecnologias-verdes-e-propriedade-intelectual-03082020

Schwab, K. (2016). *La Cuarta Revolución Industrial*. Debate.

Silva, J. (2011). *Rol de las ecotecnologías, tecnologías limpias y de tratamiento, en el control de la contaminación generada por las curtiembres de Trujillo-Perú*. Obtenido de Tesis doctoral: https://dspace.unitru.edu.pe/handle/UNITRU/8191

Suasana, I. G., & Ekawati, N. (2018). Environmental commitment and green innovation reaching success new products of creative industry in Bali. *Journal of Business and Retail Management Research, 12*(4), 246-250. Obtenido de https://www.jbrmr.com/cdn/article_file/content_90455_18-07-04-02-02-40.pdf

Tang, M., Walsh, G., Lerner, D., Fitza, M. A., & Li, Q. (2017). Green Innovation, Managerial Concern and Firm Performance: An Empirical Study. *Business Strategy and the Environment*.

UNCTAD. (2023). *Informe sobre Tecnología e Innovación 2023*. UNCTAD.

Useche, A. J., Juarez, F., & Ramirez, A. (2022). *Tecnologías de la cuarta revolución industrial y su aplicación en la Armada Nacional de Colombia*. Universidad del Rosario. Obtenido de., doi:10.12804/urosario9789587848786

Villa de Prado, R. (2019). Consecuencias económicas y sociales de la cuarta revolución industrial y estrategias pensadas para la adaptación de la actividad económica. *Revista Aportes de la Comunicación y la Cultura*. Obtenido de http://www.scielo.org.bo/scielo.php?script=sci_arttext&pid=S2306-86712019000100010&lng=es&tlng=es

Villagaray, S., & Bautista, E. (2011). Sistemas agroforestales con tecnología limpia en los suelos del VRAEM, Perú. *Acta Nova, 2*(5), 289-311. Obtenido de http://scielo.iics.una.py/scielo.php?script=sci_arttext&pid=S2414-89382020000100097#B63

Vivanco, E. (2020). *Energías renovables y no renovables*. Obtenido de Biblioteca del Congreso Nacional de Chile: https://www.bcn.cl/obtienearchivo?id=repositorio/10221/29102/1/BCN_Energia_renovable_y_no_renovable_ventajas_y_desventajas_final.pdf

Wood, A., & Berge, K. (1997). Exporting manufactures: Human resources, natural resources,and trade policy. *The Journal of Development Studies*, *34*(1), 35–59. doi:10.1080/00220389708422502

Wossen, T., Muller, A., & Nichterlein, K. (2019). Eco-efficiency and agricultural innovation systems in developing countries: Evidence from macro-level analysis. *PLoS One*, *14*(4), e0214115. Advance online publication. doi:10.1371/journal.pone.0214115 PMID:30951548

Wu, J., & Strezov, V. (2023). *Green technologies and sustainability: A new trend*. Green Technologies and Sustainability.

Chapter 6
Creating a Sustainable Win-Win Situation:
Addressing Gender Disparity in Sustainable Technological Innovation - Advancing Inclusion and Diversity

Maria Mansab
Quaid i Azam University, Pakistan

ABSTRACT

Concerning gender diversity, the IT industry faces both a challenge and an opportunity. As in many other disciplines, women are underrepresented in the leadership of the technology sector. However, this situation also presents an opportunity: tech companies that can expand their ranks of female executives stand to gain significantly. It is critical to promote and support more women and girls to enter STEM (science, technology, engineering, and mathematics) sectors for the sake of global economies and the global transition to sustainability. There were many ways in which technological advancements aided women's capability and resource development. Women's adoption, usage, and pervasiveness of ICTs span the broad areas of outreach (e.g., health promotion), education (e.g., health literacy opportunities), lifestyle (e.g., peer coaching and planning), prevention (e.g., screening opportunities), health challenges (e.g., intimate partner violence apps), and perceptions of barriers.

INTRODUCTION

The chapter commences by emphasizing the historical backdrop, including the proclamation by the United Nations General Assembly designating February 11 as the International Day of Women and Girls in Science. The paper also examines the progression of gender ratios in STEM disciplines over multiple decades, specifically from 1955 to 2010. The inclusion of a historical perspective facilitates readers' comprehension of the enduring character of the matter at hand. The chapter presents a comprehensive perspective on the gender gap in STEM across many locations and fields by referencing studies from

DOI: 10.4018/978-1-6684-9979-5.ch006

multiple sources, such as the Proceedings of the National Academy of Sciences (PNAS), UNESCO, the STEM Equity Monitor, and the National Science Foundation. The utilization of a wide range of data sources enhances the reliability and trustworthiness of the information being delivered. The platform offers current statistical data about the disparity between genders, encompassing the proportions of women engaged in scientific, technological, engineering, and mathematical (STEM) fields, as well as their presence within academic and industrial environments. An analysis of the present condition is necessary to comprehend the current terrain. The chapter also addresses the concept of intersectionality by examining the inequalities experienced by marginalized populations, including African-Americans, Hispanics, Native Americans, and those with disabilities. This contribution enhances the breadth of the discourse and recognizes that the gender disparity in STEM is not the sole facet of inequality. The chapter encompasses regional viewpoints, emphasizing disparities in female representation across different nations and continents. This statement acknowledges the importance of considering the contextual variations in the gender gap across different locations, introducing a nuanced perspective to the ongoing topic. The chapter provides an analysis of gender inequality within the academic sphere, with a particular focus on the obstacles encountered by women in the fields of science, technology, engineering, and mathematics (STEM) about their research and publication endeavors. This analysis offers a comprehensive perspective on the gender discrepancies observed across various domains within the field of STEM.

Concerning gender diversity, the IT industry faces both a challenge and an opportunity. As in many other disciplines, women are underrepresented in the leadership of the technology sector. However, this situation also presents an opportunity: tech companies that can expand their ranks of female executives stand to gain significantly. It is critical to promote and support more women and girls to enter STEM (science, technology, engineering, and mathematics) sectors for the sake of global economies and the global transition to sustainability. There were many ways in which technological advancements aided women's capability and resource development. Women's adoption, usage, and pervasiveness of ICTs, which span the broad areas of outreach (e.g., health promotion), education (e.g., health literacy opportunities), lifestyle (e.g., peer coaching and planning), prevention (e.g., screening opportunities), health challenges (e.g., intimate partner violence apps), and perceptions of barriers (i.e., women's health).

Despite women's widespread usage of technological solutions, scholars disagreed on how best to define and apply the concept of empowerment. Empowerment is hard to assess since different research use different and often insufficient measures of empowerment. Young women and girls all across the world perform vital roles as activists and protectors of human rights. Since they are often shut out of official decision-making processes, they participate in public life through nontraditional channels like internet discussion boards. They have restricted access to resources including education, career development, and employment. Their engagement is hindered by a variety of circumstances, including climate change, ongoing conflicts, authoritarian regimes, gender and age-based discrimination, and online abuse and harassment.

Adopting a holistic human rights-based approach that is centered on child rights, gender-responsive, and intersectional is vital to ensuring a sustainable climate for girls' and young women's engagement. Authentic participation from women human rights defenders is essential, as is the provision of a safe, welcoming environment in which to plan joint initiatives to advance gender parity. calling on the global community to work together to stop the backlash against women and girls and defend their human rights. Women continue to be underrepresented in the STEM fields that are driving the digital revolution, despite the widespread skills gap that is stunting economic expansion. In 2018, women made up 33% of researchers, representing parity in the biological sciences in a number of nations. Women are underrepresented in STEM fields, with only 28% of engineering graduates and 40% of computer science graduates being female. According to UNESCO, women make up a relatively small percentage of the AI workforce (22% or less globally).

Women are underrepresented in STEM (science, technology, engineering, math) fields, as noted by the UNESCO Science Report 2021. In a technologically driven future, UNESCO predicts that advances in technology like artificial intelligence will break down previously insurmountable barriers between men and women. It demands that COVID-19-inspired improvements in work-life balance be codified into law to prevent women from spending an inordinate amount of time in unpaid caregiving, teaching, and housework. UNESCO asserts that women need equal access to education and information to compete with men for future employment.

These days, it's impossible to imagine life without some form of digital technology. online education, access to formal and informal learning, access to health and well-being information and support, the ability to engage in one's own creative and cultural practices, the expression of one's ideas and opinions, the opportunity to relax, play, and connect with peers, and the pursuit of employment, career information, and entrepreneurial opportunities. Girls are at a disadvantage when it comes to digital adoption; they do not have as much access to or use digital technology as males do, and they do not reap the same benefits from it. Women's needs can only be met with digital products and services that are created by and for women. When developing digital solutions, goods, and content, it is common practice to assume a 'default' user, one who has the same internet connection, the same device, the same digital platform, the same level of digital literacy, and the same interests and preferences as the majority of the population. Despite their best intentions, many design teams only consider men while creating their products. Fewer women than males work on the development, design, and testing of products. Therefore, women are less likely to actively pursue and implement digital solutions due to a lack of access to and/or interest in such products and services. This in turn further disadvantages women and widens the digital gender gap.

Efforts to encourage more women to pursue careers in STEM (science, technology, engineering, and mathematics) have been ongoing for as long as I can remember. However, a recent study showed that the gender gap for women in technology is currently higher than it was in 1984, despite my hope that our combined efforts have contributed to improving the share of women in leadership roles. The study, performed by Accenture, found that compared to men, women are 45% more likely to leave their careers in technology by the age of 35. Only 21% of women and 8% of women of color said they saw a future for themselves in technology, according to a survey (Accenture, 2020). A worldwide epidemic makes things worse because working women already carry a disproportionate share of the load in terms of juggling work, school, and caring for children and elderly relatives. The number of women who left the labor force in August and September of 2020 was 865,000, while the number of men was 216,000. According to new data compiled by Women in the Workplace, one-third of mothers are considering leaving the workforce entirely or significantly reducing their hours. Work schedule adjustments, career changes to something less taxing, leaves of absence, and complete retirement are all viable alternatives. The lack of services like daycare and in-person schooling has been blamed by experts for the large exodus.

What does this indicate about the IT industry's endeavors to become more inclusive and diverse? We must do much more to not only recruit but also retain outstanding female employees. And there is much more work to be done if we are to close the gender disparity and diversify the sector as a whole. Although there are numerous ways to increase inclusion and diversity in technology, I would like to see us focus on three areas. The first step is to foster an inclusive and diverse culture and then monitor progress. I am working diligently to establish a flexible work environment that allows our employees to work according to their schedules. It is impossible to exaggerate the significance of support networks, training, and mentoring. What we do to make technology more inclusive for women will help us become more inclusive as a whole.

Understanding Gender Disparity in Sustainable Technological Innovation

To advance gender parity in STEM fields, the United Nations General Assembly declared February 11 as the International Day of Women and Girls in Science. On this day, people throughout the world are encouraged to reflect on why diversity in STEM is so important and how it may be improved. This section details the historical and contemporary gender gap in STEM fields of study and professional fields. At the start of 2020, the Proceedings of the National Academy of Sciences (PNAS) published an analysis of the gender gap in scientific fields across 83 countries and 13 fields of study (Dossi, 2022). The following patterns were found in the authors' data when they looked at the publication records of scientists from 1955 to 2010:

73% of authors are men, and 27% are women, up from 12% in 1955 to 35% in 2005 across all disciplines and countries. With 18% female authors in engineering and 15% in physics and mathematics, STEM disciplines have the lowest gender ratios. Male authors published an average of 13.2 articles over the course of a career, compared to 9.6 articles for female authors, but both sexes published at the same average annual rate (1.3 articles). The disparity between the average career lengths of male and female authors, which are 9.3 years and 1.7 years, respectively, may help to explain this (Huang, Gates &Barabasi, 2020). The number of citations received after 10 years of publication per manuscript was used to calculate the academic career impact, which was found to vary substantially by gender. 30% more male authors were cited than female authors.

The 2020 UNESCO Institute of Statistics fact document titled 'Women in Science' provides additional information on how this gender disparity varies by country. The percentages for each continent are shown in the following graph. Central Asia and Latin America and the Caribbean had the highest proportions of female researchers, at 49% and 46%, respectively. In 2016, Venezuela had the highest national percentage at 61.4%. In Europe, the aggregate proportion of female researchers is lower, with the Netherlands having the lowest national average in 2017 at 26.4% (UNESCO, 2020).

The workforce WISE programs utilize Office for National Statistics labor Force data to provide statistics on women in the STEM workforce of the United Kingdom. The percentage of female STEM workers has increased from 21% in 2016 to 24% in 2019, indicating a minority of female STEM workers. In 2019, 10% of engineering professionals are female, while 46% of scientific professionals are female (Dossi, 2022). The gender gap varies by field. These numbers shed light on the current and historical gender disparities in STEM academics and industry. Although gender disparities in academic participation, especially in STEM fields, have decreased over time, the disparity still exists.

Men remain overrepresented in STEM (science, technology, engineering, and mathematics) occupations. A little over a quarter of STEM workers across all sectors is women. Women are significantly underrepresented in STEM fields among college students. According to the STEM Equity Monitor, in 2018, just 21% of STEM course enrollments and 23% of STEM course completions were made up of female students at universities and VET institutions (Ananthram, Bennett, and Bawa (2021) found that in 2018, women enrolled in and completed 61% of all non-STEM courses.

The most recent and comprehensive analysis of diversity trends in STEM employment and education was released by the National Centre for Science and Engineering Statistics (NCSES) in Diversity and

Figure 1. Global average percentage of female researchers across disciplines and regions in 2017
Source: Imperial College London

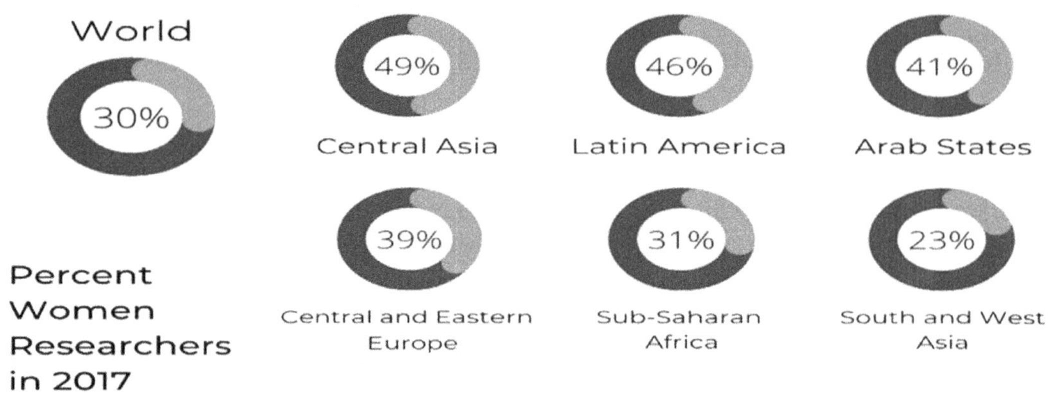

Figure 2. A significant gap in STEM
Source: STEM Equity Monitor

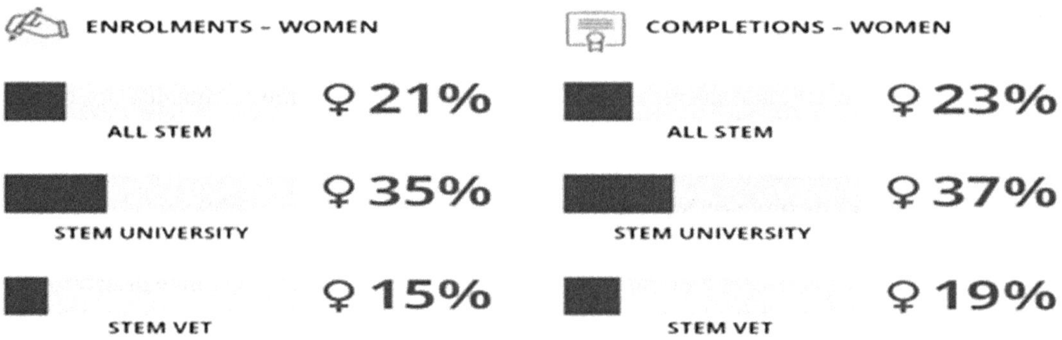

STEM: Women, Minorities, and Disabled People 2023. More women, African-Americans, Hispanics, Americans of Indian and Alaska Native descent, and Asian-Americans have worked in STEM fields and earned degrees in science and engineering at all levels, according to new studies. These groups, including people with disabilities, continue to be underrepresented in STEM fields compared to their total distribution in the U.S. population, reflecting larger equity challenges in our country. Since STEM occupations tend to pay better and have lower unemployment rates, they should be available to people of all backgrounds and abilities.

Women earned 50 percent of all bachelor's degrees in science and engineering and nearly half of all associate's degrees. 35% of the STEM workforce consisted of women, and their pay was consistently lower than men's. During the COVID-19 epidemic, more women, Hispanics, and African-Americans pursued graduate degrees in science and engineering. In 2021, minorities such as Hispanics, African-Americans, Native Americans, and Alaska Natives made up 31% of the US population but held only 24% of STEM positions. They were more likely to work in STEM jobs that required certification or technical training than in those that required a four-year degree. Employees in the STEM fields who identify as Hispanic,

Black, American Indian, or Alaska Native earn lower wages than their white and Asian counterparts. The number of Hispanic students earning associate's degrees in STEM fields has increased dramatically. Hispanic students earned 32% of all associate's degrees in STEM fields between 2011 and 2020, according to the National Science Foundation (NSF) (2023). Sixty-three percent of Hispanics in STEM fields (or 24%) are employed in jobs that don't require a bachelor's degree.

Unemployment rates for people of color in STEM fields (6.6% in 2021) and Hispanics (5.7% in 2021) will be higher than for whites (2.9%) and Asians (2.3%) that same year. Less of the United States will exist by 2020. Both in terms of raw numbers and as a share of all bachelor's degrees, a disproportionate percentage of Indian and Alaska Native students got degrees in science and engineering. Twenty-one percent of workers in 2021 with at least one impairment had STEM jobs, while disabled people made up three percent of the STEM workforce. In 2021, 11% of scientific and engineering Ph.D. recipients reported having a disability of some kind.

Women make up only 18.5% of the academic workforce in South and West Asia, while they make up 23.4% of the academic workforce in East and Southeast Asia (Ugwuegbula, 2020). These dismal statistics are indicative of a wider problem of gender inequality in academia, especially in the STEM

Figure 3. US STEM workforce characteristics
Source: National Science foundation

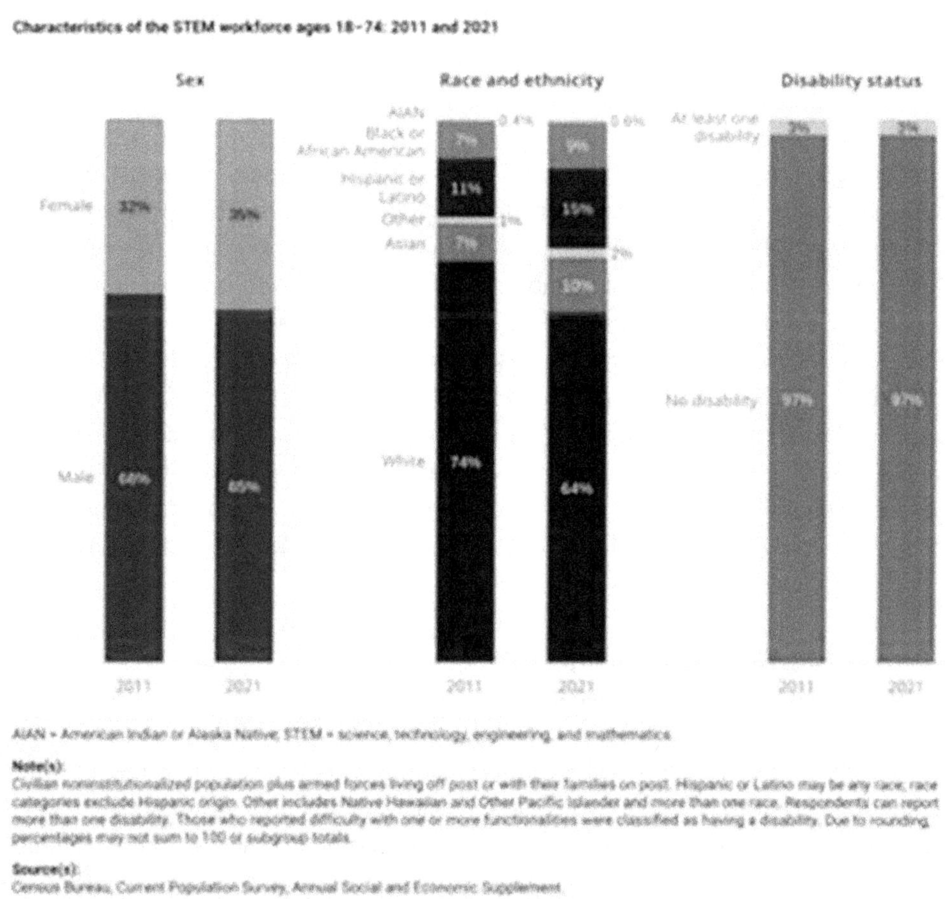

fields. Due to the increasing demand for STEM professionals, it is imperative that Asia cultivate a big talent pool in these areas. Up to 80% of Southeast Asian jobs will necessitate both basic digital literacy and applied ICT skills by the year 2030. To achieve this goal, it is necessary to take advantage of the full spectrum of the workforce, not just the male half. By 2030, countries like Indonesia would need an additional 600,000 to 1.2 million people trained in IT, media, and telecom. The potential $21.8 billion in lost GDP due to a talent shortage has been estimated. Companies and economies lose a lot of money because of the lack of women working in STEM fields. When it comes to closing the gender gap in STEM fields, the Philippines is the only Asian country in the top 10. In the Philippines, women make up 52% of the STEM academic community, but in Indonesia and Singapore they only make up 33% (TAF, 2021). Even in technologically advanced nations like Japan and Korea, women make up only 15% and 18% of researchers respectively.

Only 14% of the 280,000 scientists, engineers, and technologists working in Indian research organizations are women, according to a UN estimate from 2020. Compared to the worldwide average of about 30% of women in scientific research, this is significantly lower. The United States (34%), the United Kingdom (38%), Germany (27%), and France (32%), all rich countries, all have fewer women in STEM fields than India (43%), which is counterintuitive. The Higher Education Commission (HEC) of Pakistan reports that out of the total of 3 million students enrolled in universities and degree institutions in the country (1.4 million women and 1.6 million males), over a million are majoring in science, technology, engineering, and mathematics.

Approximately 415,008 students are enrolled in STEM related majors, 276,659 students enrolled in ICT programs, 178,260 students enrolled in health science programs, and 166,448 students enrolled in engineering programs. Pakistan had 157,102 STEM graduates last year, putting it in the top twelve countries worldwide. According to Haq (2023), over 43,000 of these grads work in the IT sector. In Pakistani universities, women make up roughly 46% of the STEM student population. There are currently about 500,000 Pakistani women enrolling in baccalaureate and graduate-level STEM programs in Pakistan (Saleh, 2022). In terms of STEM and IT graduates from universities, Pakistan is expected to be in the top dozen or so countries. However, when compared to countries with a similar level of middle income, the country's university enrollment rate is significantly lower than average.

Evidence of gender discrimination in academia is extensive, albeit patchy, with women underrepresented in the majority of scientific fields, publishing fewer articles throughout their careers, and having their work referenced less frequently. Gender differences have been affected by the surprising fact that the past 60 years have seen a surge in the number of women working in science. This all-encompassing depiction of gender inequality in academia has the potential to recast the discussion about the future of women's academic careers, which has major ramifications for both academic institutions and government policy.

Factors Contributing to Gender Disparity in Sustainable Technological Innovation

This research highlights the importance of STEM networks in reducing obstacles to women's contribution in and advancement within STEM areas. In preparation for the subsequent section on "Effective Network Approaches," this one presents an overview of the major barriers that women and girl's encounter.

Women and girls encounter numerous, overlapping challenges as they pursue STEM fields of study and jobs. Disabled persons and members of other minority groups in Asia and around the world, for example, find these limitations particularly difficult to bear. Human and structural elements, both official and informal, contribute to these challenges. Gendered societal norms state that males and boys are 'naturally' better at STEM and more suitable to professions in STEM fields than girls and women. Parents, institutions, and management are all examples of gatekeepers and influencers who can either increase or decrease the severity of these pressures. Gender bias, preconceptions, and a scarcity of female role models make it more difficult for girls to imagine a future in which they pursue and thrive in STEM careers. Girls' interest in STEM areas begins to wane around the age of 10, when they begin to internalize the assumptions that these fields are dominated by males and that girls are inherently less intelligent in these topics. The absence of inspiring women working in STEM only served to reinforce these preconceived notions (Ugwuegbula, 2020). When added to gender-blind workplaces or government policies or casual and systemic discrimination, such barriers tend to diminish a woman's ability to remain in STEM and grow in her career.

The "leaky pipeline" theory is commonly used to explain why fewer and fewer women pursue careers in science, technology, engineering, and mathematics (STEM). Despite its aesthetic appeal, this metaphor has serious flaws. According to this model, the "pipeline problem" of the STEM gender gap is caused by a dearth of qualified or interested female students in these disciplines. It disregards the many institutional, social, and cultural barriers that keep women from pursuing jobs in STEM fields throughout their lives. A single-entry pipeline also fails to account for the reality that many women return to STEM occupations after a brief hiatus. If a girl doesn't get inspired, encouraged, and given the chance to excel in STEM fields early on in her education, she has little chance of ever having a career in these fields (PBCWE, 2019). Women who did not initially pursue careers in the STEM fields sometimes end up creating their opportunities in these fields. This is especially true as national and multinational tech enterprises grow, giving rise to new economic sectors and job prospects.

It is challenging for many companies to retain women working in STEM fields due to strict laws and standards. Potential barriers to women's leadership in STEM include a lack of family leave policies or childcare assistance, a lack of flexibility in work hours, and a dearth of opportunities for sector-specific upskilling. Cultural traditions across Asia (and the world) place women at the center of caregiving roles. To better balance their personal and professional lives, some women may look for flexible work arrangements. A crucial turning point for women's careers occurs between the ages of 30 and 35, when males in STEM may focus on the papers, patents, and project leadership necessary for promotion but women must juggle work and care responsibilities.

Lack of flexible policies or enough support for childcare and other mundane job responsibilities may prevent women from advancing in their careers if they are unable to take on leadership roles, travel for business, or work long hours. When women reach this age, they may decide to retire from their careers. According to a UNESCO 2015 report, 53.1% of women in Korea who work in STEM disciplines left their employment because they wanted to start a family. Workplace bias, both subtle and overt, can limit women's opportunities and achievements. Implicit bias in STEM fields shows that women may be passed over for promotion because "people are not used to looking at women as competent." That there is "no policy barrier, but there is a perception barrier" is not always the case. There may be more overt opposition to women's success in STEM disciplines.

Gender prejudices in society and culture can discourage women from pursuing careers in STEM. The stress of not wanting to abandon their children can prevent women from traveling to work on bridge-building projects with her. This is especially difficult for women working in industries like mining and construction where employees are sometimes required to spend long amounts of time in remote areas.

Pressure from family members to conform to traditional roles for wives and daughters might also discourage women from pursuing jobs in science, technology, engineering, and mathematics (STEM) later in life. Some individuals believe that women shouldn't follow their passions and pursue professions in data or technology because their loved ones could find it strange or unattractive or that doing so in the middle of one's career is too risky, even if they enjoy it.

Women's underrepresentation in STEM is often attributed to a lack of network support; however, not all women have access to traditional networking possibilities. If joining a network requires you to be an "insider," such as a graduate of a specific school or an employee of a specific company, then that network is likely to be exclusive. Cultural and gendered barriers to networking include the tendency to socialize during non-work hours or over alcoholic beverages, for example. Networks that rely on face-to-face connection and participation can be hampered by physical distance. Some women still have trouble integrating into online networks, even if the number of virtual interactions has increased throughout the Covid-19 pandemic.

Problems can arise from a lack of familiarity with virtual networks, poor Internet access, or inadequate technical skills. Women's access to online communities may be hindered by cultural biases and gender conventions in some countries. The "Zoom fatigue" experienced by some network executives during the pandemic led them to abandon online networking altogether (TAF, 2021). Women in Asia have less time than men to participate in networking events and build professional relationships outside of the immediate workplace because of conventional family commitments. This also implies that people have fewer chances to practice networking skills like introducing themselves, talking about themselves, and asking for help.

Benefits of Gender Diversity in Sustainable Technological Innovation

Several empirical studies have shown that having a more diverse workforce, especially one that includes women, boosts innovation and creativity by providing more perspectives and perspectives. Also, different perspectives, value systems, and behavioral patterns can be brought into a company thanks to gender diversity (Xie, Zhou, Zong, & Lu, 2020). It's difficult to predict the future, but we're positive that technology, especially digital technology, and sustainability will play major roles in shaping the world in the years to come. According to the European Union, these sectors will get the bulk of the €724 billion ($810 billion) COVID-19 recovery package. However, they need a third component—women—to fulfill their full potential (Sultan & Scharf, 2022). This is not merely a social justice problem. The participation of women (who make up half of the population) is crucial if businesses are to attract the expertise they need in underserved fields.

According to Eurostat, just 17% of highly technical occupations (including coding, systems analysis, and software development) are held by women. While women make up over half of the environmental science workforce, they make up only 20% of the electrical, mechanical, and civil engineering fields, all of which are crucial to ensuring a sustainable future. There is data to suggest that businesses that effectively utilize female talent also excel in other key metrics. When compared to companies without any women on their boards, those with at least 30% women had much higher environmental scores and R&D spending (1.5x and 4x, respectively). According to McKinsey's research, companies with a higher percentage of women on their executive teams tend to perform better financially (ESCAP, 2021). The issue is the lack of women in leadership roles and the scarcity of available workers in two major sectors. Fortunately, there are ways to deal with the issue.

The status quo won't cut it if we want more women in the workforce, but expanding the employment pipeline could help. Think about hiring for experience instead of education. One way to increase the number of qualified applicants is to partner with educational institutions that provide courses like coding boot camps (Sultan & Scharf, 2022). The representation of women must be prioritized and actively pursued to improve. Executives could take one approach by promising to work toward gender parity in all future ventures and projects. Increasing the representation of women in positions of power is another priority. Progress is doable, and we know that. But efforts like this either don't exist or aren't very successful. Women need to play a substantial part in all economic areas if we are to fully achieve the possibilities of technology and sustainability. If we can make life better for women, we can make life better for everyone.

Strategies for Advancing Inclusion and Diversity for Women in Sustainable Technological innovation

Networks throughout the region are taking several different steps to encourage more girls and women to pursue STEM fields of study and jobs and to provide tailored support to keep them there and help them succeed. Supporting inclusive leadership and institutional change to make STEM fields more accessible, creating safe spaces for women in STEM to discuss their needs, helping women develop confidence in their abilities and a career path vision, making STEM affordable, approachable, and accessible to women, and facilitating mentorship, sponsorship, and network building are all effective strategies.

Disparities in women's participation and leadership in STEM, particularly between countries, STEM fields, and degrees of seniority within fields, make it impossible to implement a universal solution to this problem. Instead, as will become clear in the succeeding section, networks are employing a wide range of effective approaches to overcome these challenges. Multifaceted approaches that target women's distinct and particular requirements are most effective, even when government-supported or other large-scale networks reach thousands of women in STEM.

The following methods are examples of network-based initiatives that have shown consistent success in attracting women to STEM areas, retaining them there, and/or advancing them. These are repeatable procedures that can be modified to fit specific locales. They are meant to serve as examples of what has worked successfully in Asia and elsewhere in the world.

We can no longer imagine our lives without digital technology. employment, career information, and entrepreneurship opportunities; online education, access to formal and informal learning; access to health and well-being information and support; the ability to engage in one's own creative and cultural practices; the expression of one's ideas and opinions; for leisure, play, and connect with peers; and the employment search. Women often do not reap the same benefits from digital technology as men do, and they face additional barriers to full digital inclusion. Digital products and services that are tailored to the needs of women are those that are created in collaboration with and by women. Many digital solutions, products, and content are created with a 'default' user in mind, rather than taking into account girls' unique needs and circumstances, such as the devices they have access to, the digital platforms they use, their levels of digital literacy, or the content they find relevant and wish to see (UNICEF, 2022). Teams, despite their best intentions, frequently cater to a largely male audience when designing products. Fewer women than males work on the development, design, and testing of products. Consequently, women are less likely to actively pursue and implement digital solutions because they either lack the resources to do so or the inclination to do so. This in turn further disadvantages women and widens the digital gender gap. An essential part of closing the gender digital divide and providing equal access for girls and young women is the Gender toolkit, which will be used by creators, designers, and implementers of digital products and services.

Creating a secure environment for female employees is crucial to attracting and retaining them in STEM fields. While workplace safety and security should be a top priority for all women, it may be more of an issue for women in the STEM areas because of the gender imbalance in those fields and the fact that many women in industries like mining and energy work in isolated, male-dominated settings. Women need to be physically safe and free from harassment if we are to create safe workplaces and inclusive communities. It's also important to foster an environment where employees feel safe asking questions, gaining knowledge, and enhancing their abilities.

Networks and mentorship circles give a safe place for women to ask tough questions, learn new technical and soft skills (including problem-solving, negotiation, and public speaking), and boost their self-assurance through practice. After that, you can take your newly acquired skills to the workplace and hone them there. Women's networks in the STEM fields emphasize networking as a crucial skill. Zendesk's Wendy Johnstone attributes much of her success to the support she received from the women's networks she joined. It took a lot of courage for me to put myself out there, ask for help, and network when I initially started my job. Women's networks were the first to show me the efficacy of women working together; they provided a safe space in which to explore the possibilities of making new relationships and making the most of genuine ones. In a welcoming and encouraging environment, women in STEM networks help their members expand their professional and personal networks beyond their immediate technical and functional spheres.

Organizations and institutions have established training and support networks for women in STEM to encourage more women and girls to enter these disciplines. Motivating more women to work in STEM disciplines can be accomplished in several ways, including providing a return on investment, exhibiting realistic role models, and assisting women in overcoming personal challenges related to confidence in STEM (see the Building Confidence section above). By removing the following obstacles, networks attempt to make STEM fields more welcoming to women of all backgrounds and ages. Free or discounted service provision is one strategy.

For mentorship to be effective in attracting and keeping women in STEM disciplines and boosting their retention once they have entered these industries, networks are essential. Mentoring is especially important for students continuing their studies in the STEM professions and at higher levels. Mentoring programs are an essential strategy for networks to address the "leaky pipeline" problem in regions where many women quit STEM fields. WISET is a South Korean organization that takes a "lifecycle" approach to mentoring and other forms of concentrated intervention to help women overcome the particular challenges they experience at different stages of their educational and career paths. As reported by Murphy (2019)," WISET, for example, has developed a mentoring program that pairs mid-career employees with senior-level experts to stop the "leak" between the middle and upper levels of leadership. Career mentors assist mentees by listening to and empathizing with their concerns, sharing anecdotes from their own experiences, offering advice on how to maintain a work-life balance, and offering various forms of guidance and support.

The visibility of women in STEM subjects can be increased through the establishment of networks that highlight the achievements of women scientists to decision-makers and the public. To do this, we must provide the women in our network with the tools they need to become more visible in social and professional settings. The advancement of women presently working in STEM and the recruitment of new women to the field depend on initiatives like these, which showcase the achievements of women in STEM. Superstars of STEM initiatives give women public recognition for their achievements in STEM professions, which can inspire more women to follow in their footsteps (TAF, 2021). Networks help women achieve greater external attention by highlighting their accomplishments. This "third-party" recognition, in which the network provides public visibility on behalf of female scientists, is especially valuable because it opens up more avenues of collaboration and access in cultural contexts where women may be discouraged from drawing attention to themselves or their accomplishments in the pursuit of career advancement.

One of the most important ways in which networks can help women advance in STEM fields is through altering governmental and private sector institutions. For this purpose, groups are formed and kept active that promote workplace diversity, prepare women for positions of leadership, and recruit male and female company executives to strive toward gender equality. Public and private sector managers and executives can play a crucial role in boosting women's advancement by amplifying women's views in informal networks and formal management conversations. Women and men can greatly boost the number of women and members of underrepresented groups who hear about job openings and other professional opportunities by having casual conversations about them. Similar to how nominating many competent women for promotions, awards, and leadership positions can help improve women's participation in high-profile jobs, people in powerful positions within an organization can also contribute. If more women are visible in the applicant pools for these positions, there is a better chance that they will be filled by women.

RECOMMENDATIONS

The study has the following recommendations:

- Create a long-term plan that accounts for all relevant factors, such as finances and staffing.
- Women need men to lead by example and meet them where they are. Establish rapport with influential people.
- Reduce the cost of STEM fields. Encourage a broad range of people to join the network. Make sure your network is secure. Emphasize the achievements of women in STEM fields.
- Educate those in the network who have been assigned the role of mentor, manager, or leader.
- Encourage the growth of a network of female professionals in STEM fields and help it focus on its most important goals by facilitating internships for students in these fields.
- Promote DEI (diversity, equity, and inclusion) by strengthening in-house connections. Make formal and informal channels for employee involvement in both internal and external networks. Involve STEM-related networks in staff training.
- Join forces with other groups to multiply the results of government programs and policies. Fix connectivity problems with the internet. Take action to increase the number of women working in STEM fields.
- Working together with existing networks to develop STEM-related curricula can aid in the creation of a "network of networks for women in STEM" and the crucial initiatives that will emerge from these networks.

CONCLUSION

Important opportunities for women and girls to connect, study, and advance in STEM fields are being created via women's networks across Asia. Their success depends on their ability to overcome the challenges that prevent women from pursuing professions in STEM or that cause them to decide that such careers are not for them. A lack of female role models in STEM disciplines, cultural norms and gatekeepers who limit participation in these fields, and outdated policies in the workplace are all things that networks try to change. Some groups focus on removing a specific barrier, while others try to help women, in general, succeed in the workplace.

This study revealed several novel strategies being used by Asian networks to entice more girls and women into STEM and to encourage their success and leadership in these fields. By sharing content that is specifically geared toward women, networks improve girls' access to STEM education and encourage their online participation in science. For free and in convenient places, networks offer opportunities for women who are interested in data science but lack the relevant experience to spend a day learning about data. Women are gaining confidence in their abilities and direction in their jobs as a result of these groups.

And others are trying to make sure women of all levels in STEM get the attention and support they need to succeed from those in positions of authority. There needs to be support and participation from the private sector, the government, and civil society for this initiative to succeed. We commit to building a "network of networks" to support women in STEM fields as part of this initiative. Collaborative efforts that bring together government training resources with those of industry associations knowledgeable in cutting-edge science, network outreach to women in STEM, and business support for work placement, among other things, could significantly increase the number of women working in technology. When conducting research that bridges the gap between the academic world and the commercial world, networks are an invaluable resource. A significant contributor to the current stagnation of women in STEM is the lack of networks that can advise businesses and governments on how to progress women in STEM through institutional and legislative adjustments.

To further DEI initiatives, the private sector can draw on resources both within and beyond its ranks. An organization's internal networks can be strong partners in promoting diversity, equity, and inclusion when they share the same priorities as the company. One respondent said that the best method to encourage more women to enter STEM disciplines is to have a female employee who is already working in a STEM sector talk about her work at the company while smiling. Businesses can facilitate the growth of internal networks by recognizing their importance and providing them with adequate resources including money, meeting space, and time away from regular duties. Improving information exchange between nodes can benefit networks. Researchers in Southeast and East Asia discovered over 70 groups dedicated to empowering women to achieve professional success in STEM professions. Some of these networks are already talking to one another and sharing valuable insights on topics of common concern.

The purpose of setting up networks for these communities is to promote ongoing communication and the exchange of knowledge. Cross-pollination and collaboration can be particularly effective in addressing the systemic and structural hurdles that women in STEM face, especially when paired with financial capital injections that strengthen networks and the women they serve. Collectively, the network, business, and government sector proposals provide a road map for increasing the effectiveness of women-in-STEM networks and reinforcing their efforts to close the gender gap in STEM.

REFERENCES

A Complex formula: girls and women in science, technology, engineering and mathematics in Asia. (2015). UNESCO Digital Library.

Accelerating Women's Advancement in STEM. (2021). *The Asia Foundation*, https://asiafoundation.org/wp-content/uploads/2021/06/Accelerating-Womens-Advancement-in-STEM_Report_update7.26.21.pdf

Baruah, D. (2022). STEMming Gender Divide in India. *The British Councel*. https://education-services.britishcouncil.org/insights-blog/stemming-gender-divide-india

Dossi, S. (2022). How is the current STEM gender bias? *Imperial London College*, https://www.imperial.ac.uk/news/233901/how-current-stem-gender-bias/

Haq, R. (2023). Over A Million Pakistani University Students Enrolled in STEM Fields. *Haq's Musings*, http://www.riazhaq.com/2023/02/over-million-pakistani-university.html

Huang, J. A. J.-L. (2020). Historical comparison of gender inequality in scientific careers across countries and disciplines. *PNAS*, https://www.pnas.org/doi/10.1073/pnas.1914221117

Innovation and Technology for Gender Equality. (2022). *UNICEF*. https://www.unicef.org/eap/innovation-and-technology-gender-equality

Luqun Xie, J. Z. (2020, February). Gender diversity in R&D teams and innovation efficiency: Role of the innovation context. *Research Policy*, *49*(1), 103885. Advance online publication. doi:10.1016/j.respol.2019.103885

Murphy, W. (2019). Advice for Men Who Are Nervous About Mentoring Women. *Harvard Business Review*. https://hbr.org/2019/03/advice-for-men-who-are-nervous-about-mentoring-women

New report on diversity trends in STEM workforce and education. (2023). *National Science Foundation*. https://phys.org/news/2023-01-diversity-trends-stem-workforce.html

Report, F. (2019). Women in STEM – A Baseline Study. *PBCWE*. https://pbcwe.com.ph/2019/11/29/full-report-women-in-stem-a-baseline-study/

Resetting tech culture. (2020). *Accenture*. https://www.accenture.com/content/dam/accenture/final/a-com-migration/pdf/pdf-134/accenture-a4-gwc-report-final1.pdf#zoo
m=50

Sandra Sancier-Sultan, S. S. (2022). Why bringing women into tech roles is good for society. *World Economic Forum*. https://www.weforum.org/agenda/2022/02/women-gender-technology-economy-representation-equality/

Subramaniam Ananthram, D. B. (2021). The gender gap in STEM is not due to women's lack of confidence. *World Economic Forum*. https://www.weforum.org/agenda/2021/03/gender-gap-in-stem-not-due-to-womens-lack-of-confidence/

The Future is equal: Gender Equality in the Technology Industry. (2021). *ESCAP*. https://www.unescap.org/kp/2021/report-gender-equality-technology-industry

Ugwuegbula, L. (2020). TEM Gender Bias Cripples Asia-Pacific Region. *The Diplomat.* https://thediplomat.com/2020/02/stem-gender-bias-cripples-asia-pacific-region/

women and digital revolution. (2021). *UNESCO science report.* https://www.unesco.org/reports/science/2021/en/women-digital-revolution

Women in Science. (2020). *UNESCO Fact Sheet No. 60.* https://uis.unesco.org/sites/default/files/documents/fs60-women-in-science-2020-en.pdf

Women Role in STEM Education in Pakistan. (2022). *Stem Education.* https://stemeducation.org.pk/2022/05/09/women-role-in-stem-education/

Chapter 7
Digital Economy:
The Correlation Between Digital Technology and Economic Growth – An Emperical Application

Dilek Dinç
Çankaya University, Turkey

Aytaç Gökmen
Çankaya University, Turkey

ABSTRACT

The aim of this study is to explain the concept of digital economy and to examine the effect of digital technology on economic growth. In the study, the causality relationship between digital technology and economic growth in the Turkish economy for the period 2004-2021 was investigated with the Toda-Yamamoto test. As a result of the study, according to 10% significance level, bidirectional causality between internet usage-economic growth, computer usage-economic growth, and one-way causality between mobile line usage and economic growth from mobile line usage to economic growth was determined. No causal relationship was found between website ownership and economic growth in enterprises. According to the 5% significance level, a one-way causality relationship was determined from only computer and internet use to economic growth.

INTRODUCTION

Technological developments have continuously created changes in economic and social life in the historical process. Today's world, on the other hand, is living in an era where digital technologies are in a race. While digitalization affects all business functions at the micro level, from production to marketing, it also affects issues such as growth, employment and investment at the macro level (Akkaya & Gerçek, 2019). Recently, large investments in the information and communication technologies (ICT) sector have greatly affected the economic relations among countries (Öztürk, 2002). In the economic literature, the

DOI: 10.4018/978-1-6684-9979-5.ch007

widespread use of digital technology has led to the emergence of concepts such as "New Economy" or "Digital Economy" (Özcan, 2016; Barışık & Yirmibeşcik, 2006). It seems that the world economy has evolved towards a new economic system, especially after the 1990s. While some of the countries adapt to this transformation easily and quickly, the adaptation of others progresses slowly. For example, the United States of America (USA) is leading this movement, followed by China.

1. THE CONCEPT OF DIGITAL ECONOMY

The main driving force of digitalization is actually digital technologies. In this respect, it is very important to examine the development of digital technology sectors in that country when evaluating the transition process of a country to the digital economy. Digital technology sectors include all products, technologies, services and systems that read, store, transmit and process digital information, the components of these products and technologies, the creation of digital information, all products, technologies, services and systems related to digital information (Taymaz, 2018).

The aim of this study is to make a general assessment about the concept of digital economy and to examine the effect of digital technology, which is the most important part of the digital economy, on economic growth. In the second part, following the introduction of the study, a general information is given about the concept of digital economy, its scope and features. In the third chapter, the relationship between digital technology and economic growth is presented with a literature review. In the fourth chapter, an econometric application is made on Turkey and the findings obtained from the analyzes are evaluated. The last part is the conclusion part.

The concept of digital economy has been defined in many ways since the mid-1990s and its scope has expanded day by day. This concept, which used to express technological change and the use of technology by producers and consumers, has gained a different dimension with the development of the internet. The concept of digital economy was first introduced to the literature by Tapscott in 1995. According to Tapscott (1995), information stored in computers moves very quickly between networks and information becomes digital in all its aspects in this way. Later, the concept of digital economy was defined by many people. Carlsson (2004) defines the digital economy as the economy based on digital technologies, Dinana (2019) as the global network of economic and social activities acquired by digital technologies, and OECD (2019) defines the change that makes technology more competent and encourages innovation in all sectors of the economy. The work of Bukth and Heeks (2017) in this area is very important for the conceptualization of the digital economy. In the study, the concept of digital economy is defined in three ways as core, narrow-scoped and broad-scoped. Using this study, the scope of the digital economy can be explained with the help of Table 1.

According to Table 1, the core scope of the digital economy is hardware manufacturing, telecommunications, information services, software and information technology sectors. The narrow scope of the digital economy is achieved by incorporating the core scope elements, the digital service, the platform economy, the sharing economy, and the business economy. The broad scope of the digital economy, which is used extensively today, is reached by including the core and narrow scope elements, as well as e-commerce, industry 4.0, precision agriculture and algorithmic economy. Therefore, the digital economy emerges as a concept that constantly improves its meaning with its content extending from core informatics activities to broader activities.

Table 1. The scope of digital economy

Core Scope	Narrow Scope	Broad Scope
Manufacturing of Hardware	Core Scope	Narrow Scope
Telecommunication	+	+
İnformation Services	Digital Service	E-Commerce
Softwhere & Information Technologies	Platform Economics	Industry 4.0
	Economics of Sharing	Delicate Agriculture
	Business Economics	Algorithmic Economics

Recently, digitalization has gone far beyond the concept of digital technology and has become very effective on society. For example, today, the archives of many public institutions have been removed, and all information has begun to be stored on computers. In addition, banking has shifted to digital banking, and the number of people making transactions from internet banking has increased very rapidly with the effect of the pandemic. Digitalization has radically changed the media, government institutions, education and health services, thus increasing the speed of customer satisfaction, efficiency and service availability. Again, when the digital economy is mentioned, a communication network has been formed between devices with the internet of things, which is defined as existing objects accessing the internet in a way and communicating with other devices, and social life has started to become easier thanks to information exchange.

It is seen that the scope of the digital economy has expanded in the past period and many subjects such as robots, cloud computing, digital platforms as well as the Internet of things have been included in the definition of digital economy. The digital economy is also called the new economy in the literature. In this economy, all kinds of information, sounds, texts, images, moving objects are transmitted by computer networks. Large amounts of information reach their recipients in an extremely fast, cheap and reliable way (Yüce & Akbulut, 2018). The main features of the digital economy are summarized in Table 2.

Table 2. The characteristics of digital economy

1.	The digital economy is an information economy. The basic factor of production is information.
2.	In the digital economy, information in digital form provided by digital devices can easily circulate among individuals in different parts of the world in the shortest possible time.
3.	There is virtual world reality in the digital economy.
4.	The digital economy is a network economy.
5.	The digital economy is a global economy.
6.	The digital economy is a new economic system that emerged as a result of digitalization and changing production factors and business processes. With digitalization, speed and efficiency increase and low costs in production are very important.
7.	With digitalization, the media sector has come to a very successful and important position in the digital economy.
8.	In the digital economy, traditional business processes are turning into a faster and more flexible environment.
9.	Intermediaries are disappearing in the digital economy.
10.	Thanks to digital information technology, the time between ordering a product and its production and distribution is reduced.
11.	With digitization, the gap between the technologically literate and those without access to technology is widening.

The digital economy does not only have positive effects on society and the economy. There are also some problems in this new economic system. For example, as a result of the development of technology, the need for human beings decreased with the introduction of robots and machines in many areas, and as a result, the problem of unemployment arose. The national security problem emerges as another problem (Karaçor et al., 2020).

2. THE CORRELATION BETWEEN DIGITAL ECONOMY & ECONOMIC DEVELOPMENT: LITERATURE REVIEW

Studies investigating the effect of the digital economy on economic growth have gained intensity in recent years, especially with the spread of digitalization. In most of these studies, the relationship between digital economy and economic growth is discussed in terms of the relationship between digital technology and economic growth.

The relationship between digital technology and economic growth can be examined from two perspectives: digital technologies as a production factor and digital technologies as consumer goods. As a factor of production, digital technologies create two direct and indirect effects on production. In the direct effect, the increase in the share of digital technology sectors in the country causes the per capita income to increase faster and therefore the economic growth to occur faster, as this sector employs more qualified labor than other sectors. In the indirect effect, increased productivity as a result of the use of digital technology is mentioned. It shows that the use of digital technology in many areas reduces production costs, increases production, and increases in production reduce prices and increase competitiveness.

In addition, digital technologies not only affect their own sector, but also lead to an increase in productivity in all other sectors where they are used and formed the infrastructure, thus providing an advantage in international competition (WEF, 2016). When considered as consumer goods, digital technologies provide the emergence of many new products and sectors. Therefore, new products based on digital technology have a positive impact on economic growth by creating new markets and new demand.

In this part of the study, the main studies investigating the relationship between digital technology and economic growth are included. In these studies, it is seen that variables such as telecommunication infrastructure (such as the number of fixed and mobile lines, fixed and mobile line usage), Internet diffusion and Internet usage are used to represent digital technology.

Dutta (2001) found a positive relationship between telecommunication infrastructure and growth by applying a causality test to 15 developed and 15 developing countries in the period 1970-1993.

Roller & Waverman (2001) found a positive and strong relationship between telecommunications infrastructure and GDP in their study using simultaneous equation models for OECD countries in the 1970-1990 period.

Pazarlıoğlu & Gürler (2007) used the number of fixed and mobile lines per 1000 people as the telecommunication indicator in the model they applied within the framework of the growth function to 30 countries, including European Union countries and candidate countries, in the 1994-2004 period. As a result of the study, it was concluded that there is a positive and significant relationship between telecommunications and real GDP per capita.

Yapraklı & Sağlam (2010) examined the relationship between information and communication technologies (ICT) and economic growth for the Turkish economy between 1980 and 2008, and concluded that the progress in information and communication technologies increased economic growth in a positive and meaningful way.

Lee et al. (2012) examined the effect of mobile phones on growth in Sub-Saharan Africa in their study using the GMM estimator. As a result of their research, they observed that mobile phones contributed significantly to economic growth in Sub-Saharan Africa.

Yıldız (2012), in his study that he applied to OECD countries for the period 1990-2009, determined that telecommunications increased economic growth as a result of panel data analysis.

Chavula (2013), in his study based on the endogenous growth model applied to 49 developing countries for the period 1990-2007, examined the effect of the number of fixed lines, mobile lines and Internet usage on economic growth and emphasized that the variable that increased the growth the most was mobile lines.

Aytun & Akın (2014) investigated the direction of causality between telecommunications infrastructure and economic growth for 21 OECD countries during the 1975-2012 period. Considering the cross-sectional dependency and heterogeneity between countries, they determined that there is a two-way causal relationship between telecommunications infrastructure and growth.

David (2019) examined the causal relationship between telecommunications infrastructure, economic growth and development for some African countries using panel data analysis for the period 2000-2015. As a result of the study, it was determined that there is a long-term causality relationship between the related variables.

Rezny et al. (2019) stated in their study that the information economy is the most important way of social transformation for a sustainable economic growth and a way out of the problems of increasing resource scarcity and climate degradation.

Haftu (2019) examined the effect of mobile phone and internet use on economic growth for the period 2006-2015 in 40 Sub-Saharan African (SSA) countries with the help of panel data analysis. As a result of the analysis, it was found that a 10% increase in mobile phone usage leads to a 1.2% increase in GDP per capita, while the Internet does not have a positive effect on GDP per capita.

Haini (2019) investigated the impact of human capital development and internet penetration on economic growth in ASEAN countries using panel data analysis. As a result of the study, it was concluded that human capital affects economic growth positively, and the increase in Internet penetration has a positive effect on economic growth at a statistically significant 5% level.

Gözüçuk (2020) examines the impact of the digital economy on economic growth in OECD countries using panel data analysis for the period 2005-2017. As a result of the study, it was found that increases in broadband subscription and individual Internet use increase economic growth. In addition, it was found that the increase in fixed telephone subscriptions affected economic growth negatively, while mobile telephone subscription did not have an effect on economic growth.

Yılmaz & Kırışkan (2020) examined the effect of telecommunication infrastructure on economic growth in Turkey in their study. In the study, in which cointegration analysis and vector error correction models were used, it was concluded that the increases in the telecommunication stock had a positive and significant effect on the GDP in the long run.

Topçu & Gedik (2021) examined the relationship between the telecommunication sector and the growth of the Turkish economy for the period of 2008-2020. They used XDSL (digital subscriber line, number of subscribers using the internet), number of subscribers, Number of Mobile Subscribers, Fixed Investments and Mobile investments as telecommunication indicators. As a result of their studies using the Johansen cointegration approach, ARDL model and Granger causality analysis, a two-way relationship between fixed investments and GDP, and a one-way relationship from GDP to XDSL has been determined. No relationship was found between GDP, mobile investments and the number of mobile subscribers.

Single et al. (2022) examined the relationship between telecommunications and economic growth for 44 countries in different regions and different income levels, using the data for the period of 2000-2015 and a dynamic panel data model. As a result of the study, they found that there is a positive and significant relationship between economic growth and investments in the telecommunications sector, foreign trade volume and telecommunications consumption.

Considering the studies in the literature, it is noteworthy that there is a positive relationship between digital technology and economic growth. When this general result is evaluated, it is seen that the development of telecommunication infrastructure investments by the state and the encouragement of the private sector are of great importance in terms of increasing economic efficiency.

3. AMPIRICAL ANALSIS

3.1. Model and Data Set

In this study, the causality relationship between digital technology and economic growth in the Turkish economy for the period 2004-2021 was investigated using the Toda-Yamamoto (1995) causality test. Representing digital technology, mobile line use, internet use, computer use, website ownership in enterprises; GDP is used to represent economic growth. The reason for the start of the study in 2004 is the lack of data. All data were obtained from the Turkish Statistical Institute (TUIK). The model used in the study is given by Equation 1, and the logarithmic forms of the variables are included in the model.

$$lnGSYH = c1 + lnMOBILit + lnINTERNETit + lnBILGISAYARit + lnWEBit + u1 \quad (1)$$

MOBIL: Mobil Hat Kullanımı

INTERNET: İnternet Kullanımı
BILGISAYAR: Bilgisayar Kullanımı
WEB: Girişimlerde Web Sitesi Sahipliği
GSYH: Gayri Safi Yurtiçi Hasıla

3.2. Methodology

3.2.1. Unit Root Tests

In this study, Extended Dickey Fuller (Augmented Dickey Fuller-ADF) (1981), Phillips-Perron (PP) (1988) and Kwiatkowski, Phillips, Schmidt, Shin (KPSS) (1992) unit root tests were used to analyze the stationarity of the series. ADF and PP unit root test results of the variables are presented in Table 3.

According to Table 3, at the 5% significance level, the ADF and PP test results show that the series used in the study are not stationary at the level; however, it shows that they become stationary by taking the difference.

Table 3. ADF ve PP unit root test findings

Variable	ADF Probability		Stationary (S), Trend (T)	Decision
MOBIL	-3.250222 (2)	0.1124	S, T	Birim kök mevcut
INTERNET	-3.375116 (3)	0.0953	S, T	Birim kök mevcut
BILGISAYAR	-1.825319 (0)	0.3565	S	Birim kök mevcut
WEB	-2.644203 (1)	0.1051	S, T	Birim kök mevcut
GSYH	-1.738380 (0)	0.0779	S	Birim kök mevcut
DMOBIL	-10.03272 (2)	0.0000	S, T	Birim kök mevcut değil
DINTERNET	-3.529095 (0)	0.0211	S	Birim kök mevcut değil
DBILGISAYAR	-4.765010 (0)	0.0084	S	Birim kök mevcut değil
DWEB	-3.352672 (0)	0.0294	S	Birim kök mevcut değil
DGSYH	-4.003398 (2)	0.0006	yok	Birim kök mevcut değil
Variable	**PP Probability**		**Stationary (S), Trend (T)**	**Decision**
MOBIL	-3.93512 (2)	0.1946	S, T	Birim kök mevcut
INTERNET	-2.331856 (0)	0.3971	S, T	Birim kök mevcut
BILGISAYAR	-2.368174 (0)	0.1643	S	Birim kök mevcut
WEB	-2.738339 (2)	0.0882	S	Birim kök mevcut
GSYH	-1.738380 (0)	0.0779	S	Birim kök mevcut
DMOBIL	-13.64341 (1)	0.0000	S	Birim kök mevcut değil
DINTERNET	-3.529095 (0)	0.0211	S	Birim kök mevcut değil
DBILGISAYAR	-3.794775 (0)	0.0127	S	Birim kök mevcut değil
DWEB	-3.526447 (2)	0.0016	yok	Birim kök mevcut değil
DGSYH	-6.816304 (2)	0.0000	yok	Birim kök mevcut değil

Note: Probability > 0.05 unit root, Probability < 0.05 no unit root, D: first difference of the series. The numbers in parentheses represent the lag lengths determined by the Schwarz (SC) information criterion.

Therefore, according to the ADF and PP unit root tests, the maximum degree of integration (dmax) of the variables was determined as 1. KPSS unit root test was performed to support ADF and PP unit root tests. KPSS test results are presented in Table 4.

According to Table 4, since the LM test statistics are absolutely larger than the KPSS critical value at the 5% significance level, it was concluded that they contain a unit root. Findings obtained from the first degree difference of the variables indicate that the variables are difference stationary. These results support the results obtained from the ADF and PP tests, and the maximum degree of integration (dmax) of the variables was determined as 1.

3.2.2. Lag-Length

Another necessary procedure for the Toda-Yamamoto test is to determine the appropriate lag length. The appropriate lag length can be determined using information criteria and descriptive tests. In this study, the lag length was determined by VAR analysis. Analysis results are shown in Table 5.

Table 4. KPSS unit root test findings

Variable	LM	Stationary (S), Trend (T)	Asymptotic Critic Value (%5)	Decision
MOBIL	0.160377	S, T	0.146000	Birim kök mevcut
INTERNET	0.181625	S, T	0.146000	Birim kök mevcut
BILGISAYAR	0.162578	S, T	0.146000	Birim kök mevcut
WEB	0.226268	S, T	0.146000	Birim kök mevcut
GSYH	0.160632	S, T	0.146000	Birim kök mevcut
DMOBIL	0.437369	S	0.463000	Birim kök mevcut değil
DINTERNET	0.095868	S	0.463000	Birim kök mevcut değil
DBILGISAYAR	0.308399	S	0.463000	Birim kök mevcut değil
DWEB	0.215203	S	0.463000	Birim kök mevcut değil
DGSYH	0.163736	S	0.463000	Birim kök mevcut değil

The lag length was determined as 2 according to the SC, FPE, AIC and HQ information criteria. The LM test was applied to determine whether there is an autocorrelation problem in the determined number of delays. According to the results given in Table 6, the null hypothesis that there is no autocorrelation problem in the 2nd lag is accepted.

In the study, the White Variance test was used to determine whether there was a variable variance problem. The test result is reported in Table 7.

According to this result, it was accepted that there was no variance problem between the error terms at the 5% significance level.

In addition to the identification tests, in the next step, the stability of the 2-delay VAR model was investigated.

Table 5. The determination of optimum lag-length in the VAR model

Lag-Length	FPE (Final Prediction Error)	AIC (Akaike)	SC (Schwarz)	HQ(Hannan-Quinn)
0	32999475	31.50057	31.74201	31.51294
1	291997.6	26.58160	28.03020	26.65578
2	**67578.69***	**23.75344***	**26.40921***	**23.88944***

*Denotes the determined lag length at 5% significance level.

Table 6. Autocorrelation LM tests results

Lag-length	LM	Probability
1	17.04889	0.8800
2	23.51799	0.5473
3	27.07284	0.3523
4	26.59170	0.3766

Table 7. White variable variance test

Lag-length	Test Statictics	Probability
2	154.8110	0.3770

Not: H_0: No variance change.

Figure 1. AR characteristic polinomium inverse units

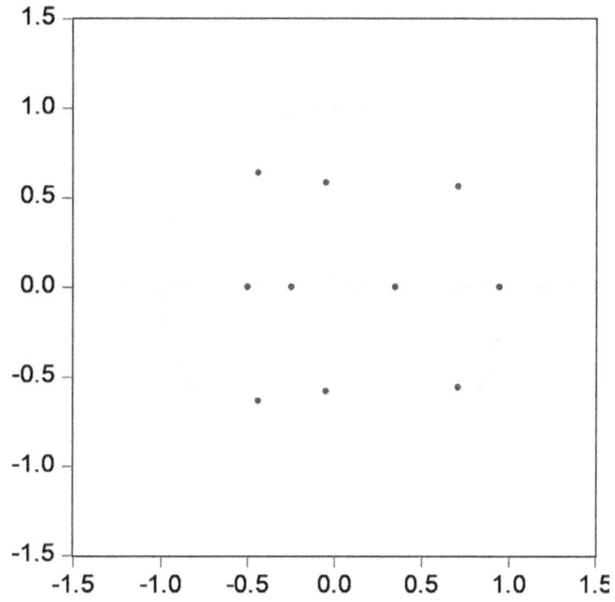

Şekil 1, AR karakteristik polinomunun ters köklerinin hiçbirinin birim çember dışında yer almadığını, dolayısıyla VAR modelinin istikrarlı olduğunu ifade etmektedir. Tablo 5, 6, 7 ve Şekil 1'den elde edilen bulgulara göre, optimal gecikme uzunluğu (*k*) 2 olarak belirlenmiştir.

3.2.3. Toda-Yamamoto Causality Test

The required (k + dmax) level for the Toda-Yamamoto causality test was determined as 3. The causality test results are reported in Table 8.

As a result of the Toda-Yamamoto causality test, according to 10% significance level, a bidirectional causality relationship was found between Internet use-economic growth, computer use-economic growth, and a unidirectional causality relationship between mobile line usage and economic growth from mobile line usage to economic growth.

No causal relationship was found between website ownership and economic growth in enterprises. According to the 5% significance level, a one-way causality relationship was found from only computer and Internet use to economic growth. These findings support the literature and indicate the existence of a causal relationship between digital technology and economic growth.

Table 8. Toda-Yamamoto causality test results

Direction	X^2	Probability	Result
BILGISAYAR →GSYH	21.8472	0.0004	There is a causal relationship from computer use to GDP*.
GSYH →BILGISAYAR	3.80669	0.0714	There is a causal relationship from GDP to computer use**.
INTERNET →GSYH	22.8797	0.0003	There is a causal relationship from Internet use to GDP*.
GSYH →INTERNET	3.49707	0.0825	There is a causal relationship from GDP to Internet usage**.
MOBIL→GSYH	3.73866	0.0736	There is a causal relationship from mobile line usage to GDP**.
GSYH →MOBİL	0.00090	0.9765	There is no causal relationship from GDP to mobile line usage*, **.
WEB→GSYH	0.46577	0.5061	There is no causal relationship from website ownership to GDP in enterprises*, **.
GSYH → WEB	0.11632	0.7381	There is no causal relationship from GDP to website ownership in enterprises*, **.

(* at 5% significance level, ** at 10% significance level.)

Table 9. Results obtained

%10 Anlamlılık Düzeyine Göre	Karar	%5 Anlamlılık Düzeyine Göre	Karar
BILGISAYAR ↔GSYH	**Çift Yönlü**	**BILGISAYAR →GSYH**	**Tek Yönlü**
INTERNET ↔GSYH	**Çift Yönlü**	**INTERNET →GSYH**	**Tek Yönlü**
MOBIL→GSYH	**Tek Yönlü**	MOBIL & GSYH	Nedensellik Mevcut Değil
WEB & GSYH	Nedensellik Mevcut Değil	WEB & GSYH	Nedensellik Mevcut Değil

When the 5% and 10% significance levels are evaluated together, it is seen that the direction of this causality relationship is from computer and Internet usage used to represent digital technology to GDP used to represent economic growth. Therefore, it can be concluded that the increase in digital technologies positively affects economic growth.

RESULT

In this study, the causality relationship between digital technology and economic growth in the Turkish economy for the period 2004-2021 was investigated with the Toda-Yamamoto causality test. Representing digital technology, mobile line use, internet use, computer use, website ownership in enterprises; GDP is used to represent economic growth.

According to the results of this study, there is a causal relationship between digital technology and economic growth. Therefore, it can be said that the use of digital technology at every stage of economic activities is very important in terms of economic growth.

It is the most important deduction that more importance is given to the field of digital technology and more effective studies in this field will contribute to Turkey's economic growth and positively affect the Turkish economy. Increasing the use of digital technology can be made possible by providing the necessary technological infrastructure by the state and by encouraging the private sector.

REFERENCES

Akkaya, H., & Gerçek, A. (2019). The Taxation of Digital Economy in OECD and some Seleceted Countries: Inferences for Turkey. *International Journal of Public Finance*, *4*(2), 166–188. doi:10.30927/ijpf.634050

Aytun, C., & Akın, C. S. (2014). Telecommunications infrastructure and economic growth in OECD countries: Cross-Section Dependent Heterogeneous Panel Causality Analysis. *Economics. Business & Finance*, *29*(340), 69–94.

Barışık, S., & Yirmibeşcik, O. (2006). Harmonization efforts aimed at accelerating the formation process of the new economy in Turkey. *ZKU Journal of Social Sciences*, *2*(4), 39–62.

Bekar, M., Tüfekci, A., Yalçın, Y., & Emirmahmutoğlu, F. (2022). Panel data analysis of the relationship between telecommunications and economic growth. *Journal of Economics. Policy & Finance Studies*, *7*(2), 366–385.

Bukth, R., & Heeks, R. (2017). *Defining, conceptualising and measuring the digital economy.* Working Paper Series, Paper No. 68.

Carlson, B. (2004). The digital economy: What is new and what is not? *Structural Change and Economic Dynamics*, *15*(3), 245–264. doi:10.1016/j.strueco.2004.02.001

Chavula, H. K. (2013). Telecommunications development and economic growth in Africa. *Information Technology for Development*, *19*(1), 5–23. doi:10.1080/02681102.2012.694794

David, O. O. (2019). Nexus between telecommunication infrastructures, economic growth and development in Africa: Panel vector autoregression (PVAR) analysis. *Telecommunications Policy*, *43*(8), 1–17. doi:10.1016/j.telpol.2019.03.005

Dickey, D. A., & Fuller, W. A. (1981). Likelihood ratio statistics for autoregressive time series with a unit root. *Econometrica*, *49*(4), 1057–1072. doi:10.2307/1912517

Dinana, H. O. (2019). The impact of information & communication technologies (ICT) on sustainable development goals (SDGs) progression in the era of digital economy. *Sustainable Development Conference*, 1-13.

Dutta, A. (2001). Telecommunications and economic activity: An analysis of Granger causality. *Journal of Management Information Systems*, *17*(4), 71–95. doi:10.1080/07421222.2001.11045658

Gözüküçük, M. F. (2020). *Digital transformation and economic growth* [Unpublished Master Thesis]. Istanbul Commerce University Institute of Social Sciences.

Kaitkowski, D., Philips, P. C. B., Schmidt, P., & Shin, Y. (1992). Testing the null hypothesis of stationarity against the alternative of a unit root. *Journal of Econometrics*, *54*(3), 159–178. doi:10.1016/0304-4076(92)90104-Y

Karaçor, Z., Güvenek, B., & Karadoğan, E. (2020). Ekonominin değişen maskesi: Dijital ekonomi. *International Journal of Social. Humanities and Administrative Sciences*, *6*(34), 2140–2149.

Kırışkan, I. & Yılmaz, D. H. (2020). Telecommunications infrastructure and economic growth in Turkey. *Bilig / Journal of Turkish World Social Sciences, 92*, 55 – 84.

Lee, S. H., Levendis, J., & Gutierrez, L. (2012). Telecommunications and economic growth: An empirical analysis of Sub-Saharan Africa. *Applied Economics*, *44*(4), 461–469. doi:10.1080/00036846.2010.508730

OECD. (2019). Tax challenges arising from digitalisation-interim report 2019 (Inclusive Framework on BEPS, OECD/G20 Base Erosion and Profit Shifting Project). OECD.

Özcan, M. P. (2016). Problems encountered in taxation of the digital economy: An evaluation within the scope of BEPS action plan 1. *Electronic Journal of Vocational Colleges*, *6*(2), 73–82.

Öztürk, L. (2002). The global dimensions of the digital divide. *Ege Academic Review*, *2*(1), 1–10.

Pazarlıoğlu, M., & Kiren Gürler, Ö. (2007). Telecommunications investments and economic growth: Panel data approach. *Finance. Political & Economic Comments*, *44*(508), 35–43.

Phillips, C.B. P. & Perron, P. (1988). Testing for a unit root in time series regressions. *Biometrika*, *75*(2), 335-346.

Rezny, L., White, J. B., & Maresova, P. (2019). The knowledge economy: Key to sustainable development? *Structural Change and Economic Dynamics*, *51*(C), 291–300. doi:10.1016/j.strueco.2019.02.003

Roller, L. H., & Waverman, L. (2001). Telecommunications infrastructure and economic development: A simultaneous approach. *The American Economic Review*, *91*(4), 909–923. doi:10.1257/aer.91.4.909

Tapscott, D. (1995). *The Digital Economy: Promise and Peril in the Age of Networked Intelligence*. McGraw-Hill Publisher.

Taymaz, E. (2018). *Digital Technologies and Economic Growth: Turkey's Position, Opportunities and Options in Digital Technology Sectors*. TÜSİAD.

Toda, H. Y., & Yamamoto, T. (1995). Statistical inference in vector auto regressions with possibly integrated processes. *Journal of Econometrics*, *66*(1-2), 225–250. doi:10.1016/0304-4076(94)01616-8

Topçu, H., & Gedik, A. (2021). The effect of telecommunication investments on economic growth in Turkey. *Journal of Academic Value Studies*, *7*(3), 316–343.

World Economic Forum (WEF). (2018). *The future of jobs report 2018*. https://www3.weforum.org/docs/WEF_Future_of_Jobs_2018.pdf

Yapraklı, S., & Sağlam, T. (2010). Information and communication technologies and economic growth in Turkey: An econometric analysis (1980-2008). *Ege Academic Review*, *10*(2), 575–596. doi:10.21121/eab.2010219637

Yıldız, F. (2012). The impact of telecommunications investments on economic growth: An empirical study on OECD countries. *Süleyman Demirel University Journal of the Faculty of FEAS*, *17*(3), 233–258.

Yüce, M., & Akbulut, N. (2018). Measures taken regarding the taxation of the digital economy within the scope of the BEPS action. *Academic Perspective International Refereed Journal of Social Sciences*, (68), 105–123.

ENDNOTE

[1] Corresponding author.

Chapter 8
Assessing the Role of Trade Openness, FDI, and Political Stability on Sustainable Development:
Evidence From Developed and Developing Economies

Sathish Pachiyappan
CHRIST University (Deemed), India

Jabez J. P.
CHRIST University (Deemed), India

Shylaja H. N.
Department of Commerce, Manipal Academy of Higher Education, Manipal, India

John Paul Raj V
CHRIST University (Deemed), India

Saravanan Vellaiyan
CHRIST University (Deemed), India

ABSTRACT

The study tries to investigate the long run and short run relationship between trade openness (TO), political stability (PO), and FDI on sustainable development of select developed and developing nations. Time series data from 1995 to 2021 of about 25 economies—10 developed economies and 15 developing economies—was collected and analyzed using Phillips Perron Fisher panel unit root test, panel auto regressive distributed lag (PARDL) model, and panel fully modified least squares/fully modified OLS. From the result, it found that FDI and TO are positively contributing to sustainability development index (SDI) in developing countries rather than the developed countries in the long run. In addition to this, changes in the SDI score is significantly influenced by the present and past import and export activities in developed as well as developing economies in the short run.

DOI: 10.4018/978-1-6684-9979-5.ch008

1. INTRODUCTION

Globalization has rapidly changed the world over the past few decades. It is a major economic factor that is synonymous to imports and exports (Gurgul & Lach, 2014). With increased industrialization across nations, impetuous production and consumption, the non-renewable natural resources have been under enormous pressure. The repercussion of such rapid growth has been on the environment which has led to dismaying problems like global warming, ozone layer depletion, biodiversity loss, unequal distribution of wealth etc., which if not checked now, would result in irreparable damages for the future generations. This has raised the importance of sustainable development goals as postulated by the United Nations. Sustainable Development Goals have become imperative with the agenda of sustainable development getting penetrated into not just the government initiatives but also every business, institutions, and other such bodies both public and private. Thus, sustainable development goals have become the centre of policy actions.

With nations across the world, both developed and developing, in their pursuit of growth are getting increasingly industrialized which in turn results in long term implications on the environment. Trade policies across the nations remain a focal point as it has both direct and indirect implications on sustainable development. Sustainable development being the objective of WTO at the Doha Round, focuses on opening world trade while focusing on removing environmentally harmful trade distortionary measures and promoting greater access to environment friendly goods and services. Also, as per both WTO and the World Bank, there is a considerable association between international trade and sustainable development. Many initiatives and programs have been launched by both these organizations that aim to devise trade strategies that could result in economic growth at the same time safeguard the environment for future generations (Balassa, 1986; Weiss, 1992; Chandia et al 2018). With increased international trades opening up, there also has been an increased flow of capital to both developed and developing nations in the form of the FDI.

FDI often results in win-win situations to both the host and the foreign country as it results in innovation, technology transfer, and a host of other economic and non-economic benefits. Both developed and the developing economies have adopted carefully crafted economic reforms and restructured their economies to attract more FDI. There are various factors that have an impact on how much FDI a country would draw. Changing global economic situations, policy changes, political environment, trade openness, political stability, exchange rate, market potential, etc., are few of the factors which have an impact on FDI decisions.

Trade openness and FDI are regarded as the catalyst for rapid economic growth (Mohammed and Ruslee, 2015). As per the international economic theory and policy, the trade openness and foreign direct investment leads to economic growth (Kong et al., 2021; Cheng et al 2021). There are several definitions for trade openness and one of the most common and convenient definitions is that it is the sum of imports and exports normalized by the Gross Domestic Product (Adow and Tahmad 2018; Zaman et al. 2018; Ho et al. 2013). Trade openness and FDI are advantageous to host countries as they expose them to innovative ideas that result in resource utilization which in turn results in optimum returns in the form of economic growth (Ciobanu, 2021).

Political risk is considered a major limiting factor for capital flows (Lucas, 1990). Investments in many developing countries are often subject to large political factors. Political risks and the conditions under which a business operates often are important determinants of FDI for countries (Singh & Jun, 1995). It is found that in countries that have corrupt governments and low levels of democracy, there has been more FDI inflows and also are characterized by higher FDI inward performance (Kim 2010). In this study the focus is on trade openness, foreign direct investment (FDI) and political stability and its influence on sustainable development of a country. The study makes an attempt to understand the long run and short run impact trade openness, FDI and political stability have on the sustainable development.

The first part of the study is the review of existing literature followed by the research methodology followed by theoretical framework and discussions.

2. LITERATURE REVIEW AND HYPOTHESIS DEVELOPMENT

The following section presents the studies that has been previously conducted by scholars in this area of research. The literature review has been presented under three heads: Trade openness and sustainable development, Political Stability and sustainable development, FDI and sustainable development.

2.1 Trade Openness and Sustainability Development

The existing literature prominently points out that outward trade policies have a significant effect on improving the standard of living and economic performance (Kim, 2011). TO has been a means of achieving socioeconomic and environmental prosperity (Frankle & Romer, 1999; Antweiler et al., 2001; Cole & Elliot, 2003; Boulatoff & Jenkins, 2010; Shabaz et al., 2013) Trade openness is found to be a significant element of economic growth and it is argued that a country's growth relies on trade openness and energy consumption. It is even argued that increased trade openness results in improvement in management systems, human capital and also provides access to eco-friendly resources that leads to sustainable economic development (Chien et al, 2023). In an attempt to investigate the relationship between trade & environment using Kuznetz curve hypothesis, Le et al (2016) concluded that TO was beneficial to developed countries while it deteriorates the developing countries as they have to increasingly depend on primary products. It is also concluded that the trade openness and industrialization are increasingly becoming the reasons to vitiate the quality of the environment (Wang et al., 2022). Trade Openness, energy consumption and foreign direct investment are found to be extremely detrimental for sustainable development (Sheikh et al., 2021). The key elements that work as a catalyst to ensure sustainable economic growth of any country are IT exports, capital formation, FDI and trade openness (Zaman et al., 2021). In the study to assess the role of energy consumption and trade openness towards environmental sustainability in Turkey, the authors found that primary energy use, trade openness and financial development damage the environmental quality (Akhayere, 2023). In a panel data analysis of 105 countries belonging to different categories, it was found that the trade openness hampers environmental quality. However, the impact is different for different categories (Shahbaz et al., 2017). A study on 176 countries wherein trade openness, FDI and renewable energy consumption and their impact on environmental sustainability assessed using OLS regressed, fixed effect and generalized method of moments, it was observed that the above three variables were negatively associated with carbon emission (Khan et al., 2021).

2.2. Political Stability and Sustainability Development

Though there are multiple studies on the impact of political stability on economic growth, there are not many studies on political stability and its impact on sustainability development. Sustainable Development is essentially the development that focuses on current use of resources to ensure growth without resulting in any harm to the future use of resources (Güney, 2017). In a study on the political stability and its impact on sustainable growth in Romania, it was found political stability has an important role in a country's economic growth and that a stable political environment helps in creating a coherent and continuous path for sustainable development (Radu, 2015).

2.3 Foreign Direct Investment and Sustainability Development

There have been multiple studies which have highlighted the inter relationship between FDI and sustainability. FDI can promote green growth and sustainable innovation (Amendolagine et al., 2021; Melane-Lavado et al., 2018; Kardos, 2014), which would reduce welfare costs associated with dirty growth, reduce pollution related mortalities. Environmentally sustainable production and consumption practices can be encouraged employing the efficiency seeking and market seeking FDI (Buchner et al., 2011; United Nations Conference on Trade and Development [UNCTAD], 2011). Environmental conservation and coastal development can be induced with the flow of FDI into the telecommunications and hospitality sector (Falk, 2016; Osinubi et al., 2022). There are also certain studies that argue that FDI can set off ecological footprint and accelerate acidifying gasses, ozone precursor gasses and even air pollutants (Doytch, 2020). An economically sustainable system not only produces goods and services for the residents on a continuing basis but also ensures that the practices they follow doesn't create any sectorial imbalances that leads to agricultural and industrial damage (Harris et al., 2001).

Based on the above discussion, the study address the solution for the following research question to fulfil the research gap. To what extent the FDI, Political Stability and Trade Openness influence on Sustainability Development in both developed as well as developing nations in short run as well as long run.

3. RESEARCH OBJECTIVES:

i. To investigate the relationship between FDI, Political Stability, Trade Openness and sustainable development in both developed and developing economies.
ii. To assess the short run and long run impact of political stability, trade openness, FDI on Sustainable development index in the both developing and developed countries.

4. RESEARCH HYPOTHESES

The following hypotheses are framed based on the research objectives in an attempt to answer the research questions.

H_1-There is a positive relationship between FDI, Political Stability and Trade Openness and sustainable development in developed and developing economies.

H_2-There is a significant short run and long run impact of FDI, Political Stability and Trade Openness on sustainable development in developed and developing economies.

5. DATA COLLECTION AND METHODOLOGY

The use of econometric model in studying the determinants of carbon emissions, including energy consumption, economic growth (Wang et al., 2018), research and development and globalization (Wang and Zhang, 2020) are very well examined in the existing studies. Therefore, the use of econometric models to the current research would provide better results to analyze the objective of the study. The following subsections discuss the data, selection of variables, econometric model and its applications.

5.1 Data

According to (World Economic Situation and Prospects, 2020), the world consists of 36 developed and 136 developing economies. By adopting convenience sampling under a non-probability method, this study uses data for developed (10 countries) and developing (15 countries) economies over a period of 27 years from 1995 to 2021. The selected continents include Europe, Asia, Africa, and American Countries. The developed countries includes United States, Canada, United Kingdom, Germany, Norway, Japan, Sweden, Switzerland, Australia, New Zealand while India, China, Brazil, South Africa, Mexico, Indonesia, United Arab Emirates, Turkey, Egypt, Philippines, Bangladesh, Vietnam, Argentina, Iran and Sri Lanka are added in the developing countries.

5.2 Selection of the Variables

5.2.1 Variables of the Study

5.2.1.1 Sustainable Development Index (SDI)

The Sustainable Development Index (SDI) has been formulated by the author and the same has been considered as the outcome variable. SDI is created from five crucial variables such as Gross Domestic Product (GDP), CO_2 Emission per Capita, Human Development Index (HDI), Sustainable Development Goals (SDG) and Energy Consumption per Capita from the selected counties. The following table presents the detailed description of crucial variables which helps to create the SDI.

SDI was formulated with the help of Data Envelopment Analysis (DEA) which is used for efficiency analysis and performance measurement. To formulate the SDI, the goal should be specified behind the Index, wherein here, the environmental and social aspects are signified in the formulation of this Index. This was computed based on efficiency measurement, whereas the two outputs have been divided as desirable and undesirable, along with three different inputs. After selecting the orientation of these variables, they need to be normalized for further use. The outputs, GDP, and CO_2 Emission per Capital, have been normalized by dividing their values by the minimum of that particular reference period. In contrast, the SDG and HDI score so to normalize them, they all have been divided by a hundred for easy calculation. The other input variable, Energy consumption per capita, has been normalized by dividing their data values with the maximum of that reference period as it goes under input. Now, weights have been assigned equally for input and output variables. As the output has two variables, and they are equally responsible for SDI, the weights have been assigned as 0.5 for each. The input variables HDI, SDG, and Energy consumption per Capita had weights as 0.4,0.4, and 0.2 because the Index has been focused on environmental and social aspects of an economy. Hence the formula becomes, the values of

Table 1. Description of the variables to create SDI

Variables	Definition	Nature	Position
GDP	Gross Domestic Product – A macroeconomic Indicator	Desirable Output – Measured in USD Million	As it directly explains the economic position of a country, this is a desirable output for our Index. If GDP is higher, the SDI will be higher.
CO_2 Emission per Capita	One of the Environment Performance of a nation.	Undesirable output – Measured in tons per annum.	This shows the environmental performance of a nation, as its getting lower, the Index will get better.
HDI	Human Development Index	Input Variable – Measured as score.	This variable has been taken as human capital Input, whereas our desired output needs the status of people (societal) to achieve sustainability.
SDG	Sustainable Development Goals Score – Combination of multiple aspects (17 Universal Goals)	Input Variable – Measured in score	This variable explains the environmental aspects of the innovation and industrial input.
Energy Consumption per Capita	A minor energy indicator of a nation.	Input variable – Measured in kWh/capita per annum	This acts as energy input for our efficiency index, as this should be reduced for a sustainable development goal of SDI.

the Sustainable Development Index (SDI) range from positive to negative values, as the economy may positively or negatively be developed in a sustainable aspect during the reference period.

5.2.1.2 Foreign Direct Investment

Foreign Direct Investment (FDI) is considered as an explanatory variable in order to check the influence of international capital flows on sustainability development index of both developed and developing economies.

5.2.1 3 Trade Openness

Trade Openness (TO) is considered as an explanatory variable. This would help in identifying to what extent the export and import support to the sustainability development index of both developed and developing economies.

5.2.1.4 Political Stability

The third explanatory variable is considered as political Stability (PO) which represents the economy's stability towards implementing and controlling national policies. This percentile provides an overall estimation of non-violence and anti-corruption measurement of that economy.

Table 2. Variable descriptions

	Variable	Abbreviation	Measurement	Source
Dependent	Sustainable Development Index	SDI	Index ranges positively and negatively	Formulated by Author/Bloomberg
Independent	Foreign Direct Investment	FDI	USD Millions (Net FDI Inflows)	Bloomberg
	Trade Openness	TO	Percentage (% of GDP on Trade)	Bloomberg
	Political Stability	PO	Percentile (Estimation over Non-Violence and Anti-Corruption)	Bloomberg

5.3. Structure of Data and Econometric Model

The study uses panel data structure since the four variable data has been collected for 25 (developed and developing) countries over a period of 27 years. Descriptive statistics was used to describe the characteristics of each variable given in the study. Followed by unit root test has been done with the help of Fisher ADF and Phillips Perron since the structure of data is unbalanced panel data (Gozgor, 2011). An analyze was done to know the relationship between the variables through correlation matrix. Further, Panel Autoregressive Distributed Lag (PARDL) model have been used to check if the short and long run does exist. In addition, to support the result of PARDL, the Panel Fully Modified Least Squares (FMOLS) method was used.

6. RESULTS AND DISCUSSION

Based on the analysis, the following section provides the results and discussions.

6.1 Descriptive Statistics

The basic statistical characteristics has been presented in the following section.

Table 3 shows the summary of descriptive statistics of actual values of the variables both for developed as well as developing countries. The highest mean score of sustainability development index with more volatility (Mean = 0.163, S.D=0.519) was observed in developing countries than the SDI score in developed countries (Mean =0.039, S.D=0.175). Also, the average investment of FDI is more with least volatility (Mean =1358.897, SD=12597.760) in developing countries whereas the mean of FDI investment is less but high volatility in developed countries. Similar mean values was found in the trade openness in both economies but however the volatility was little higher in developing countries than the developed countries. Political stability mean value is high with more variation (Mean =0.030, S.D= 0.134) in developed countries than the developing countries. Apart from this, minimum and maximum value of each variable has been given in the table. Probability of Jarque Berra test provides that all the variables are not normally distributed. Overall, the actual mean score of SDI, FDI and TO are relatively more in emerging countries than the developed economies.

Table 3. Summary of descriptive statistics

	Developed Countries				Developing Countries			
Statistics	**DSDI**	**DFDI**	**DTO**	**DPS**	**DSDI**	**DFDI**	**DTO**	**DPS**
Mean	0.039	2191.039	-0.01	0.030	0.163	1358.897	-0.01	0.011
Maximum	0.535	299412.000	0.101	0.994	3.634	112646.400	0.899	0.810
Minimum	-0.66	-303623.80	-1.16	-0.994	-1.649	-67739.750	-2.09	-0.804
Std. Dev	0.175	65573.500	0.134	0.470	0.519	12597.760	0.174	0.204
Probability	0.000	0.000	0.000	0.042	0.000	0.000	0.000	0.000
Observations	260	260	260	260	390	390	390	390

6.2 Stationarity Tests

Table 4 provides the results of Phillips perron fisher panel unit root test to check the presence of unit root problems or not in the variables. It is evidenced that foreign direct investment and political stability are stationarity at level but the other two variables such as sustainability index and trade openness has unit root problem. Variables which are having unit root problem cannot be processed for further analysis. Hence, all the variables are converted into the first difference and becomes stationarity at one percent significant level.

6.3 Correlation Matrix

The correlation matrix shows the correlation coefficients between pairs of variables in two panels of data both for developing as well as developed countries which is depicted in Table 5. There was an inverse relationship found in the TO ($r=-0.20$, $p<0.05$) and PS ($r=-0.20$, $p<0.05$) with SDI in developed countries. Similar relationship exists in the TO ($r=-0.18$, $p<0.05$ and PS ($r=-0.11$, $p<0.01$) with SDI in

Table 4. Phillips Perron Fisher panel unit root test results

	Developed Economies		Developing Economies	
Variables	Level	First Diff.	Level	First Diff.
SDI	0.9897	0.0000*	1.0000	0.0000*
FDI	0.0000*	0.0000*	0.9538	0.0000*
TO	0.2945	0.0000*	0.2452	0.0000*
PS	0.0732***	0.0000*	0.0001*	0.0000*

*&*** indicates 1 & 10 percent significance level

Table 5. Correlation matrix

Developed Economies Panel				
Variables	DSDI	DFDI	DTO	DPS
DSDI	1	-	-	-
DFDI	0.0050	1	-	-
DTO	-0.2019**	-0.0764	1	-
DPS	-0.2058**	0.0799	0.0187	1
Developing Economies Panel				
Variables	DSDI	DFDI	DTO	DPS
DSDI	1	-	-	-
DFDI	0.2938**	1.0000	-	-
DTO	-0.1831**	-0.0307	1.0000	-
DPS	-0.1147*	-0.0503	-0.0164	1.0000

** & * indicates 5 & 1 percent significance level

developing countries also. However, FDI does not have any relationship exist with SDI while FDI have positive and significant relationship with SDI by 29 percent at 5 percent significant level. Overall, the contribution of foreign direct investment is relatively more towards the improvement of SDI score in developing countries than developed countries.

6.4 Auto Regressive Distributed Lag (ARDL) Model

This model was initiated by Pesaran and Shin (1999) and further it was enlarged by Pesaran, Shin and Smith (2001). They stated that the main assumption of ARDL approach is selected variables does not require the same order of integration and this model takes into the account of a different order of integration i.e I*(0)* or I*(1)* to estimate the short and long run relationship. Hence, the panel ARDL model has been adopted to find short and long run exist.

Table 6 shows the output of panel ARDL model in the long run. It is observed that DFDI and DTO does not have long run impact on SDI in developed economies whereas long run influence has been identified at one percent significant level in the developing countries. On the other hand, for developed economies, DPS does have significant influence on SDI in the long run while there is no significant influence on SDI in the developing economies. The result of this study is aligned with the study of (Chien et al, 2023) stated that increased trade openness results in improvement in management systems, human capital and also provides access to eco-friendly resources that leads to sustainable economic development. On the other hand, similar results found by (Amendolagine et al., 2021; Melane-Lavado et al., 2018; Kardos, 2014) mentioned that FDI can promote green growth and sustainable innovation which would reduce welfare costs associated with dirty growth, reduce pollution related mortalities. Overall, foreign direct investment and trade openness has major contributing for the sustainability development index in developing countries than the developed economies.

Table 7 depicts the short run influence of political stability, trade openness and foreign direct investment on sustainability index in both developed and developing economies. It is observed that changes in SDI score does not have significant influence by previous months score of SDI in developing economies but however, changes in SDI score has significantly influenced by the second month of SDI score in developed economies in the short run. On the other hand, present and past month FDI investment does not majorly contribute towards the improvement of SDI score in developed as well as developing economies. Changes in the SDI score does significantly influence by the present and past import and export activities in developed as well as developing economies in the short run. However, the magnitude of the effect is larger for developing economies. Political stability has negative and significant influence on SDI score in the developed economies than the developing countries. However, SDI score does not

Table 6. Panel ARDL long run estimation

Variables	Developed Economies			Developing Economies		
	Co-Efficient	Std. Error	Probability	Co-Efficient	Std. Error	Probability
DFDI	-8.62E-07	6.79E-07	0.2071	2.86E-05	4.47E-06	0.0000*
DTO	1.3035	0.8156	0.1132	1.4351	0.4947	0.0043*
DPS	2.4422	0.5020	0.0000*	-0.0331	0.7717	0.6687

** and * indicates 5 & 1 percent significance level

Table 7. Panel ARDL short run estimation

Variables	Developed Economies			Developing Economies		
	Co-Efficient	Std. Error	Probability	Co-Efficient	Std. Error	P Probability
ECT_1	-0.3899	0.0836	0.0000*	-0.5234	0.1155	0.0000*
D(DSDI(1))	0.0572	0.1011	0.5729	-0.0027	0.1167	0.9811
D(DSDI(-2)	-0.5145	0.1113	0.0000*	-0.1122	0.0906	0.2176
D(DSDI(-3)	-0.0431	0.1156	0.7099	0.0044	0.1288	0.9727
D(DFDI)	8.47E-06	7.10E-06	0.2360	-1.60E-06	3.36E-05	0.9622
D(DFDI(-1))	8.72E-06	6.26E-06	0.1672	-5.06E-06	3.79E-05	0.8827
D(DFDI(-2))	5.15E-06	4.59E-06	0.2642	1.85E-06	4.27E-05	0.9655
D(DFDI(-3))	1.41E-06	2.06E-06	0.4965	-4.11E-05	6.15E-05	0.5051
D(DTO)	-0.6867	0.2215	0.0025*	-2.6208	0.6293	0.0001*
D(DTO(-1))	-1.3445	0.8340	0.1102	-3.3192	0.7648	0.0000*
D(DTO(-2))	-0.0253	0.3644	0.9448	-2.2272	0.8003	0.0061*
D(DTO(-3))	0.3966	0.3205	0.2190	-1.0351	0.8967	0.2502
D(DPS)	-1.0351	0.2693	0.0002*	0.2566	0.4348	0.5560
D(DPS(-1))	-0.5617	0.2665	0.0377**	0.0671	0.4971	0.8928
D(DPS(-2))	-0.1863	0.1931	0.2844	0.5286	0.5428	0.3317
D(DPS(-3))	0.0140	0.0856	0.8704	0.9070	0.4449	0.0447**

** and * indicates 5 & 1 percent significance level

influence majorly by past political events. The coefficient of error correction term has negative and significant at one percent level. It indicates the short run deviation gets corrected or adjusted by 0.38 percent towards the long run equilibrium for every month in the developed economies while 0.52 percent in developing countries.

6.5 Panel Fully Modified Least Squares/Fully Modified OLS

Fully Modified OLS (FMOLS) can be used to support the results of panel data from ARDL method. FMOLS is a method used to estimate co-integrating relationships, which is similar to the ARDL method. Both methods can be used to test for the existence of a long-run relationship between variables and estimate the parameters of this relationship. While the ARDL method is particularly useful for panel data, FMOLS can be used for both panel data and time series data. FMOLS can also provide more accurate estimates of the long-run parameters when the regressors are endogenous or exhibit serial correlation. Therefore, FMOLS can be a useful complement to ARDL when analyzing panel data, particularly when dealing with issues related to endogeneity and serial correlation.

This Panel Dynamic Least square method is a variation of the Ordinary Least Squares (OLS) regression but with the addition of lagged dependent variables and/or lagged regressors to capture the dynamic relationship between variables over time. This technique has been used to support the long and short-run relationships between dependent and independent variables which is given in Table 8. As, the R-square value for developed economies resulted in 5 percent, which gives changes in the SDI score is explained

Table 8. Panel fully modified OLS: Cointegration regression

	Developed Economies			Developing Economies		
Variables	**Co-efficient**	**Std. Error**	**Probability**	**Co-efficient**	**Std. Error**	**Probability**
DFDI	2.46E-08	1.77E-07	0.8893	1.13E-05	2.28E-06	0.0000*
DTO	-0.2784	0.0857	0.0013*	-0.5864	0.1615	0.0003*
DPS	-0.0826	0.0264	0.0020*	0.2927	0.1478	0.0485**
Dependent Variable	DSDI			DSDI		
Cross Sections	10			15		
Observations	250			375		
R-squared	0.0982			0.2055		
Adjusted R-squared	0.05275			0.1677		
Long Run Variance	0.0322			0.2898		

** and * indicates 5 & 1 percent significance level

by changes in the PO, FDI and TO. However, the variable of FDI does not have significant, which implies that for developed economies can be explained by more undisclosed variables. But Political stability is statistically significant, and it explains SDI Index crucially in developed economies. In developing economies, the R-square value reveals that model is very well valid and dependent variables is nearly 16 percent affected by FDI, TO and PO. All the chosen explanatory variables are contributing for changes in the Sustainability index score in developing economies.

7. CONCLUSION AND POLICY IMPLICATIONS

The study scrutinized the influence of Foreign Direct Investment (FDI), Trade Openness (TO) and Political Stability (PO) on Sustainability Development Index (SDI) in both developed and developing nations. The study assessed the influence level each factor explanatory variables on SDI by including 25 economies (10 developed and 15 developing nations) from 1995 to 2021. Furthermore, the new contribution made by the authors in this paper is formulated the new index is known as Sustainability Development Index (SDI). This has been formulated by considering of five variables such as GDP, CO2 emissions per capita, HDI score, SDG score, and energy consumption per capita. To properly deal with panel structure data, the study used Panel Autoregressive Distributed Lag Model (PARDL) to check the short run and long run influence. In addition, fully modified OLS has been done to support the result of PARDL model.

The findings of the study have been summarized in three ways. Firstly, the empirical results of correlation analysis shows that negative relationship found between TO and SDI followed by PO and SDI in both developed and developing countries. PO and TO increases, the sustainability development index decreases. This is reflected in the review of literature as well as multiple studies have established that TO has differential affects on both developed and developing economies. It indicates, one rationale is that greater trade openness would result in more intense economic activity, which would consume more resources and worsen the environment. Additionally, nations with greater trade openness can put more emphasis on economic development than on sustainability issues, which would result in lower sustainability index scores. On the other hand, political instability frequently interferes with governance, policy making, and the adoption of sustainable practices, which could be one explanation for this negative link. Thus the findings are in sync with the limited literature review available where it has been established that political stability influences sustainable development. In times of political unrest, governments may find it difficult to put long-term sustainability goals first and may choose to prioritize short-term stability or other urgent matters. In wealthy nations, there is no connection between FDI and the sustainability score. This suggests that changes in FDI levels do not correspond to changes in sustainability performance, and that the level of FDI has no bearing on the sustainability index in industrialized countries. But, however, the contradictory result found in the developing countries. Developed and developing economies have behaved differently to FDI as per the observations made by previous available studies reflected in the literature section. Also it has been found that sustainability score rises as FDI rises. FDI can help developing nations via capital investments, knowledge transfer, and technological advancements. Through the promotion of more environmentally friendly production methods, environmental preservation, and social advancement and these aspects can support sustainable development. Foreign direct investment (FDI) inflows can also spur economic growth, which, if supported by efficient governance and regulations, may open doors for sustainable development.

Second, the estimated the result of PARDL model indicates that there is no co-integration between TO and SDI as well as FDI and SDI in developed countries. This suggests that shifts in FDI levels and TO do not correspond to long-term adjustments or gains in the sustainability development index in industrialized nations. The fact that developed nations frequently have strong institutional frameworks, regulatory frameworks, and governance structures in place to enable sustainable growth regardless of FDI levels is one rationale. It's possible that these nations have already attained a certain level of sustainability and have built policies and practices that influence sustainability outcomes, which will eventually reduce the influence of FDI. On the other hand, The fact that industrialized nations frequently have social norms, environmental rules, and sustainable practises in place regardless of the degree of trade openness is one explanation for this argument. The level of sustainability that these nations have already attained and the implementation of practises and policies that promote sustainability may have lessened the long-term importance of trade openness. However, FDI and TO has co-integrated with SDI in developing countries. This suggests that adjustments in the level of FDI and TO are related to long-term adjustments or enhancements in the sustainability development index in emerging nations. The ability of FDI to provide developing nations with access to finance, technology, knowledge, and markets is one explanation for this inference. By encouraging economic growth, job creation, infrastructure development, and social advancement, these variables can support sustainable development. The transfer of sustainable practises, environmental norms, and governance concepts from industrialized to developing nations is another benefit of FDI inflows. With regard to TO, the assumption can be justified by the idea that trade openings might help developing nations gain access to global markets while also promoting economic expansion and job creation. Investing in sustainability initiatives like environmental preservation, social advancement, and capacity building can be made possible by this economic development, which can also give the necessary resources and opportunities. Additionally, trade openness may encourage the adoption of new technologies, best practises, and information from more sustainable economies, all of which may improve sustainability over the long run.

Furthermore, Political stability and the sustainability development index in industrialized nations are related over the long term. This suggests that improvements or sustained changes in the sustainability development index over time in industrialized countries are correlated with changes in the degree of political stability. This conclusion can have as its justification political stability, which creates a favourable climate for programmes promoting sustainable development. It makes it possible to put long-term plans into action, to invest in social welfare and environmental preservation, and to build efficient governance institutions. Economic growth, social cohesion, and international and domestic investment can all be boosted by political stability, which will ultimately lead to sustainable development outcomes. However, there is no co integration between PS and SDI in developing countries. The assumption could be explained by the fact that developing nations have a difficult time converting political stability into results in terms of sustainable development. The successful implementation of sustainable development projects may be hampered by elements including shoddy governance frameworks, poor institutional capability, social injustices, and economic weaknesses. Furthermore, developing nations can put short-term sustainability last in favour of quick economic growth or other pressing issues, which over time would result in a lower association between political stability and the sustainability development index.

Finally, the current SDI score in developing economies is not greatly influenced by the score from the previous month. However, in developed economies, the score from the second month has a major impact on changes in the SDI score. This suggests that the SDI score and its variations in these two types of economies follow various patterns or correlations. The lack of a discernible correlation between the SDI score from the previous month and the current SDI score in developing economies shows that other variables or causes may have a bigger influence on the sustainability development index. This would suggest that emerging markets encounter particular difficulties, such as a lack of infrastructure, resources, or institutional restraints, which might have a more significant impact on the present SDI score than the past values whereas the strong impact of the SDI score from the second month on changes in the SDI score in developed economies implies that short-term alterations or revisions to the sustainable development index have a discernible effect. This may be a sign that developed economies have more established and stable institutions, enabling them to be more adaptable to modifications in sustainability practises, policies, or performance. On the other hand, level of FDI in the present and past months does not significantly contribute to the improvement of the SDI score in both developed and developing economies. This implies that other factors or variables have a greater influence on the SDI score than FDI investments in these economies in the short run. The volume of import and export activity over the most recent two months has a big impact on how quickly the SDI score changes in both developed and developing nations. Through a number of ways, import and export activity might affect the SDI score. Increased export activity may be a sign of better economic output, employment, and income production, all of which are beneficial for sustainable development. By facilitating access to resources, technologies, and knowledge that support efforts towards sustainable development, import operations can also have an impact on the SDI score. Additionally, trade operations could have an impact on the SDI score due to environmental factors including transportation-related carbon emissions. In developed economies, political stability has a short-term, significant negative impact on the SDI score. This suggests that in industrialized economies, the SDI score declines in the short run as political stability rises. Political stability on a high level could cause complacency or a lack of urgency in solving sustainability issues. It can mean that political stability permits the status quo to be maintained, which could make it more difficult to execute critical policy reforms or sustainability efforts. In developed economies, other factors and variables may also have an impact on the SDI score, and there may be a complex and context-specific relationship between political stability and the SDI score.

The main policy implications arising from the study can be presented as follows: Firstly, By fostering economic growth, knowledge transfer, job creation, and access to new markets, engaging in international trade and luring foreign investment can support attempts to achieve sustainable development in emerging economies. Second, the significance of including environmental factors in trade and FDI activity. Developing nations can gain from adopting sustainable practises to make sure that trade and investment operations contribute to long-term sustainability. Examples of such practises include supporting renewable energy, sustainable resource management, and environmental regulations. Third, underline the necessity for developing nations to improve their institutional ability in order to manage trade and investment flows properly. To guarantee that trade openness and FDI are in line with sustainable development goals, this includes creating strong regulatory frameworks, enhancing governance structures, and improving monitoring and enforcement systems. Fourth, To maximize the benefits of trade openness and foreign direct investment on sustainable development, policymakers in developing nations should establish strategies. This may entail putting in place laws that encourage ethical investment, maintaining a good business climate, encouraging knowledge transfer, and incorporating sustainability standards into trade agreements.

8. SCOPE AND LIMITATIONS OF THE STUDY

The current study assessing the role of trade openness, foreign direct investment and political stability on sustainability development index in both developed and developing nations, and the availability of data on variables used in the study restricts its scope. Future research may take into account including the other factor such as economic factors, social factors, environmental factors to find the influence on SDI. The study is only restricts to 25 countries which includes 10 developed and 15 developing nations. Future research can broaden the data sources used in the analysis to include other nations and areas in order to overcome this constraint, improving the conclusions' robustness and reliability.

REFERENCES

Akhayere, E., Kartal, M. T., Adebayo, T. S., & Kavaz, D. (2023). Role of energy consumption and trade openness towards environmental sustainability in Turkey. *Environmental Science and Pollution Research International*, *30*(8), 21156–21168. doi:10.100711356-022-23639-9 PMID:36261639

Antweiler, W., Copeland, B. R., & Taylor, M. S. (2001). Is Free Trade Good for the Environment? *The American Economic Review*, *91*(4), 877–908. doi:10.1257/aer.91.4.877

Balassa, B. (1986). Policy responses to exogenous shocks in developing countries. *The American Economic Review*, *5*, 75–78.

Boulatoff, C., & Jenkins, M. (2010). Long-term Nexus Between Openness, Income, and Environmental Quality. *International Advances in Economic Research*, *16*(4), 410–418. doi:10.100711294-010-9283-y

Buchner, B., Falconer, A., Hervé-Mignucci, M., Trabacchi, C., & Brinkman, M. (2011). *The landscape of climate finance*. Climate Policy Initiative.

Chandia, K. E., Gul, I., Aziz, S., Sarwar, B., & Zulfiqar, S. (2018). An analysis of the association among carbon dioxide emissions, energy consumption and economic performance: An econometric model. [CrossRef]. *Carbon Management*, *9*(3), 227–241. doi:10.1080/17583004.2018.1457930

Cheng, C. Y., Chien, M. S., & Lee, C. C. (2021). ICT diffusion, financial development, and economic growth: An international cross-country analysis. *Economic Modelling*, *94*, 662–671. doi:10.1016/j.econmod.2020.02.008

Chien, F., Paramaiah, C., Thuy, N. T. T., Phan, T. T. H., & Ngo, T. Q. (2023). The impact of eco-innovation, trade openness, financial development, green energy and government governance on sustainable development in ASEAN countries. *Renewable Energy*, *211*, 259–268. doi:10.1016/j.renene.2023.04.109

Ciobanu, A. M. (2021). The impact of FDI on economic growth in case of Romania. *International Journal of Economics and Finance*, *12*(12), 1–81.

Cole, M. A., & Elliott, R. J. R. (2003). Determining the trade–environment composition effect: The role of capital, labor and environmental regulations. *Journal of Environmental Economics and Management*, *46*(3), 363–383. doi:10.1016/S0095-0696(03)00021-4

Doytch, N. (2020). The impact of foreign direct investment on the ecological footprints of nations. *Environmental and Sustainability Indicators*, *8*, 100085. doi:10.1016/j.indic.2020.100085

Falk, M. (2016). A gravity model of foreign direct investment in the hospitality industry. *Tourism Management*, *55*, 225–237. doi:10.1016/j.tourman.2016.02.012

Frankel, J. A., & Romer, D. (1999). Does trade cause growth? *The American Economic Review*, *89*(3), 379–399. doi:10.1257/aer.89.3.379

Gozgor, G. (2011). *Munich Personal RePEc Archive Panel unit root tests of purchasing power parity hypothesis: Evidence from Turkey Panel Unit Root Tests of Purchasing Power Parity Hypothesis: Evidence from Turkey*. Academic Press.

Güney, T. (2016). *Governance and sustainable development: How effective is governance?* doi:10.1080/09638199.2016.1249391

Gurgul, H., & Lach, Ł. (2014). Globalization and economic growth: Evidence from two decades of transition in CEE. *Economic Modelling*, *36*, 99–107. doi:10.1016/j.econmod.2013.09.022

Haksoon, K. (2010). Political Stability and Foreign Direct Investment. *International Journal of Economics and Finance*, *2*(3). Advance online publication. doi:10.5539/ijef.v2n3p59

Harris, Wise, Gallagher, & Goodwin. (Eds.). (2001). A Survey of Sustainable Development: Social and Economic Dimensions. Island Press.

Ho, C. Y., Wang, W., & Yu, J. (2013). Growth spillover through trade: A spatial dynamic panel data approach. *Economics Letters*, *120*(3), 450–453. doi:10.1016/j.econlet.2013.05.027

Kardos, M. (2014). The relevance of Foreign Direct Investment for sustainable development. Empirical evidence from European Union. *Procedia Economics and Finance*, *15*, 1349–1354. doi:10.1016/S2212-5671(14)00598-X

Khan, H., Weili, L., & Khan, I. (2021). Environmental innovation, trade openness and quality institutions: An integrated investigation about environmental sustainability. *Environment, Development and Sustainability*, 1–31.

Kim, H. (2010). Political stability and foreign direct investment. *International Journal of Economics and Finance*, *2*(3), 59–71. doi:10.5539/ijef.v2n3p59

Kong, Q., Peng, D., Ni, Y., Jiang, X., & Wang, Z. (2021). Trade openness and economic growth quality of China: Empirical analysis using ARDL model. *Finance Research Letters*, *38*, 101488. doi:10.1016/j.frl.2020.101488

Le, T. H., Chang, Y., & Park, D. (2016). Trade openness and environmental quality: International evidence. *Energy Policy*, *92*, 45–55. doi:10.1016/j.enpol.2016.01.030

Lucas, R. E. Jr. (1990). Liquidity and interest rates. *Journal of Economic Theory*, *50*(2), 237–264. doi:10.1016/0022-0531(90)90001-Z

Melane-Lavado, A. (2018). Green foreign direct investments and the deepening of capabilities for sustainable innovation in multinationals: Insights from renewable energy. *Journal of Cleaner Production*.

Muhammad Shahbaz, S. K. (2013). The dynamic links between energy consumption, economic growth, financial development and trade in China: Fresh evidence from multivariate framework analysis. *Energy Economics*, *40*, 8–21. doi:10.1016/j.eneco.2013.06.006

Osinubi, T. T., Ajide, F. M., & Osinubi, O. B. (2022). Governance, FDI and tourism in top 10 tourist destinations in Africa. *Transnational Corporations Review*, *14*(2), 140–152. doi:10.1080/19186444.2022.2040345

Radu, M. (2015). Political stability-a condition for sustainable growth in Romania? *Procedia Economics and Finance*, *30*, 751–757. doi:10.1016/S2212-5671(15)01324-6

Shahbaz, M., Nasreen, S., Ahmed, K., & Hammoudeh, S. (2017). Trade openness–carbon emissions nexus: The importance of turning points of trade openness for country panels. *Energy Economics*, *61*, 221–232. doi:10.1016/j.eneco.2016.11.008

Sheikh, M., Malik, M., & Masood, R. (2021). Dynamic linkage between trade openness and sustainable development: Evidence from the BRICS countries. *BRICS Journal of Economics*, *2*(1), 23–39. doi:10.38050/2712-7508-2021-27

Singh, H., & Jun, K. W. (1995). Some new evidence on determinants of foreign direct investment in developing countries (No. 1531). World Bank Publications.

Tahmad & Adow. (2018). The impact of trade openness on foreign direct investment in Sudan by sector in the 1990-2017 period: An empirical analysis. *Economic Analysis, 7*(8), 14–21.

Wang, Q., Su, M., & Li, R. (2018). Toward to economic growth without emission growth: the role of urbanization and industrialization in China and India. *J Clean Prod, 205*, 499-511. https://doi.org/. doi:10.1016/j.jclepro.2018.09.034

Wang, Q., & Zhang, F. (2020). The effects of trade openness on decoupling carbon emissions from economic growtheevidence from 182 countries. *Journal of Cleaner Production*, *279*, 123838. doi:10.1016/j.jclepro.2020.123838 PMID:32863606

Wang, W., Rehman, M. A., & Fahad, S. (2022). The dynamic influence of renewable energy, trade openness, and industrialization on the sustainable environment in G-7 economies. *Renewable Energy*, *198*, 484–491. doi:10.1016/j.renene.2022.08.067

Weiss, E. B. (1992). Environment and trade as partners in sustainable development: A commentary. *The American Journal of International Law*, *86*(4), 728–735. doi:10.2307/2203789

World Economic Situation and Prospects. (2020). https://www.un.org/development/desa/dpad/wpcontent/uploads/sites/45/WESP2020_Annex.pdf

Yusoff, M. B., & Nuh, R. (2015). Foreign direct investment, trade openness and economic growth: Empirical evidence from Thailand. *Foreign Trade Review*, *50*(2), 73–84. doi:10.1177/0015732515572055

Zaman, M., Pinglu, C., Hussain, S. I., Ullah, A., & Qian, N. (2021). Does regional integration matter for sustainable economic growth? Fostering the role of FDI, trade openness, IT exports, and capital formation in BRI countries. *Heliyon*, *7*(12), e08559. doi:10.1016/j.heliyon.2021.e08559 PMID:34917822

Zhang, C., & Zhou, X. (2016). Does foreign direct investment lead to lower CO2 emissions? Evidence from a regional analysis in China. *Renewable & Sustainable Energy Reviews*, *58*, 943–951. doi:10.1016/j.rser.2015.12.226

Chapter 9
Factors Affecting the Implementation of Business Analytics in the Healthcare Sector in MENA Countries

Aftab Ara
https://orcid.org/0000-0002-2239-5897
Independent Researcher, Saudi Arabia

ABSTRACT

This chapter delves into the factors influencing the implementation of business analytics in the healthcare sector of MENA (Middle East and North Africa) countries. A comprehensive theoretical model has been developed to encompass various organizational, technological, human, and external factors. The data analysis plan involves both descriptive and inferential statistical methods to analyze the collected data. Regression analysis, factor analysis, and correlation analysis will be used to examine the relationships between the identified factors and the success of business analytics implementation. Additionally, qualitative analysis will be conducted to identify common strategies and approaches that have been effective in overcoming barriers to the adoption of business analytics in healthcare. The findings of this study will contribute to the existing body of knowledge on business analytics adoption in healthcare and provide insights for managers and policymakers in MENA countries to overcome barriers and promote the effective utilization of business analytics in the healthcare sector.

1. INTRODUCTION

In this study, we aim to investigate the factors influencing the successful implementation of business analytics in the healthcare sector in MENA countries. To achieve this objective, we have developed a theoretical model that incorporates various organizational, technological, human, and external factors that may impact the adoption and utilization of business analytics in healthcare. This theoretical model

DOI: 10.4018/978-1-6684-9979-5.ch009

serves as a framework for understanding the complex dynamics involved in implementing business analytics in the healthcare context.

1.1. Theoretical Model Development

The theoretical model developed in this study is based on a comprehensive review of the existing literature on business analytics adoption in healthcare (Saleh et al., 2009). We have synthesized and integrated key concepts and variables from previous studies to create a model that captures the multifaceted nature of the implementation process. The model includes variables such as organizational culture, leadership support, resources availability, technological infrastructure, data quality, data literacy, change readiness, regulatory environment, and industry collaboration.

By developing this theoretical model, we aim to provide a structured framework for analyzing the factors that influence the successful implementation of business analytics in the healthcare sector. This model will guide our data collection, analysis, and interpretation, enabling us to gain a deeper understanding of the challenges and opportunities associated with adopting business analytics in MENA healthcare.

The healthcare industry is increasingly recognizing the potential of business analytics to improve decision-making, enhance operational efficiency, and drive innovation. Several studies have explored the adoption and implementation of business analytics in healthcare settings, shedding light on the challenges and opportunities associated with this emerging field. However, there is a notable gap in research when it comes to understanding the specific factors and challenges that impede the successful adoption of business analytics in the healthcare sector within the Middle East and North Africa (MENA) region. While there have been studies conducted on the adoption of business analytics in healthcare in the MENA region, they often focus on broader challenges or do not provide a comprehensive understanding of the barriers specific to the healthcare sector (Shahbaz et al., 2020; Alkraiji et al., 2013; Ward et al., 2014). This study aims to fill this gap by providing a detailed analysis of the factors and challenges specific to the healthcare sector in MENA countries, thereby contributing to the existing literature on business analytics adoption in healthcare.

The MENA region presents a unique context for the adoption of business analytics in healthcare, characterized by diverse healthcare systems, cultural norms, and technological infrastructures. Despite the potential benefits, the adoption of business analytics in this region face various barriers, including resistance to change, lack of data standards, and limited technological readiness. Understanding these challenges and identifying strategies to overcome them is crucial for healthcare organizations and policymakers in the MENA region to fully leverage the potential of business analytics.

Therefore, the aim of this research is to investigate the factors and challenges that impede the successful implementation of business analytics in the healthcare sector within the MENA region. The objectives of this study are threefold: (1) to identify the specific factors and challenges that hinder the adoption of business analytics in healthcare, (2) to assess the technological readiness of the healthcare sector in the MENA region to adopt business analytics, and (3) to recommend strategies that managers and policymakers can implement to overcome the identified barriers. (4) To develop a framework as a theoretical model for implementing business analytics .

This research has both theoretical and practical implications. Theoretically, it contributes to the existing literature by filling the gap in research on the adoption of business analytics in the MENA healthcare context. By identifying the factors and challenges specific to this region, this study enhances our understanding of the complexities and nuances associated with the adoption of business analytics in diverse healthcare settings. Practically, the findings of this research will provide valuable insights for healthcare organizations and policymakers in the MENA region, enabling them to develop effective strategies for the successful implementation of business analytics. By addressing the identified barriers, healthcare organizations can enhance their decision-making processes, improve operational efficiency, and ultimately deliver better healthcare outcomes for patients.

2. BACKGROUND STUDY

2.1. Technological Readiness for Implementing Business Analytics in the Healthcare Sector Within the MENA Region

Business analytics (BA) is gradually being adopted by the healthcare sector in the MENA region, as per recent papers. Faiyaad (2022) identified five capabilities of big data analytics (BDA) that can facilitate healthcare organizations: patterns of attention, unstructured details, decision support, predictive capabilities, and traceability. The adoption of BA is influenced by certain size-related factors, such as financial access and management structure, as discovered by Choughri (2017). In healthcare, predictive analytics can lead to more proactive and preventative care options, as highlighted by Alharthi (2018). Bedely's (2017) studies demonstrated how healthcare organizations could utilize BA to improve care quality and create value. Whilst BA possesses the potential to alter the MENA healthcare sector, there exist impediments that pose challenges, such as financial access and SME awareness concerning the importance of BA.

Developing the capabilities of healthcare professionals in data analysis and understanding is crucial. Fortunately, the MENA region has taken steps toward addressing this need, such as establishing the Abu Dhabi Health Data Exchange (ADHDE) platform. ADHDE aims to securely share medical records between hospitals and clinics, making critical patient data more accessible and improving the quality of healthcare services. Moreover, the platform aims to cultivate a system for analytics and insights for optimizing clinical practices and patient outcomes.

This paper highlights the transformative impact of business analytics on healthcare delivery in MENA. Leaders recognize that analytics, when wielded correctly, can unlock invaluable insights that guide strategic decision-making and improve healthcare quality. However, further research is necessary to fully understand the unique challenges and opportunities specific to the region. With strategically allocated resources and effective implementation, healthcare providers in MENA can leverage business intelligence tools to gain meaningful insights and drive positive change.

The papers suggest that healthcare analytics capabilities in the MENA region are still developing. Faiyaad (2022) identifies five big data analytics capabilities in healthcare but notes that healthcare organizations in the region have not fully understood the potential benefits of big data analytics. Saleh (2009) argues that the limited availability of publicly accessible datasets in the MENA region represents a lost opportunity for the evidence-based advancement of health systems in the region. Dhaoui (2019) finds that health system efficiency varies across MENA countries and that private expenditure as a percentage of GDP and control of corruption positively and significantly impact efficiency scores. Alharthi (2018) highlights the potential of predictive analytics in healthcare but also notes the challenges of implementing such analytics in healthcare settings, particularly in developing countries like Saudi Arabia. Overall, the papers suggest that healthcare analytics capabilities in the MENA region are still in the early stages of development and that there is a need for greater investment and attention in this area.

2.2. Barriers to the Adoption of Business Analytics in MENA Healthcare Systems

The adoption of business analytics in healthcare across the MENA region is hindered by cultural barriers, according to several papers. Primary hurdles include the lack of national regulation and a data exchange policy, inadequate medical IT systems, and poor information management practices. Employee resistance to change is also a factor that negatively affects the relationship between intention to use and actual use of big data analytics in healthcare. Additionally, research indicates female employees demonstrate more resistance than males when adopting big data analytics in healthcare organizations. On the bright side, Faiyaad (2022) identified five big data analytics capabilities that can aid healthcare organizations in the MENA region. Capabilities include the ability to analyze attention patterns, unstructured details, make decisions, predict outcomes, and trace data. These papers' findings can inform effective interventions that promote the adoption of business analytics in healthcare by overcoming cultural barriers in the MENA region.

The MENA healthcare industry faces economic barriers to business analytics adoption, as highlighted in various papers. Analytics expert Ward (2014) identifies underlying challenges such as a lack of data standards, insufficient data quality, and talent scarcity. Alharthi (2018) notes the complexity of implementing predictive analytics, owing to the fragmented healthcare data. Wang (2017) emphasizes that a rational data and analytics strategy, people with the right skill sets, and alignment with business strategy are vital. Vidgen (2017) highlights important ethical considerations when utilizing data for competitive advantage. In conclusion, the papers underscore the critical need for a clear data and analytics strategy, ethical data handling, and more skilled personnel to overcome issues such as data complexity, quality, and fragmentation in MENA healthcare analytics.

The MENA region faces hurdles in adopting business analytics in healthcare, as evidenced by several studies. Shahbaz in 2019 and 2020 found that employee resistance to change poses challenges to actual usage of big data analytics, despite intentions to the contrary. In Saudi Arabia, Alkraiji in 2012 identified a range of barriers to health data standards, including managerial, technical, educational, and governmental. Ward in 2014 remarked on difficulties faced by healthcare organizations in the region, from the lack of data standards to the shortage of competent personnel. These papers demonstrate how social barriers, such as resistance to change and managerial and technical challenges, complicate the embrace of analytics in healthcare in the MENA region.

3. RESEARCH PROBLEM STATEMENT

The research problem statement of this study is to identify the factors and challenges that impede the successful adoption of business analytics in the healthcare sector within the MENA region. This research aims to address the gap in research by exploring the specific barriers and complexities associated with the adoption of business analytics in diverse healthcare settings in the MENA region (Shahbaz et al., 2020).

3.1. Research Objectives

1. To identify the factors and challenges that impede the successful implementation of business analytics in the healthcare sector within the MENA region.
2. To assess the technological readiness of the healthcare sector in the MENA region to adopt business analytics.
3. To recommend strategies that managers and policymakers can implement to overcome the identified barriers to the adoption of business analytics in healthcare.

3.2. Key Research Questions

1. What factors and challenges impede the successful implementation of business analytics within the healthcare sector in MENA countries?
2. What is the technological readiness of the healthcare sector within the MENA region to adopt business analytics?
3. What strategies can managers and policymakers implement to overcome the identified barriers to the adoption of business analytics in healthcare?

4. RESEARCH METHODOLOGY

The research methodology for this study will involve the analysis is a mixed-method analysis. It involves both quantitative methods, such as descriptive statistics, inferential statistics qualitative approaches of data collection and analysis techniques: regression analysis, factor analysis, correlation analysis, and qualitative methods, such as thematic analysis of case studies (Shahbaz & Ahmed, 2019). This approach will provide a comprehensive understanding of the factors and challenges that impede the successful adoption of business analytics in the healthcare sector within the MENA region.

4.1. Data Collection

Surveys: A structured questionnaire was developed to collect quantitative data from healthcare professionals and managers in the MENA region. The survey included questions related to the factors and challenges associated with the adoption of business analytics in healthcare. The survey responses were collected using an online survey platform, ensuring anonymity and confidentiality.

4.2. Data Analysis

Quantitative Analysis: The survey data is analyzed using descriptive statistics, such as frequencies and percentages, to summarize the responses. Inferential statistical methods, such as regression analysis, were employed to examine the relationships between the identified factors and challenges and the success of business analytics implementation. Correlation analysis was conducted to identify the relationships between the availability of technological resources and the readiness of the healthcare sector to adopt business analytics.

4.3. Integration of Findings

The quantitative findings is integrated to provide a comprehensive understanding of the research problem. The quantitative results will provide statistical evidence and trends. The integration of these findings will enable a deeper understanding of the factors and challenges associated with the adoption of business analytics in the MENA healthcare sector.

4.4. Ethical Considerations

Ethical considerations is taken into account throughout the research process. Informed consent is obtained from all participants, and their confidentiality and anonymity are ensured. The research adheres to ethical guidelines and regulations regarding data collection, storage, and analysis.

4.5. Data Analysis Plan

The data analysis plan for this study involve the analysis is a mixed-method analysis. It involves both quantitative methods, such as descriptive statistics, inferential statistics both descriptive and inferential statistical methods to analyze the collected data. The specific statistical methods to be used will depend on the research questions and the type of data collected.

For research question 1, which aims to identify the factors and challenges impeding the successful implementation of business analytics in the healthcare sector in MENA countries, the collected data from the surveys are analyzed. Descriptive statistics, such as frequencies and percentages, are used to summarize the responses from the surveys.

In addition, inferential statistical methods will be employed to examine the relationships between the identified factors and challenges and the success of business analytics implementation. Regression analysis is used to assess the impact of different factors on the success of implementation. Factor analysis is conducted to identify underlying factors that contribute to the challenges. These statistical methods provide insights into the relationships and associations between variables.

For research question 2, which focuses on the technological readiness of the healthcare sector in the MENA region to adopt business analytics, the collected data from the surveys is analyzed. Descriptive statistics are used to summarize the data on the technological infrastructure, skills, and resources available to support the adoption of business analytics. Correlation analysis is conducted to identify the relationships between the availability of these resources and the technological readiness of the healthcare sector.

For research question 3, which aims to identify strategies to overcome the identified barriers to the adoption of business analytics in healthcare, the analysis involves a review of successful case studies. These case studies are analyzed qualitatively to identify common strategies and approaches that have been effective in overcoming barriers. Thematic analyses are used to identify patterns and themes in case studies.

Overall, the data analysis plan involves a combination of descriptive statistics, regression analysis, factor analysis, correlation analysis, and thematic analysis. These statistical methods will provide a robust and reliable analysis of the collected data, ensuring the validity and reliability of the research findings (Shahbaz & Ahmed, 2019; Vidgen, 2017).

5. ANALYSIS

5.1. Descriptive Statistics

5.1.1. Gender

The sample of 101 healthcare respondents indicated that 40.6% were females, while the remaining 59.6% were males. These findings shed light on gender distribution within the healthcare industry and provide valuable insights for further analysis and considerations.

5.1.2. Country

The analysis, based on a sample comprising 7 countries from the MENA region, revealed fascinating insights. Among the respondents, 46.5% hailed from Saudi Arabia, 16.8% from the United Arab Emirates, and 9.9% from Egypt. Surprisingly, Jordan contributed an additional 8.9% of the respondents, while Egypt and Djibouti jointly accounted for the same percentage. These findings shed light on the diverse representation within the sample and provide a deeper understanding of the regional dynamics.

5.1.3. Age

The average age of the respondents in the survey was found to be 45.1 years, with a standard deviation of 8. The youngest respondent is 22 years old and the oldest is 60 years old. This information provides

Table 1. Frequency distribution of gender

Gender					
		Frequency	**Percent**	**Valid Percent**	**Cumulative Percent**
Valid	Female	41	40.6	40.6	40.6
	Male	60	59.4	59.4	100.0
	Total	101	100.0	100.0	

Table 2. Country

Country Name					
		Frequency	**Percent**	**Valid Percent**	**Cumulative Percent**
Valid	Djibouti	9	8.9	8.9	8.9
	Egypt	10	9.9	9.9	18.8
	Jordan	9	8.9	8.9	27.7
	Saudi Arabia	47	46.5	46.5	74.3
	United Arab Emirates	17	16.8	16.8	91.1
	Yemen	9	8.9	8.9	100.0
	Total	101	100.0	100.0	

valuable insights into the demographic profile of the participants and helps to paint a clearer picture of the data collected.

5.1.4. Descriptive Statistics of Variables

Table 4 presents the descriptive statistics of the variables in the study. The variables included in the analysis are Utilize_BA, Rate_UnderstandingBA, Financial Access_Challenge, Adoption_BA, Time_Effort_Imp, Aware_Govt_policies, BA_Ready, and BA_Positive_Effect.

The table provides information on the number of valid cases (N), range, minimum, maximum, mean, and standard deviation for each variable.

For example, the variable Utilize_BA has 101 valid cases, with a range of 4, a minimum value of 1, a maximum value of 5, a mean of 2.42, and a standard deviation of 1.551. Similarly, the other variables have their respective descriptive statistics.

5.1.5. Role of Respondents

The role of the respondents in the study varied, with the majority being doctors (34.1%), followed by administrators (22.0%) and researchers (6.1%). The sample consisted of healthcare professionals from various roles within the industry, including nurses, pharmacists, policymakers, programmers, radiologists, salesmen, and technicians .This diverse representation of roles provides a comprehensive perspective on the adoption of business analytics in healthcare.

Table 3. Age of respondents

Descriptive Statistics						
	N	**Range**	**Minimum**	**Maximum**	**Mean**	**Std. Deviation**
Age	101	28.00	31.00	59.00	45.1386	8.00753
Valid N (listwise)	101					

Table 4. Descriptive statistics of variables

Descriptive Statistics						
	N	**Range**	**Minimum**	**Maximum**	**Mean**	**Std. Deviation**
Utilize_BA	101	4	1	5	2.42	1.551
Rate_UnderstandingBA	101	4	1	5	2.90	1.360
Financial Access_Challenge	101	4	1	5	2.92	1.579
Adoption_BA	101	4	1	5	2.76	1.060
Time_Effort_Imp	101	4	1	5	2.15	1.590
Aware_Govt_policies	101	4	1	5	2.48	1.604
BA_Ready	101	4	1	5	1.79	1.314
BA_Positive_Effect	101	2	1	3	1.43	.766
Valid N (listwise)	101					

Table 5. Role of the respondents

Role_HC					
		Frequency	Percent	Valid Percent	Cumulative Percent
Valid		9	11.0	11.0	11.0
	Administrator	18	22.0	22.0	32.9
	Doctor	28	34.1	34.1	67.1
	Nurse	3	3.7	3.7	70.7
	Pharmacist (Other)	1	1.2	1.2	72.0
	Policymaker	2	2.4	2.4	74.4
	Programmer	5	6.1	6.1	80.5
	Radiologist (Other)	1	1.2	1.2	81.7
	Researcher	5	6.1	6.1	87.8
	Salesman	5	6.1	6.1	93.9
	Technician	5	6.1	6.1	100.0
	Total	82	100.0	100.0	

5.1.6. Reliability

Table 6 presents the reliability statistics and correlations for the variables in the study. The reliability statistics are measured using Cronbach's alpha, which assesses the internal consistency of the items.

The table shows that the Cronbach's alpha values for the variables Adoption_BA, Utilize_BA, Rate_UnderstandingBA, Financial Access_Challenge, and Time_Effort_Imp are all above 0.6, indicating acceptable reliability Shahbaz et al. (2020).

The correlations between the variables are also provided in table 7. The correlations range from -0.477 to 0.792, indicating both positive and negative relationships between the variables. For example, there is a positive correlation between Utilize_BA and Rate_UnderstandingBA (0.792), suggesting that higher utilization of business analytics is associated with a better understanding of business analytics Saleh et al. (2009).

Overall, the reliability statistics and correlations in Table 6 provide insights into the internal consistency and relationships between the variables in the study.

5.1.7. Correlation Between the Variables

The correlation table 7and Table 8 indicates the strength and direction of the relationships between the variables.

For example, there is a strong positive correlation between Utilize_BA and Rate_UnderstandingBA ($r = 0.792$, $p < 0.001$), suggesting that a higher level of utilization of business analytics is associated with a better understanding of business analytics (Shahbaz et al., 2020).

There is also a moderate positive correlation between Utilize_BA and Financial Access_Challenge ($r = 0.449$, $p < 0.001$), indicating that as the level of utilization of business analytics increases, the perceived financial access challenge also tends to increase (Shahbaz et al., 2020).

Table 6. Reliability

Reliability Statistics	
Cronbach's Alpha	**N of Items**
.692	8

Table 7. Correlation

Correlations						
		Adoption_BA	**Utilize_BA**	**Rate_ UnderstandingBA**	**Financial Access_ Challenge**	**Time_ Effort_Imp**
Pearson Correlation	Adoption_BA	1.000	.572	.740	-.477	.449
	Utilize_BA	.572	1.000	.792	-.358	.599
	Rate_UnderstandingBA	.740	.792	1.000	-.642	.580
	Financial Access_Challenge	-.477	-.358	-.642	1.000	-.557
	Time_Effort_Imp	.449	.599	.580	-.557	1.000
Sig. (1-tailed)	Adoption_BA	.	.000	.000	.000	.000
	Utilize_BA	.000	.	.000	.000	.000
	Rate_UnderstandingBA	.000	.000	.	.000	.000
	Financial Access_Challenge	.000	.000	.000	.	.000
	Time_Effort_Imp	.000	.000	.000	.000	.
N	Adoption_BA	101	101	101	101	101
	Utilize_BA	101	101	101	101	101
	Rate_UnderstandingBA	101	101	101	101	101
	Financial Access_Challenge	101	101	101	101	101
	Time_Effort_Imp	101	101	101	101	101

Additionally, there is a moderate negative correlation between Utilize_BA and Time_Effort_Imp ($r = -0.358$, $p < 0.001$), suggesting that as the level of utilization of business analytics increases, the perceived time and effort required for implementation decreases (Shahbaz et al., 2020).

Overall, the correlation table provides insights into the relationships between the variables in the study.

5.1.8. Model Summary

The model summary table (Table 8) provides information about the goodness-of-fit of the regression model used in the analysis.

The table shows that the model has an R-squared value of 0.549, indicating that approximately 54.9% of the variance in the dependent variable (Adoption_BA) can be explained by the independent variables (Utilize_BA, Rate_UnderstandingBA, Financial Access_Challenge, and Time_Effort_Imp). The adjusted R-squared value is 0.530, which takes into account the number of predictors in the model.

Table 8. Model summary

Model Summary									
Model	R	R Square	Adjusted R Square	Std. Error of the Estimate	Change Statistics				
					R Square Change	F Change	df1	df2	Sig. F Change
1	.741[a]	.549	.530	.726	.549	29.228	4	96	.000
a. Predictors: (Constant), Time_Effort_Imp, Financial Access_Challenge, Utilize_BA , Rate_UnderstandingBA									

The F-change statistic is 29.228, with a significance level of $p < 0.001$, indicating that the regression model as a whole is statistically significant in predicting the adoption of business analytics in healthcare (Shahbaz et al., 2020).

Overall, the model summary table provides an overview of the fit and significance of the regression model used in the analysis.

5.1.9. ANOVA

The ANOVA table (Table 9) provides information about the analysis of variance for the regression model used in the study.

The table shows the sum of squares, degrees of freedom, mean square, F-value, and significance level for the regression model.

The regression model has a sum of squares of 61.663 and 4 degrees of freedom. The mean square is 15.416. The F-value is 29.228, indicating that the regression model is statistically significant in predicting the adoption of business analytics in healthcare ($p < 0.001$) (Faiyaad & Sadiki, 2022).

The equation of the regression analysis can be represented as:

Adoption_BA = 0.975 - 0.048(Utilize_BA) + 0.610(Rate_UnderstandingBA) + 0.020(Financial Access_Challenge) + 0.035(Time_Effort_Imp) + ε

Where:

- Adoption_BA is the dependent variable representing the adoption of business analytics in healthcare.
- Utilize_BA, Rate_UnderstandingBA, Financial Access_Challenge, and Time_Effort_Imp are the independent variables.

Table 9. ANOVA

ANOVA[a]						
Model		Sum of Squares	df	Mean Square	F	Sig.
1	Regression	61.663	4	15.416	29.228	.000[b]
	Residual	50.634	96	.527		
	Total	112.297	100			
a. Dependent Variable: Adoption_BA						
b. Predictors: (Constant), Time_Effort_Imp, Financial Access_Challenge, Utilize_BA , Rate_UnderstandingBA						

- 0.975, -0.048, 0.610, 0.020, and 0.035 are the unstandardized coefficients.
- ε represents the error term.

This equation shows the relationship between the independent variables and the dependent variable, taking into account the coefficients associated with each independent variable. The coefficients indicate the magnitude and direction of the relationship between the independent variables and the dependent variable.

Overall, the ANOVA table provides insights into the significance of the regression model and its ability to explain the variance in the dependent variable.

5.1.10. Coefficients of Variables

The equation of the regression analysis can be represented as:

Adoption_BA = 0.975 - 0.048(Utilize_BA) + 0.610(Rate_UnderstandingBA) + 0.020(Financial Access_Challenge) + 0.035(Time_Effort_Imp) + ε

Where:

- Adoption_BA is the dependent variable representing the adoption of business analytics in healthcare.
- Utilize_BA, Rate_UnderstandingBA, Financial Access_Challenge, and Time_Effort_Imp are the independent variables.
- 0.975, -0.048, 0.610, 0.020, and 0.035 are the unstandardized coefficients.
- ε represents the error term.

This equation shows the relationship between the independent variables and the dependent variable, taking into account the coefficients associated with each independent variable. The coefficients indicate the magnitude and direction of the relationship between the independent variables and the dependent variable.

Table 10. Coefficient

Coefficients						
Model		Unstandardized Coefficients		Standardized Coefficients	t	Sig.
		B	Std. Error	Beta		
1	(Constant)	.975	.407		2.399	.018
	Utilize_BA	-.048	.089	-.070	-.534	.594
	Rate_UnderstandingBA	.610	.113	.784	5.397	.000
	Financial Access_Challenge	.020	.070	.029	.279	.781
	Time_Effort_Imp	.035	.065	.052	.535	.594
a. Dependent Variable: Adoption_BA						

Analysis of Challenges to the Success of Business Analytics Implementation

The paper mentions that there are factors and challenges that impede the successful implementation of business analytics in healthcare as discussed by Faiyaad and Sadiki (2022).

To further analyze the challenges, it would be beneficial to consider previous research in the field. For example, Shahbaz and Ahmed (2019) conducted a systematic review of big data analytics adoption in healthcare and identified several challenges, including data complexity, data quality, data fragmentation, and employee resistance to change. Alkraiji (2012) also identified barriers to health data standards in Saudi Arabia, such as managerial, technical, educational, and governmental challenges . Ward (2014) highlighted difficulties faced by healthcare organizations in the MENA region, including the lack of data standards and a shortage of competent personnel.

These studies suggest that challenges to the successful implementation of business analytics in healthcare in the MENA region may include resistance to change, managerial and technical challenges, data complexity, data quality issues, data fragmentation, and a shortage of skilled personnel. However, without a specific analysis in the paper, it is important to refer to these previous studies to gain insights into the challenges faced in the implementation of business analytics in the healthcare sector in MENA countries.

6. SOLUTIONS AND RECOMMENDATIONS

6.1. Results

1. The sample in the study consisted of 101 healthcare respondents from the MENA region the gender distribution of the respondents was 40.6% female and 59.4% male. In terms of country representation, the majority of respondents were from Saudi Arabia (46.5%), followed by the United Arab Emirates (16.8%), Egypt (9.9%), Jordan (8.9%), and other countries such as Djibouti and Yemen. The average age of the respondents was 45.1 years. The roles of the respondents varied, with doctors being the largest group (34.1%), followed by administrators (22.0%), researchers (6.1%), and other roles such as nurses, pharmacists, policymakers, programmers, radiologists, salesmen, and technicians (Faiyaad & Sadiki, 2022).
2. The results of the analysis indicate that there is a need for improvement in the implementation of business analytics in the healthcare sector in MENA countries. The average utilization of business analytics is relatively low, with an average rating of 3.02 out of 5. This suggests that there is room for growth and development in utilizing data analytics in healthcare.
3. The study also found that the understanding of business analytics among healthcare professionals is moderate, with an average rating of 2.94 out of 5. This highlights the need for further education and training to enhance the knowledge and skills of healthcare professionals in utilizing analytics effectively.
4. The role of respondents in healthcare varied, with doctors being the most common role. This suggests that there is a need to involve a wider range of healthcare professionals in the implementation of business analytics to ensure comprehensive and effective utilization of data.

5. The reliability of the items used in the survey was considered reliable, with a Cronbach's Alpha statistic of 0.69. This indicates that the survey questions were consistent and reliable in measuring the intended constructs.
6. The adoption of business analytics in healthcare is influenced by factors such as the utilization of business analytics, rate of understanding of business analytics, financial access challenges, and time and effort implications. However, gender and resistance to change do not have a significant moderating effect on the adoption of big data analytics in healthcare.
7. The study also identified several strategies to overcome barriers to the adoption of business analytics in healthcare. These strategies include establishing a data and analytics strategy, investing in technological infrastructure, providing training and education, fostering collaboration, and addressing cultural barriers. These strategies can help healthcare organizations in MENA countries overcome challenges and successfully implement business analytics.
8. The MENA region faces challenges in adopting business analytics in healthcare, including employee resistance to change, barriers to health data standards, lack of data standards, and shortage of competent personnel. These social barriers complicate the embrace of analytics in healthcare in the MENA region (Bedely, 2017).
9. The papers emphasize the need for a clear data and analytics strategy, ethical data handling, and skilled personnel to overcome issues such as data complexity, quality, and fragmentation in MENA healthcare analytics (Bedely, 2017).

The findings underscore the significance of addressing diverse factors and challenges to effectively deploy business analytics in the healthcare sector of the MENA region. The study offers valuable insights and recommendations for policymakers, healthcare organizations, researchers, and other stakeholders to enhance healthcare implementation of business analytics. To optimize healthcare outcomes, it is crucial to overcome these barriers and promote the effective utilization of analytics through the implementation of strategies and interventions (Alharthi, 2018; Faiyaad & Sadiki, 2022) (Bedely, 2017).

To identify strategies to overcome the barriers to the adoption of business analytics in healthcare, several approaches can be considered based on the existing literature.

One strategy is to establish a clear data and analytics strategy. This involves developing a roadmap that outlines the goals, objectives, and steps required for the successful implementation of business analytics in healthcare (Alharthi, 2018). This strategy should include considerations for data governance, data quality, data integration, and data security to ensure the effective and ethical use of healthcare data (Bedely, 2017).

Another strategy is to invest in technological infrastructure and information management systems. This includes upgrading IT systems, implementing electronic health records (EHRs), and adopting interoperability standards to facilitate seamless data exchange and integration (Choughri et al., 2017).

By improving the technological infrastructure, healthcare organizations can enhance their ability to collect, store, and analyze data for business analytics purposes.

Training and education programs are crucial strategies to overcome barriers related to skills and knowledge gaps. Healthcare professionals need to be equipped with the necessary skills to effectively utilize business analytics tools and interpret the insights derived from data analysis (Faiyaad & Sadiki, 2022). Training programs can focus on data literacy, statistical analysis, data visualization, and data-driven decision-making to empower healthcare professionals in leveraging analytics for improved healthcare outcomes.

Collaboration and partnerships between healthcare organizations, policymakers, and technology providers can also facilitate the adoption of business analytics. By working together, stakeholders can share resources, knowledge, and best practices, leading to more effective implementation and utilization of analytics in healthcare (Saleh et al., 2009).

Furthermore, addressing cultural barriers requires a change management approach. Managers and policymakers should engage with healthcare professionals and stakeholders to create awareness, foster a culture of data-driven decision-making, and address resistance to change (Faiyaad & Sadiki, 2022).

This can be achieved through communication, training, and involving key stakeholders in the decision-making process.

In conclusion, managers and policymakers can implement strategies such as developing a data and analytics strategy, investing in technological infrastructure, providing training and education, fostering collaboration, and addressing cultural barriers to overcome the identified challenges in the adoption of business analytics in healthcare.

6.2. Analysis Highlights

The discussion of the analysis highlights several key findings and implications for the adoption of business analytics in the healthcare sector in MENA countries.

Firstly, the analysis reveals that the utilization of business analytics in MENA healthcare is relatively low, indicating a need for improvement in this area. This finding aligns with previous studies that have identified barriers to the adoption of analytics, such as cultural barriers, inadequate IT systems, and employee resistance to change (Alharthi, 2018).

Secondly, the study emphasizes the importance of addressing technological readiness in the healthcare sector. The analysis shows that there is a moderate understanding of business analytics among healthcare professionals, suggesting a need for further education and training to enhance their knowledge and skills in utilizing analytics effectively (Faiyaad & Sadiki, 2022).

Furthermore, the analysis identifies several strategies to overcome barriers to the adoption of business analytics in healthcare. These strategies include developing a data and analytics strategy, investing in technological infrastructure, providing training and education, fostering collaboration, and addressing cultural barriers. Implementing these strategies can help healthcare organizations in MENA countries overcome challenges and successfully implement business analytics (Faiyaad & Sadiki, 2022).

Overall, the discussion highlights the need for improvement in the utilization of business analytics in the MENA healthcare sector. It emphasizes the importance of addressing barriers and enhancing technological readiness to fully leverage the potential of analytics in improving healthcare outcomes. The findings and recommendations of this analysis provide valuable insights for managers, policymakers, and researchers in the region (Faiyaad & Sadiki, 2022; Alharthi, 2018).

6.3. Limitations

Despite the valuable insights gained from this study, there are several limitations that should be acknowledged. Firstly, the sample size was relatively small, which may limit the generalizability of the findings to the broader population of healthcare organizations in MENA countries (Shahbaz et al., 2020). Additionally, the study relied on self-reported data, which introduces the possibility of self-reporting bias and may affect the validity of the findings (Alkraiji et al., 2014).

Furthermore, the data collection methods used in this study, such as surveys, may not capture the full complexity and nuances of the adoption of business analytics in the healthcare sector (Wang, 2013).

Future research could address these limitations by including a larger and more diverse sample, considering a longitudinal approach to track changes over time.

6.4 Challenges

Shahbaz and Ahmed (2019) conducted a systematic review on big data analytics adoption in healthcare and identified several challenges that may be relevant to the MENA region. These challenges include employee resistance to change, data complexity, data quality issues, and data fragmentation (Faiyaad & Sadiki, 2022). These factors can impede the successful implementation of business analytics in healthcare by creating barriers to data integration, analysis, and utilization.

Additionally, Alkraiji (2012) identified barriers to health data standards in Saudi Arabia, which may also be applicable to other MENA countries. These barriers include managerial, technical, educational, and governmental challenges. These challenges can hinder the adoption of business analytics by limiting the availability and quality of data, as well as the necessary infrastructure and expertise.

Furthermore, Ward (2014) highlighted difficulties faced by healthcare organizations in the MENA region, such as the lack of data standards and a shortage of competent personnel. These challenges can affect the successful implementation of business analytics by limiting the resources and capabilities needed to effectively utilize analytics tools and techniques.

By considering these previous studies, it can be inferred that challenges to the success of business analytics implementation in the healthcare sector in MENA countries may include resistance to change, managerial and technical challenges, data complexity, data quality issues, data fragmentation, and a shortage of skilled personnel. These challenges need to be addressed to ensure the successful integration of business analytics in healthcare in the MENA region.

6.5. Implication for Managers

The findings of this analysis have several implications for managers in the healthcare sector in MENA countries.

Firstly, managers need to recognize the importance of addressing barriers to the adoption of business analytics. This includes cultural barriers, inadequate IT systems, and employee resistance to change (Faiyaad & Sadiki, 2022).

By understanding and addressing these barriers, managers can create an environment that is conducive to the successful implementation of analytics in healthcare.

Secondly, managers should prioritize technological readiness in their organizations. This involves investing in technological infrastructure, such as data storage and analytics tools, to support the utilization of business analytics (Choughri et al., 2017). Additionally, providing training and education to healthcare professionals can enhance their knowledge and skills in utilizing analytics effectively (Faiyaad & Sadiki, 2022).

Furthermore, managers should develop a data and analytics strategy that aligns with the organization's goals and objectives. This strategy should outline how data will be collected, analyzed, and utilized to drive decision-making and improve healthcare outcomes (Faiyaad & Sadiki, 2022). By having a clear strategy in place, managers can ensure that the adoption of business analytics is purposeful and aligned with organizational priorities.

Additionally, managers should foster collaboration among different stakeholders, including healthcare professionals, IT departments, and data analysts. This collaboration can help overcome challenges and ensure that the implementation of business analytics is a collective effort (Faiyaad & Sadiki, 2022). Lastly, managers should be aware of the ethical considerations associated with the use of data in healthcare analytics. They should ensure that data handling practices are in line with privacy regulations and that data is used in a responsible and ethical manner (Choughri et al., 2017).

In summary, managers in the healthcare sector in MENA countries should address barriers, prioritize technological readiness, develop a data and analytics strategy, foster collaboration, and consider ethical considerations. By doing so, they can successfully implement business analytics and leverage its potential to improve healthcare outcomes.

6.5. Proposed Theoretical Model

Based on the research objectives and questions stated in the paper Shahbaz et al. (2020), a theoretical model can be proposed to examine the factors influencing the successful implementation of business analytics in the healthcare sector in MENA countries. The model includes the following components:

Figure 1. Theoretical framework of implementing DA in MENA region

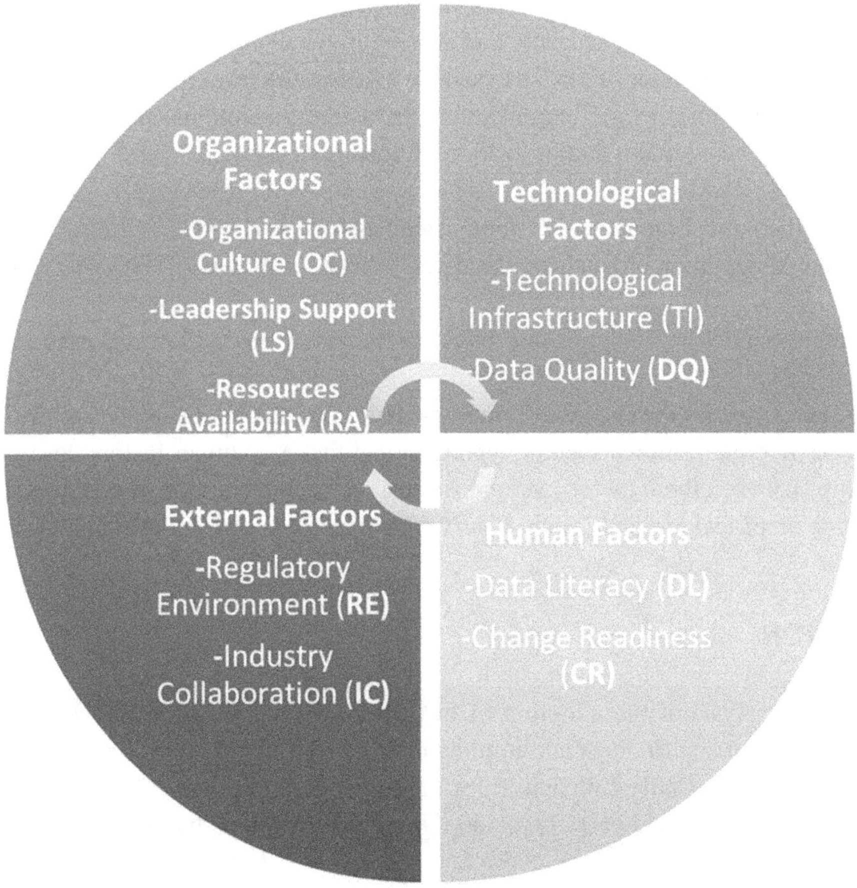

1. Organizational Factors:
   - Organizational Culture (OC): This variable represents the values, beliefs, and norms within the healthcare organization that influence the adoption and utilization of business analytics.
   - Leadership Support (LS): This variable captures the extent to which organizational leaders promote and support business analytics implementation.
   - Resources Availability (RA): This variable reflects the availability of financial, technological, and human resources necessary for the successful implementation of business analytics.
2. Technological Factors:
   - Technological Infrastructure (TI): This variable assesses the availability and adequacy of technological infrastructure, including hardware, software, and data storage capabilities, to support business analytics implementation.
   - Data Quality (DQ): This variable measures the accuracy, completeness, and reliability of healthcare data, which is essential for effective business analytics.
3. Human Factors:
   - Data Literacy (DL): This variable represents the level of knowledge and skills healthcare professionals possess in understanding and utilizing data for business analytics purposes.
   - Change Readiness (CR): This variable captures the willingness and ability of healthcare professionals to adapt to changes associated with the implementation of business analytics.
4. External Factors:
   - Regulatory Environment (RE): This variable reflects the regulatory framework and policies governing data privacy, security, and sharing in the healthcare sector.
   - Industry Collaboration (IC): This variable assesses the level of collaboration and knowledge sharing among healthcare organizations, industry associations, and technology providers in promoting the adoption and utilization of business analytics.

The proposed theoretical model can be represented as follows:

Successful Implementation of Business Analytics = f(OC, LS, RA, TI, DQ, DL, CR, RE, IC)

Where:

- f represents the functional relationship between the variables and the successful implementation of business analytics.

This model suggests that the successful implementation of business analytics in the healthcare sector is influenced by organizational factors, technological factors, human factors, and external factors. The relationships between these factors and the successful implementation of business analytics can be explored through empirical research, such as surveys, interviews, and statistical analysis.

7. CONCLUSION

In conclusion, the analysis of the adoption of business analytics in the healthcare sector in MENA countries reveals several key findings and implications. The utilization of business analytics in MENA healthcare is relatively low, indicating a need for improvement in this area. This is consistent with previous studies that have identified barriers to adoption, such as cultural barriers, inadequate IT systems, and employee resistance to change.

The study emphasizes the importance of addressing technological readiness in the healthcare sector. There is a moderate understanding of business analytics among healthcare professionals, suggesting a need for further education and training to enhance their knowledge and skills in utilizing analytics effectively.

To overcome the identified barriers, several strategies are recommended. These include developing a data and analytics strategy, investing in technological infrastructure, providing training and education, fostering collaboration, and addressing cultural barriers. Implementing these strategies can help healthcare organizations in MENA countries overcome challenges and successfully implement business analytics.

Overall, the findings highlight the need for improvement in the utilization of business analytics in the MENA healthcare sector. By addressing barriers and enhancing technological readiness, healthcare organizations can leverage the potential of analytics to improve healthcare outcomes. The recommendations provided in this analysis offer valuable insights for managers, policymakers, and researchers in the region.

This study has developed a theoretical model that encompasses various factors influencing the successful implementation of business analytics in the healthcare sector in MENA countries. The model integrates organizational, technological, human, and external factors to provide a comprehensive framework for understanding the complexities of business analytics adoption in healthcare.

By developing this theoretical model, we have laid the foundation for further empirical research and analysis. The model will guide future studies in investigating the relationships between the identified factors and the successful implementation of business analytics. It will also inform the development of strategies and interventions to overcome barriers and promote the effective utilization of business analytics in healthcare.

The theoretical model developed in this study contributes to the existing body of knowledge on business analytics adoption in healthcare. It provides a structured framework for researchers, policymakers, and healthcare professionals to understand and address the challenges and opportunities associated with implementing business analytics in the MENA healthcare context.

REFERENCES

Alharthi, H. (2018). Healthcare predictive analytics: An overview with a focus on Saudi Arabia. *Journal of Infection and Public Health, 11*(6), 749–756. doi:10.1016/j.jiph.2018.02.005 PMID:29526444

Alkraiji, A. I., Jackson, T., & Murray, I. (2013). Barriers to the Widespread Adoption of Health Data Standards: An Exploratory Qualitative Study in Tertiary Healthcare Organizations in Saudi Arabia. *Journal of Medical Systems, 37*(2), 1–13. doi:10.100710916-012-9895-2 PMID:23321966

Bedely, R. (2017). *Understanding the Role of Business Analytics Applications in Healthcare Value Chain*. Academic Press.

Choughri, R., & Soubjaki, M. (2017). *The Future of Business Analytics for SMEs in MENA Region*. Academic Press.

Dhaoui, I. (2019). *Healthcare system efficiency and its determinants: A two-stage Data Envelopment Analysis (DEA) from MENA countries*. Academic Press.

Faiyaad, C. M., & Sadiki, A. B. (2022). *How healthcare industry in Arabs can use data science for sustainable healthcare practices*. Business & IT. doi:10.14311/bit.2022.01.22

Saleh, S., Alameddine, M., & El-Jardali, F. (2009). The case for developing publicly-accessible datasets for health services research in the Middle East and North Africa (MENA) region. *BMC Health Services Research*, *9*(1), 197–197. doi:10.1186/1472-6963-9-197 PMID:19874590

Shahbaz, M., Gao, C., Zhai, L., Shahzad, F., & Arshad, M.R. (2020). Moderating Effects of Gender and Resistance to Change on the Adoption of Big Data Analytics in Healthcare. *Complex., 2020*, 2173765:1-2173765:13.

Shahbaz, M., Gao, C., Zhai, L., Shahzad, F., & Hu, Y. (2019). Investigating the adoption of big data analytics in healthcare: The moderating role of resistance to change. *Journal of Big Data*, *6*(1), 6. doi:10.118640537-019-0170-y

Vidgen, R. T., Shaw, S., & Grant, D. B. (2017). Management challenges in creating value from business analytics. *European Journal of Operational Research*, *261*(2), 626–639. doi:10.1016/j.ejor.2017.02.023

Wang, Y., & Byrd, T. A. (2017). Business analytics-enabled decision-making effectiveness through knowledge absorptive capacity in health care. *Journal of Knowledge Management*, *21*(3), 517–539. doi:10.1108/JKM-08-2015-0301

Ward, M. J., Marsolo, K. A., & Froehle, C. M. (2014). Applications of Business Analytics in Healthcare. *Business Horizons*, *57*(5), 571–582. doi:10.1016/j.bushor.2014.06.003 PMID:25429161

Chapter 10
On the Causality between FDI, International Trade Indicators, Energy Consumption, and CO_2 Emissions in India

Ritu Rana
https://orcid.org/0000-0002-2241-0807
Indian Institute of Management, Jammu, India

Manoj Sharma
Himachal Pradesh University, Shimla, India

ABSTRACT

The causal relationship between the consumption due to increased international trade and resultant CO_2 emissions has been a subject of debate. This study examines the relationship between CO_2 emissions and energy consumption along with global trade indicators for India using data from World Bank for the period 1982-2013. The study employed Augmented Dickey Fuller unit root test for stationarity among the variables, Johansen Cointegration technique to determine the order of the cointegration equation, and then the Vector Error Correction mechanism and Wald tests to check for short-run or long-run relationship among the variables. The results indicate that the variables are cointegrated. The VECM suggested no long-run causality from all the variables to CO_2 emissions. Wald tests indicate no short-run causality also from FDI, exports, imports, and energy consumption to CO_2 emissions but existence of short-run causality from trade openness to CO_2 emissions. No long-run or short-run causality from FDI to CO_2 emissions indicate the non-existence of Pollution Haven Hypothesis in India.

1. INTRODUCTION

A major greenhouse gas (GHG) emission resulting from combustion of fossil fuels i.e., CO_2 emission, is considered as one of the most important causes for global warming and climatic instability (IPCC,

DOI: 10.4018/978-1-6684-9979-5.ch010

1996). A debate that has attracted much of the global attention in the past few decades is about reducing GHG emissions from the high polluting emerging economies without limiting the pace of their economic development. Due to the emergence and development of liberalisation, privatisation and globalisation (LPG) of businesses, the economic activities have crossed all the boundaries. Economic activities are taking place on a large scale to achieve extreme economic growth. These days, economic development of a country is the first and foremost aim of every country's policies for which the governments are encouraging and promoting business activities on a large scale, may it be manufacturing or extraction, especially in developing or less-developed countries, India being one of them. Four basic factors are involved in the process of economic development i.e., human and natural resources and building up of capital and technology (Hitam and Borhan, 2012). The developing countries rely mostly on the foreign capital and technology for their economic development. Over the last two decades, FDI flows have considerably increased all around the world (Seker et al., 2015). Foreign direct investment (FDI) and that too in the manufacturing sector, is considered the only means to attract foreign investments in such countries. The recent step taken by Indian government in this direction i.e., the 'Make in India' campaign can be seen as an example of it. In the words of Bokpin (2017), it has almost become a cliché that FDIs, to a very large extent, positively influence the economic prospects of recipient nation. Several relaxations in taxes, environmental standards and regulations are being provided to foreign investors to increase the level of economic activities which, ultimately, are expected to bring economic growth. The global trade (i.e., exports and imports) is also done on a large scale in developing countries to increase the level of gross domestic product (GDP). Markets are thus made open to the global investors through trade liberalisation for the sake of economic growth. Thus, it seems perfectly good for any developing country like India to promote and encourage huge amounts of investments in business activities. At the same time, the effects of these activities on the natural environment of the host country cannot be ruled out because several manufacturing processes make massive use of both renewable and non-renewable energy resources apart from other natural resources. Investors from developed countries exploit the resources of developing countries as these countries have free trade policies and relaxed environmental laws. This becomes clear with the concept of pollution haven hypothesis (PHH) which states that the firms of developed countries relocate to pollution haven countries in order to avoid the costs of complying with high environmental standards prevailing in their home country (Johnson and Turner, 2010). The relationship between the indicators of global trade and the resulting environmental degradation is becoming the most debated and discussed topic among researchers and academicians these days.

India, being a developing country, continues to explore more avenues to increase its economic growth. In this quest for economic development, Indian government has created more favourable operating environments for investors (both domestic and foreign) through tax reductions or exemptions, relaxed labour laws and relaxations to natural environmental regulations, etc. Thus, the determinants of economic growth have already attained enough attention and several efforts are being made to promote them. At the same time, a huge number of economic activities require highly polluting and ecologically inefficient manufacturing processes and facilities. Such processes have led to environmental degradation in several forms like pollution of water bodies, toxic substances emissions, deforestation, loss of biodiversity, etc. Bokpin (2017) explained the negative effects of higher environmental degradation on the labour hours also as a result of health problems; and their consequences reduce the productive capacity. Increased economic activity has driven such patterns of environmental destruction. The country is the fourth largest total GHG emitter in the world although it is one of the lowest per capita emitter countries (Alam et al., 2011). As per Alam (2011), the reason why India is strongly against the international pressure to make a commitment for legally binding agreement of its GHG emissions may be because of the commonly held view that reduction in GHG emissions may constraint economic growth. To further understand this matter, it becomes important to understand the causal relationships between the determinants of economic growth and the environmental degradation in India. Hence, investigation of causality relationships among CO_2 emissions, energy consumption and various indicators of international trade are of utmost importance in order to formulate strategies to cope up with the global warming and climatic instability. This paper is an attempt to study the causal relationships between the determinants of international trade i.e., FDI, imports and exports along with the country's trade openness and the levels of energy consumption with the natural environment.

2. BRIEF REVIEW OF RELATED LITERATURE

The relationship between environmental pollution-energy consumption and environmental pollution-indicators of international trade and economic growth has been well documented in the literature. A wide range of literature is devoted to the analysis of environmental pollution represented by either carbon dioxide (CO_2) emissions or sulphur dioxide (SO_2) and nitrous oxide (N_2O) (Seker et al., 2015). Several studies have linked environmental pollution to economic growth (expressed in terms of GDP) and have examined the causal relationships among them (e.g., Choi et al., 2010; Hitam and Borhan, 2012; and Alam et al., 2011) alongwith examining the existence of Environmental Kuznets Curve (EKC) Hypothesis. The finance literature consists of a well debated rationale for the need of foreign investment in less developed countries due to several benefits it provides to the host country e.g., bridging the internal resource and savings gap, increasing managerial abilities, reducing foreign exchange shortage and improving balance of payment (Aliyu, 2005). While there are some studies that have linked FDI with the environmental pollution on the basis of pollution haven hypothesis and have tried to find out the causal relationship among them (e.g., Seker et al., 2015; Shahbaz et al., 2015; Jorgenson et al., 2007; and Bokpin, 2017), the indicators of global trade i.e., exports and imports, have also been used to examine such relationships (e.g., Chakraborty and Mukherjee, 2013; and Hitam and Borhan, 2012). The literature also comprises of similar studies taking openness to global trade and the role of governance and institutions as the variables to examine such relationships (e.g., Choi et al., 2010; and Bokpin 2017). A variable of much importance in the studies of causal relationships among environmental pollution and economic growth is energy consumption. Various similar studies have used energy consumption as an endogenous variable to examine these relationships (e.g., Alam et al., 2011; Choi et al., 2010; and Seker et al., 2015). Although it has been recognised that interrelationships among environmental pollution, capital accumulation and other growth parameters are of central importance in growth theory (Xepapadeas, 2005) yet, there seems to be no consensus regarding the dynamic causality relationship between foreign direct investment, energy consumption, and CO_2 emissions which also included indicators of international trade (i.e., imports and exports) and trade openness as additional variables under the same framework, especially for India.

3. THE MODEL AND THE METHODOLOGY

The study initially uses a linear regression model to examine the relationship between FDI, energy consumption, indicators of international trade and environmental degradation in India during 1982 to 2013. As Acharyya (2009) mentions CO_2 as the major global pollutant, CO_2 emissions are taken as the representative of environmental degradation. The initial model is:

$$CO_2 = f\left(FDI, EXP, IMP, OPEN, EC\right) \tag{1}$$

and the equation of the model is:

$$lnCO_2 = \beta_0 + \beta_1 lnFDI + \beta_2 lnEXP + \beta_3 lnIMP + \beta_4 OPEN + \beta_5 lnEC + u \tag{2}$$

where, CO_2 is carbon dioxide emissions (metric tons per capita); *FDI* is net inflows of foreign direct investment (current US$); *EXP* is exports (% of GDP); *IMP* is imports (% of GDP); *OPEN* is a measure of trade liberalisation; *EC* is energy consumption (kg of oil equivalent per capita) and *u* is the error term or the stochastic random variable.

Consistent with Choi et al. (2010), trade liberalisation is measured as the sum of imports and exports as a share of total GDP in a given year. All the data on the variables used for the study is sourced from the World Development Indicators (WDI) of World Bank Group (2016) Data Reports for India (https://databank.worldbank.org/data/reports.aspx?source=2&country=IND).

The double log model has been used for this study as Benoit (2011) explained logarithmic transformations as a convenient means to transform a highly skewed variable into an approximately normal one. Also, the purpose of using a log-log model is that the slope coefficients can be interpreted as elasticities and as per Gujarati (2012) an advantage of elasticities is that they are pure numbers i.e., devoid of units in which the variables are measured, such as dollars, hours etc.

At first, the study tests for stationarity of the available data using conventional time series unit root test i.e., the Augmented Dickey Fuller (ADF) unit root test. Johansen cointegration test is used to determine the order of the cointegrating equation. Vector Error Correction (VEC) mechanism is used to check for a short-run and long-run relationship among the variables. Next, the study examines whether the residuals of the final estimation regression are homoskedastic, have no serial correlation and are normally distributed. Serial Correlation Lagrange's Multiplier (LM) test and Correlogram of Q-statistics are used to test for any serial correlation among the residuals of the final estimation regression. Breusch-Pagan-Godfrey Heteroskedasticity test is used to examine the heteroskedasticity and Jarque-Bera test is used to examine the normality of the residuals. For determining the optimum lag length Akaike Information Criterion (AIC) is followed (Figure 1). Hence, optimum lag selected is 2.

4. GRAPHICAL REPRESENTATION OF DATA SERIES

Before moving further and going into rigorous statistical and econometric analysis, a graphical representation of the time series of all the concerned variables is very important. Such graphical representation is crucial to determine the direction in which the analysis should move ahead for getting accurate results. Figure 2 provides such graphical representation of all the variables of interest.

5. UNIT ROOT TESTS

As per Stock and Watson (2007), the assumption that the future will be like the past has a great importance in time series regression. If the future is like the past, then the historical relationships can be used to forecast the future. But, if the future differs fundamentally from the past, then the historical relationships might not be reliable guides to the future. This is also true that future will never be like the past. Therefore, in the context of time series regression, the idea that historical relationships can be generalised

Figure 1. Optimum lag selection

Lag	LogL	LR	FPE	AIC	SC	HQ
0	43.18400	NA	3.38e-09	-2.478933	-2.198694	-2.389282
1	232.7421	290.6557*	1.27e-13*	-12.71614	-10.75446*	-12.08858*
2	271.2956	43.69401	1.47e-13	-12.88637*	-9.243261	-11.72091

* indicates lag order selected by the criterion

Figure 2. Graphical representation of all the data series

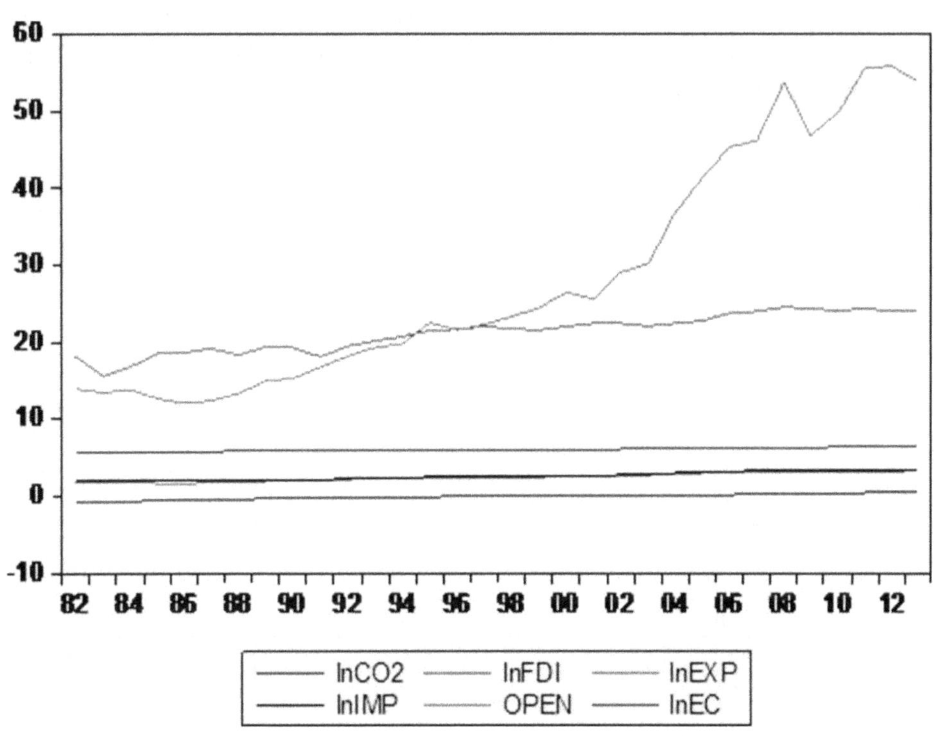

to the future is formalized by the concept of stationarity. A procedure to formally test for non-stationarity was devised by Dickey and Fuller (1979, 1981) known as the Dickey-Fuller unit root test. In order to eliminate autocorrelation in the test equation, Dickey and Fuller extended their test procedure suggesting an augmented version of the test known as Augmented Dickey Fuller (ADF) unit root test which includes extra lagged terms of the dependent variable which are determined by minimising the Akaike Information Criterion (AIC) or Schwarz Information Criterion (SIC). The ADF test equation is given as under:

$$\Delta Z_t = \gamma Z_{t-1} + \sum_{i=1}^{k} \delta_i \Delta Z_{t-i} + u_t \tag{3}$$

Here, k is the number of augmented lags. The null hypothesis is that H_0: $\gamma = 0$ i.e., a unit root exists in Z_t time series. In this study, ADF unit root test has been applied to examine the stationarity of the data series.

Through the graphical representation of the data series of all the variables, it is evident that all these time series do not evolve around zero mean, they actually fluctuate around some other mean. Due to these reasons, deterministic element i.e., a constant has been included in the ADF unit root test. Thus, the equation of ADF test now becomes:

$$\Delta Z_t = \alpha + \gamma Z_{t-1} + \sum_{i=1}^{k} \delta_i \Delta Z_{t-i} + u_t \tag{4}$$

where, α is the intercept term.

Table 1 displays the results of ADF unit root test for the data series of all the variables at level and at first difference. The results indicate that all the data series are non-stationary at level i.e. I(0) at 1% significance level but stationary at first difference i.e. I(1) at 1% significance level. Thus all the variables are integrated of the same order i.e., I(1). These results confirmed that the model meets the requirements to proceed with the cointegration test.

6. UNDERSTANDING CAUSALITY

Before doing any statistical or econometric analysis, it becomes crucial to understand the basic concepts of causality, a brief description of which is given below.

Table 1. ADF unit root test results at level and at first difference (including intercept/constant)

Sr. No.	Variable	p-Value (at Level)	p-Value (at First Difference)
1	$lnCO_2$	0.8592	0.0007***
2	lnFDI	0.7597	0.0000***
3	lnEXP	0.9536	0.0000***
4	lnIMP	0.9738	0.0002***
5	OPEN	0.9850	0.0000***
6	lnEC	0.9996	0.0024***

(Note: *** indicates significant at 1% significance level)

6.1. Vector Auto Regressive (VAR) Models

In economics, it is quite common to have models where some variables are not only explanatory for a given dependent variable but are also explained by the variables that they are used to determine which gives rise to the models of simultaneous models. Thus, a clear identification between endogenous and exogenous variables becomes necessary (Asteriou, 2007). Sims (1980) criticized any differentiation among endogenous and exogenous variables on the grounds that if there is simultaneity among a number of variables, then all these variables should be treated in the same way i.e., all variables are treated as endogenous. This means that in its general reduced form, each equation has the same set of regressors which leads to the development of Vector Auto Regressive (VAR) models. VAR is defined as a system of Auto Regressive Distributed Lag (ARDL) equations describing dynamic evolution of a set of variables from their common history, where 'vector' implies multiple variables involved. VAR allows all variables to interact with themselves (past values) and with each other in regression analysis without any theoretical structure on the estimators. VAR can explain the dynamic structure of time series better than the static Ordinary Least Squares (OLS) estimation method but at the expense of complexity in interpretation of effects (Choi et al., 2010).

Let Y_t and X_t be two series, where Y_t is affected by not only its past (or lagged) values but also by current and lagged values of X_t and vice-versa, then a simple bi-variate VAR model is given by:

$$Y_t = A_1 + B_1 X_t + \sum_{j=1}^{p} C_j Y_{t-j} + \sum_{j=1}^{p} D_j X_{t-j} + u_{1t} \tag{5}$$

$$X_t = A_2 + B_2 Y_t + \sum_{j=1}^{p} E_j Y_{t-j} + \sum_{j=1}^{p} F_j X_{t-j} + u_{2t} \tag{6}$$

where, it is assumed that u_{1t} and u_{2t} are uncorrelated error terms called impulses or shocks in the language of VAR.

If there are I(1) variables, VAR is estimated using first differences of the variables to remove common trends and if the variables are cointegrated, the Error Correction Term (ECT) has to be included in VAR.

6.2. Error Correction Mechanism

In case of a bi-variate model, the purpose and nature of the structural modelling is to discover the dynamic causal relationship between Y_t and X_t i.e., to discover the short-run slope or long-run elasticity of Y_t with respect to X_t ($\partial Y_t / \partial X_t$). For simplifying the analysis, following (ARDL) (1,1) model is considered:

$$Y_t = A_0 + A_1 Y_{t-1} + B_0 X_t + B_1 X_{t-1} + u_t \tag{7}$$

From eq. 7, both short-run and long-run effects (slopes or elasticities) can be analysed as follows for any expanded model ARDL (p,q):

a. a. Short-run or static effect:

$$\frac{\partial Y_t}{\partial X_t} = B_0 \tag{8}$$

b. b. Long-run or dynamic or equilibrium effect:

$$\frac{\partial Y_t}{\partial X_t} = \frac{B_0 + B_1 + B_2 \ldots + B_q}{1 - A_1 - A_2 \ldots - A_p} \tag{9}$$

By subtracting Y_{t-1} from both sides of eq. 7 and adding and subtracting B_0X_{t-1} to the RHS (right hand side), we get what is widely known as the Error Correction Model (ECM) which is applicable with all ARDL models as follows:

$$\Delta Y_t = B_0 \Delta X_t - \pi \left(Y_{t-1} - \alpha - \beta X_{t-1} \right) + u_t \tag{10}$$

where, $\pi = \left(1 - A_1\right)$ i.e., speed of adjustment ($0 < \pi < 1$)

and $\left(Y_{t-1} - \alpha - \beta X_{t-1}\right)$ is the error correction term (ECT) where $\alpha = \frac{A_0}{\left(1 - A_1\right)}$ and $\beta = \frac{\left(B_0 + B_1\right)}{\left(1 - A_1\right)}$

Notice the negative sign of the πECT, which signifies that the result of the coefficient π, from the model generated through system, must have a negative sign alongwith being significant.

Therefore, ECM and ARDL are basically the same if the series are integrated of the same order [often I(1)] and are cointegrated. According to Asteriou (2007), when the two variables are cointegrated, the ECM incorporates not only short-run but also long-run effects. The concept of cointegration will be dealt with in detail, in section 6.4.

6.3. Vector Error Correction (VEC) Models

Introduction of ECT in a VAR model makes the model a Vector Error Correction Model (VECM) which can be seen as a restricted VAR. VECM is an appropriate modelling strategy when the variables are cointegrated. It is useful when long-run forecast is desired; as VAR doesn't explicitly takes into account the long-run relationship (Chamalwa, 2016). VECM gives long-run structural relations plus information on adjustment, which provides better insight in economic processes.

A VAR (p) can be reformulated in a VECM. Suppose we have three variables; Y_t, X_t and Z_t, which can all be endogenous i.e., we have that (using matrix notation for $W_t = [Y_t, X_t, Z_t]$):

$$W_t = A_1 W_{t-1} + A_2 W_{t-2} + \ldots + A_p W_{t-p} + u_t \tag{11}$$

The general case of the VECM including all the various options that can possibly happen is as follows:

$$\Delta W_t = \Pi W_{t-1} + \Gamma_1 \Delta W_{t-1} + \Gamma_2 \Delta W_{t-2} + \ldots + \Gamma_{p-1} \Delta W_{t-p+1} + u_t \tag{12}$$

where, Δ is the difference operator, W_t is a vector with the 6 variables (CO_2, FDI, EXP, IMP, OPEN and EC), $\Gamma_1,\ldots,\Gamma_{p-1}$ represents the matrix of the short-run dynamics, Π represents the long-run dynamics and $\Pi = \pi\beta'$ (Johansen, 1988; and Johansen and Juselius, 1992) where π will include the speed of adjustment to equilibrium coefficients and β' will be the long-run matrix of coefficients (Mohamed et al., 2014). Therefore, the $\beta' W_{t-1}$ term is equivalent to the error correction term ($Y_{t-1} - \alpha - \beta X_{t-1}$) in the single-equation case (eq. 10), except that now $\beta' W_{t-1}$ contains upto *p-1* vectors in a multivariate framework.

The possible ranks of matrix Π and their respective implications are given in Table 2 as under (Binh, 2013):

6.4. Johansen Cointegration Test

Cointegration is an important requirement for any economic model using non-stationary time series data because if the variables do not cointegrate, a case of spurious regression may arise making the econometric work almost meaningless. Suppose that if there really is a genuine long-run relationship between two variables, then although the variables will rise overtime (because they are trended), there will be a common trend or cointegrating vector that links them together (Binh, 2013). Engle and Granger (1987) also proposed that even though all the variables in the model are non-stationary, the series may be stationary if the variables have a linear combination. For an equilibrium, or long-run relationship to exist, a linear combination of the variables (which is a stationary variable i.e., an I(0) variable) is required. In economic terms, two variables will be cointegrated if they have a long-run relationship (or equilibrium) between them. If we have more than two variables in the model, then there is a possibility of having more than one cointegrating vector (Asteriou, 2007) i.e., the variables in the model might form several equilibrium relationships. Generally, in case of *k* number of variables, only upto *k-1* cointegrating vectors can exist. Johansen's methodology is used to find out how many cointegrating relationships exist among the variables.

In order to identify the number of cointegration vectors, Johansen (1988) proposes the trace and maximum eigenvalue statistics. The trace statistic is based on a likelihood ratio test about the trace of the matrix. The trace of a matrix is a very simple operation defined only for square matrices. For any $n \times n$ matrix A, the trace of a matrix A, is the sum of its diagonal elements (Wooldridge, 2013). The maximum eigenvalue statistic is designed to test the null hypothesis that Rank (Π) = *r* i.e., *r* cointegration relationships against the alternative of *r+1* as follows:

$$\lambda_{\max s}\left(r, r+1\right) = -T\ln\left(1-\hat{\lambda}_{r+1}\right) \tag{13}$$

Table 2. Rank of matrix Π and its implications

Rank of Π	Implications
r = 0	There is no cointegration. No stable long-run relations between variables. VECM is not possible (only VAR in first differences).
0 < r < k	There are r cointegrating vectors. These vectors describe the long-run relationships between variables. VECM is introduced.
r = k	All variables are already stationary. No need to estimate the model as VECM. VAR on untransformed data can be done.

Where, T represents the number of observations and $\hat{\lambda}_i$ shows the estimated values of the roots.

The trace statistic considers whether the trace is increased by adding more eigenvalues beyond the rth eigenvalue. Here, the null hypothesis is that the number of cointegrating vectors is less than or equal to r.

$$\lambda_{trace}(r) = -T \sum_{i=r+1}^{n} \ln\left(1 - \hat{\lambda}_i\right) \quad (14)$$

6.5. The Causality Results

Once all the series are confirmed to be categorizing as stationary, and that too, of the same order i.e., I(1), the study used the Johansen Cointegration test with lag order 2 to test whether the dependent variable and all the independent variables exhibit fundamental long-run relationship among each other. The results obtained from the Johansen Cointegration test are displayed in Figure 3. The results from both the maximum eigenvalue and trace test confirm that there is at least 1 cointegrated equation indicating a long-run relationship between the variables which means that all the variables are cointegrated.

As stated earlier, if the variables are cointegrated, ECT has to be included in VAR which will make the model a VECM. The target system model is formulated with the VEC estimates as follows:

D(LNCO2) = C(1)*(LNCO2(-1) - 0.0124064334952*LNFDI(-1) - 0.244690801817*LNEXP(-1) - 0.118927504267*LNIMP(-1) + 0.0362430240936*OPEN(-1) - 3.20828757016*LNEC(-1) + 19.5399530314) + C(2)*D(LNCO2(-1)) + C(3)*D(LNCO2(-2)) + C(4)*D(LNFDI(-1)) + C(5)*D(LNFDI(-2)) + C(6)*D(LNEXP(-1)) + C(7)*D(LNEXP(-2)) + C(8)*D(LNIMP(-1)) + C(9)*D(LNIMP(-2)) + C(10)*D(OPEN(-1)) + C(11)*D(OPEN(-2)) + C(12)*D(LNEC(-1)) + C(13)*D(LNEC(-2)) + C(14) (15)

where, C(1) is the cointegration equation when $lnCO_2$ is dependent variable from where we can derive the residual of cointegration equation, C(2) to C(13) are short-run co-efficients and C(14) is the residual.

Figure 4 represents the results of the VECM model thus formulated with the respective p-values of the coefficients. As we can see in Figure 4, the coefficient C(1) does not contain a negative sign (0.417250) but is significant at 5% significance level, it indicates that there is no long-run causality running from all the variables to $lnCO_2$ (as stated in 6.2). C(1) is also the speed of adjustment towards long-run equilibrium. So, from the results, we can say that there is no long-run causality from FDI, exports, imports, trade openness and energy consumption to CO_2 emissions. This means, all the independent variables viz., FDI, exports, imports, trade openness and energy consumption do not influence the dependent variable i.e. CO_2 emissions in the long-run.

C(4) to C(13) are short-run coefficients. Short-run causality among variables is examined with the help of Wald Test. Null hypothesis in Wald test is that the coefficients of any variable are all equal to zero [i.e., C(4) = C(5) = 0, C(6) = C(7)= 0 and so on in this case] which means there is no short-run causality running from the variables to the target dependent variable. If the p-value of chi-square statistics is less than 5%, the null hypothesis is rejected and signifies that there is a short-run causality from the variables to the dependent variable. Table 3 represents the results of the Wald test to examine the short-run causality. The results indicate that there is no short-run causality running from FDI, exports, imports and energy consumption to CO_2 emissions but openness to international trade influence CO_2 emissions in the short-run.

7. STATISTICAL POINT OF VIEW

From the statistical point of view, it can be seen that the R-squared is 0.804014 which is more than 80%; Durbin-Watson statistic is 1.967244; and F-statistics is 4.733530 with p-value of 0.002688. All these are indicating a good fit of the model.

Figure 3. Johansen cointegration test results

Date: 12/10/17 Time: 17:15
Sample (adjusted): 1985 2013
Included observations: 29 after adjustments
Trend assumption: Linear deterministic trend
Series: LNCO2 LNFDI LNEXP LNIMP OPEN LNEC
Lags interval (in first differences): 1 to 2

Unrestricted Cointegration Rank Test (Trace)

Hypothesized No. of CE(s)	Eigenvalue	Trace Statistic	0.05 Critical Value	Prob.**
None *	0.917593	205.7123	95.75366	0.0000
At most 1 *	0.915163	133.3260	69.81889	0.0000
At most 2 *	0.663653	61.78248	47.85613	0.0015
At most 3 *	0.427589	30.18375	29.79707	0.0451
At most 4	0.382277	14.00471	15.49471	0.0828
At most 5	0.001205	0.034978	3.841466	0.8516

Trace test indicates 4 cointegrating eqn(s) at the 0.05 level
* denotes rejection of the hypothesis at the 0.05 level
**MacKinnon-Haug-Michelis (1999) p-values

Unrestricted Cointegration Rank Test (Maximum Eigenvalue)

Hypothesized No. of CE(s)	Eigenvalue	Max-Eigen Statistic	0.05 Critical Value	Prob.**
None *	0.917593	72.38635	40.07757	0.0000
At most 1 *	0.915163	71.54352	33.87687	0.0000
At most 2 *	0.663653	31.59872	27.58434	0.0144
At most 3	0.427589	16.17904	21.13162	0.2146
At most 4	0.382277	13.96973	14.26460	0.0556
At most 5	0.001205	0.034978	3.841466	0.8516

Max-eigenvalue test indicates 3 cointegrating eqn(s) at the 0.05 level
* denotes rejection of the hypothesis at the 0.05 level
**MacKinnon-Haug-Michelis (1999) p-values

Figure 4. Causality

Dependent Variable: D(LNCO2)
Method: Least Squares (Gauss-Newton / Marquardt steps)
Date: 12/10/17 Time: 19:09
Sample (adjusted): 1985 2013
Included observations: 29 after adjustments
D(LNCO2) = C(1)*(LNCO2(-1) - 0.0124064334952*LNFDI(-1) -
0.244690801817*LNEXP(-1) - 0.118927504267*LNIMP(-1) +
0.0362430240936*OPEN(-1) - 3.20828757016*LNEC(-1) +
19.5399530314) + C(2)*D(LNCO2(-1)) + C(3)*D(LNCO2(-2)) + C(4)
*D(LNFDI(-1)) + C(5)*D(LNFDI(-2)) + C(6)*D(LNEXP(-1)) + C(7)
*D(LNEXP(-2)) + C(8)*D(LNIMP(-1)) + C(9)*D(LNIMP(-2)) + C(10)
*D(OPEN(-1)) + C(11)*D(OPEN(-2)) + C(12)*D(LNEC(-1)) + C(13)
*D(LNEC(-2)) + C(14)

	Coefficient	Std. Error	t-Statistic	Prob.
C(1)	0.417250	0.079474	5.250136	0.0001
C(2)	-0.264303	0.198197	-1.333538	0.2023
C(3)	-0.712248	0.228571	-3.116094	0.0071
C(4)	-0.009019	0.007559	-1.193188	0.2513
C(5)	0.003325	0.005958	0.558130	0.5850
C(6)	0.134125	0.099499	1.348005	0.1977
C(7)	0.025194	0.088544	0.284531	0.7799
C(8)	0.283562	0.174190	1.627832	0.1244
C(9)	0.092763	0.119789	0.774388	0.4507
C(10)	-0.014571	0.006313	-2.308033	0.0357
C(11)	-0.009740	0.004234	-2.300481	0.0362
C(12)	0.273185	0.398067	0.686278	0.5030
C(13)	0.580158	0.354752	1.635391	0.1228
C(14)	0.069042	0.014408	4.791788	0.0002

R-squared	0.804014	Mean dependent var	0.039380
Adjusted R-squared	0.634159	S.D. dependent var	0.029389
S.E. of regression	0.017776	Akaike info criterion	-4.915692
Sum squared resid	0.004740	Schwarz criterion	-4.255618
Log likelihood	85.27754	Hannan-Quinn criter.	-4.708965
F-statistic	4.733530	Durbin-Watson stat	1.967244
Prob(F-statistic)	0.002688		

8. RESIDUAL DIAGNOSTICS

It becomes important to do a check on the residuals of the model before making any conclusions. Table 3 gives the results of residual diagnostics performed on the final estimated model. Breusch-Godfrey

Table 3. Wald test results

Short-Run Causality	p-Value
lnFDI to $lnCO_2$	0.2485
lnEXP to $lnCO_2$	0.4031
lnIMP to $lnCO_2$	0.2128
OPEN to $lnCO_2$	0.0043
lnEC to $lnCO_2$	0.1266

Serial Correlation Lagrange Multiplier (LM) test is used to examine the existence of serial correlation among the residuals of the final estimation. In case of a p-value of more than 5%, the null hypothesis of 'no serial correlation exists' cannot be rejected. Also, a correlogram of the residuals is used for double-checking the existence of serial correlation. Correlogram of Q-statistics is used for this purpose (Figure 5) where the null hypothesis is that the 'data is stationary' against the alternative of 'data has a unit root'. The results from both the tests indicate that the entire residuals are free from autocorrelation. Residuals of the final estimation are then checked for the existence of heteroskedasticity using the Breusch-Pagan-Godfrey heteroskedasticity test. In case of a p-value of more than 5%, the null hypothesis of 'no heteroskedasticity' cannot be rejected. The result indicates that the residuals of the final estimation are homoskedastic. Through the Jarque-Bera normality test, it can be seen that the residuals are normally distributed. The Jarque-Bera statistics of more than 5% indicates that the residuals of the final estimation are normally distributed.

From the results of various tests performed on the residuals of the final estimation, it becomes clear that the model is well-formulated because the residuals of the final estimation are free from autocorrelation and heteroskadasticity and are normally distributed.

9. CONCLUSIONS AND RECOMMENDATIONS

The results of the VEC Model can be summarized briefly as follows:

1). There is no long run causality running from FDI, exports, imports, trade openness and energy consumption to CO_2 emissions i.e., all the variables do not affect CO_2 emissions in the long-run.
2). There is no short run causality running from FDI, exports, imports and energy consumption to CO_2 emissions i.e., FDI, exports, imports and energy consumption do not influence CO_2 emissions in the short-run.

Table 4. Residual diagnostics results

Tests Performed	p-Values
Breusch-Godfrey Serial Correlation Lagrange Multiplier (LM) test	0.3885
Breusch-Pagan-Godfrey heteroskedasticity test	0.7688
Jarque-Bera Normality test	0.807352

Figure 5. Correlogram of residuals

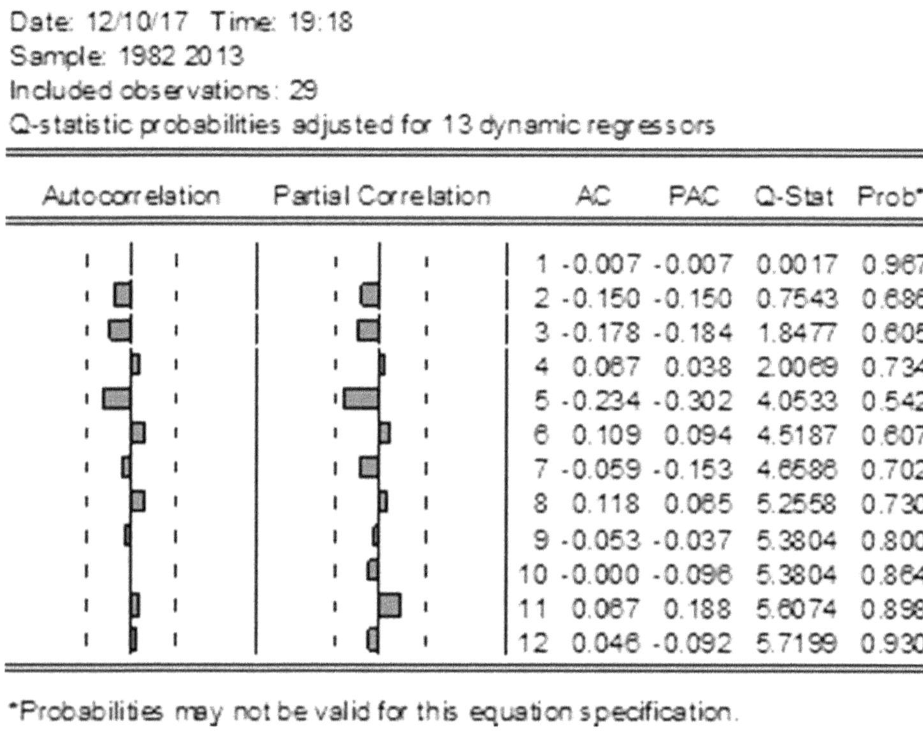

Date: 12/10/17 Time: 19:18
Sample: 1982 2013
Included observations: 29
Q-statistic probabilities adjusted for 13 dynamic regressors

Autocorrelation	Partial Correlation		AC	PAC	Q-Stat	Prob*
		1	-0.007	-0.007	0.0017	0.967
		2	-0.150	-0.150	0.7543	0.686
		3	-0.178	-0.184	1.8477	0.605
		4	0.067	0.038	2.0069	0.734
		5	-0.234	-0.302	4.0533	0.542
		6	0.109	0.094	4.5187	0.607
		7	-0.059	-0.153	4.6586	0.702
		8	0.118	0.065	5.2558	0.730
		9	-0.053	-0.037	5.3804	0.800
		10	-0.000	-0.096	5.3804	0.864
		11	0.067	0.188	5.6074	0.898
		12	0.046	-0.092	5.7199	0.930

*Probabilities may not be valid for this equation specification.

3). There is a short run causality running from trade openness to CO_2 emissions i.e., all trade liberalisation becomes one of the causes of CO_2 emissions in the short-run but not in long run.

Based on the above summary, it can be concluded that India can open its markets for investors and can continue to reduce trade barriers for a free trade environment to accumulate capital; and that too to some extent because trade openness causes CO_2 emissions in the short-run but in the long-run, it does not do so. Similarly, FDI, exports, imports and even energy consumption do not cause CO_2 emissions in long-run as well as short-run i.e., India can continue to export to and import from other countries but with a limit check by the government institutions and that too to some extent because they both do not cause environmental pollution in the short-run. Import of dirty goods may be banned, if possible, or may be reduced to a certain limit. As far as FDI is concerned, the results showing that FDI does not cause CO_2 emissions in the long-run as well as short-run indicate that Pollution Haven Hypothesis does not exist in India currently. India should welcome foreign investments for its economic growth but with regular checks so that it does not start encouraging dirty investments from developed countries to India in future. New investors should be given license only after detailed verification. Local innovation should be encouraged. Emission limits must be defined for the polluting industries and every effort should be made to make them comply with the rules.

Despite the notable findings, the study suffers data availability limitations. The data for all the variables is taken for the years 1982 to 2013 as per their availability upto 2013, whereas the 'Make in India' program of Indian government was launched in the year 2014 that increased the number of economic activities on a large scale. Hence its impacts (both negative and positive) on all the variables are not taken into account that may have both long-run and short-run effects both on economy and the natural environment of India. Future studies can be expected to take these impacts into consideration.

REFERENCES

Acharyya, J. (2009). FDI, Growth and the Environment: Evidence from India on CO_2 Emission during the Last Two Decades. *Journal of Economic Development*, *34*(1), 43–58. doi:10.35866/caujed.2009.34.1.003

Alam, M. J., Begum, I. A., Buysse, J., Rahman, S., & Huylenbroeck, G. V. (2011). Dynamic Modeling of Causal Relationship between Energy Consumption, CO_2 Emissions and Economic Growth in India. *Renewable & Sustainable Energy Reviews*, *15*(6), 3243–3251. doi:10.1016/j.rser.2011.04.029

Aliyu, M. A. (2005). Foreign Direct Investment and the Environment: Pollution Haven Hypothesis Revisited. *8th Annual Conference on Global Economic Analysis*, Lubeck, Germany.

Asteriou, D., & Hall, S. G. (2007). *Applied Econometrics: A Modern Approach Using Eviews and Microfit* (Revised Edition). Palgrave Macmillan.

Benoit, K. (2011). *Linear Regression Models with Logarithmic Transformations*. London School of Economics.

Binh, P. T. (2013). *Unit Root Tests, Cointegration, ECM, VECM and Causality Models*. Topics in Time Series Econometrics.

Bokpin, G. A. (2017). Foreign direct investment and environmental sustainability in Africa: The role of institutions and governance. *Research in International Business and Finance*, *39*, 239–247. doi:10.1016/j.ribaf.2016.07.038

Chakraborty, D., & Mukherjee, S. (2013). How do Trade and Investment Flows affect Environmental Sustainability? Evidence from Panel Data. *Environmental Development*, *6*, 34–47. doi:10.1016/j.envdev.2013.02.005

Chamalwa, H. A., & Bakari, H. R. (2016). A Vector Autoregressive (VAR) Cointegration and Vector Error Correction Model (VECM) Approach for Financial Deepening Indicators and Economic Growth in Nigeria. *American Journal of Mathematical Analysis*, *4*(1), 1–6.

Choi, E., Heshmati, A., & Cho, Y. (2010). An Empirical Study of the Relationships between CO_2 Emissions, Economic Growth and Openness. *SSRN*, (5304), 1–27. doi:10.2139srn.1708750

Dickey, D. A., & Fuller, W. A. (1979). Distribution of the Estimators for Autoregressive Time Series with a Unit Root. *Journal of the American Statistical Association*, *74*(366), 427–431. doi:10.2307/2286348

Dickey, D. A., & Fuller, W. A. (1981). Likelihood Ratio Statistics for Autoregressive Time Series with a Unit Root. *Econometrica*, *49*(4), 1063. doi:10.2307/1912517

Engle, R., & Granger, C. W. J. (1987). Cointegration and Error Correction: Representation, Estimation and Testing. *Econometrica*, *55*(2), 251–276. doi:10.2307/1913236

Gujarati, D. N. (2012). *Econometrics By Example* (2nd ed.). Palgrave Macmillan.

Hitam, M. B., & Borhan, H. B. (2012). FDI, Growth and the Environment: Impact on Quality of Life in Malaysia. *Procedia: Social and Behavioral Sciences*, *50*, 333–342. doi:10.1016/j.sbspro.2012.08.038

IPCC. (1996). Second Assessment on Climate Change. International Panel on Climate Change. Cambridge, UK: Cambridge University Press.

Johansen, S. (1988). Statistical Analysis of Cointegration Vectors. *Journal of Economic Dynamics & Control*, *12*(2), 231–254. doi:10.1016/0165-1889(88)90041-3

Johansen, S., & Juselius, K. (1992). Testing Structural Hypothesis in a Multivariate Cointegration Analysis of the PPP and the UIP for UK. *Journal of Econometrics*, *53*(1-3), 211–244. doi:10.1016/0304-4076(92)90086-7

Johnson, D., & Turner, C. (2010). *International Business: Themes and Issues in the Modern Global Economy* (2nd ed.). Routledge, Taylor & Francis Group. doi:10.4324/9780203861615

Jorgenson, A. K., Dick, C., & Mahutga, M. C. (2007). Foreign Investment Dependence and the Environment: An Ecostructural Approach. Society for the Study of Social Problems, 54(3), 371-394.

Mohamed, M. B. H., Skima, I., Saafi, S., & Farhat, A. (2014). Price modeling: Analysis with a Vector Error Correction Model. *International Conference on Innovation & Engineering Management, Proceedings*, 99-102.

Seker, F., Ertugrul, H. M., & Cetin, M. (2015). The Impact of Foreign Direct Investment on Environmental Quality: A Bounds Testing and Causality Analysis for Turkey. *Renewable & Sustainable Energy Reviews*, *52*, 347–356. doi:10.1016/j.rser.2015.07.118

Shahbaz, M., Nasreen, S., Abbas, F., & Anis, O. (2015). Does Foreign Direct Investment Impede Environmental Quality in High, Middle and Low-income Countries? *Energy Economics*, *51*, 1–35. doi:10.1016/j.eneco.2015.06.014

Sims, C. A. (1980). Macroeconomics and Reality. *Econometrica*, *48*(1), 1–48. doi:10.2307/1912017

Stock, J. H., & Watson, M. W. (2007). *Introduction to Econometrics* (2nd ed.). Pearson Education.

Wooldridge, J. M. (2013). *Introductory Econometrics: A Modern Approach* (5th ed.). South-Western Cengage Learning.

World Bank Group. (Ed.). (2016). *World Development Indicators 2016*. World Bank Publications.

Xepapadeas, A. (2005). Economic Growth and the Environment. Handbook of Environmental Economics, 1(2), 1219-1271.

Chapter 11
Mapping the Future of Sustainable Development Through Cloud-Based Solutions:
A Case Study of OpenStreetMap

Munir Ahmad
https://orcid.org/0000-0003-4836-6151
Survey of Pakistan, Pakistan

Asmat Ali
https://orcid.org/0000-0002-8804-2285
Survey of Pakistan, Pakistan

ABSTRACT

OSM plays a crucial role in advancing sustainable development across the economy, environment, and society. It can foster economic sustainability by enabling infrastructure planning, disaster response, and energy efficiency. Moreover, OSM can contribute to environmental sustainability through resource management, climate change mitigation, and biodiversity protection. Additionally, OSM can promote social sustainability by empowering communities and supporting healthcare, education, and gender equality. Spatial analysis of OSM data can help to identify areas that need essential services, addressing social inequalities. OSM's vast capabilities, coupled with cloud computing, can further enhance its role in driving sustainable progress. Continuous efforts are required to enhance data accuracy and completeness in OSM using advanced techniques like machine learning and artificial intelligence while also promoting inclusive mapping involving marginalized communities.

INTRODUCTION

Over the years, cloud computing has garnered widespread recognition as the most effective method for executing and managing high-performance applications across a diverse range of domains (Kamanga et

DOI: 10.4018/978-1-6684-9979-5.ch011

al., 2023). It is widely acknowledged for its versatility and efficiency in delivering computing resources and services to users and industries, enabling seamless and scalable performance across various sectors (Tank & Gandhi, 2023). Cloud computing, as a cutting-edge technology, has become deeply ingrained in our daily lives, much like the internet. For instance, cloud storage applications enable us to share vast amounts of data online without any geographical limitations. Likewise, cloud-based online meeting applications allow us to effortlessly share views, audio, and video online, transcending geographical boundaries (Huawei Technologies Co., Ltd., 2023).

Cloud computing indeed exhibits several key characteristics that set it apart from traditional computing models such as on-demand self-service, easy-to-access standardized mechanisms, resource pooling, rapid elasticity, auditability & certifiability, and measured service (Hamdaqa & Tahvildari, 2012). Cloud Computing has become a widely accepted approach for constructing service-centric systems. Nonetheless, Cloud Computing does have limitations when it comes to handling massive amounts of data, specifically concerning ultra-low latency, high bandwidth, security, and real-time analytics. However, Fog and Edge Computing present viable solutions to address these constraints, offering alternative avenues to overcome the limitations of Cloud Computing (Kalyani & Collier, 2021).

The integration of cloud technology solutions enhanced the sustainability and efficiency of research and development, along with teaching practices, resulting in an overall improvement in the quality of education within academic institutions. Both public and private learning institutions should seize the opportunity presented by cloud infrastructure to ensure the delivery of high-quality services, even with limited resources at their disposal (Al-Rasheedi & Khan, 2021). The integration of IoT-based agriculture systems with relevant technologies, such as cloud computing, big data storage, and analytics, is instrumental in the implementation of smart farming practices (Farooq et al., 2019). The notion of smart cities has developed through the integration of information and communication technology (ICT) into diverse subsystems and processes within the urban environment. Advancements in digital technologies, including IoT, fog/edge/cloud computing, and cyber-physical systems, have transformed smart city applications (Ahmed et al., 2022). Leveraging advanced technologies, such as IoT, AI, cloud computing, real-time data analytics, and smart cities can strengthen their disaster preparedness and response capabilities (Munawar et al., 2022).

Integration of cloud computing and big data has led to significant advancements in science discoveries and application development. Cloud computing offers crucial solutions for handling big data effectively, especially considering the influence of spatiotemporal thinking and various application domains. The inherent spatiotemporal principles in big data and geospatial sciences provide valuable insights for optimizing cloud computing and processing large datasets (C. Yang et al., 2017). OpenStreetMap (OSM) serves as a widely adopted platform for Volunteered Geographic Information (VGI). Notably, OSM also constitutes a significant example of big data, as evidenced in previous studies by Castro et al., (2019; Ma et al., (2015; and Schweizer et al., (2021). Over the past 18 years, the OSM community has contributed to the creation and modification of billions of geographical features, such as roads and buildings, continuously enriching the dataset. As the OSM community continues to grow, the dataset becomes even more valuable, necessitating advanced tools, interfaces, and methodologies to harness its full potential. In collaboration with the AWS Public Datasets program, OSM data archives are now readily accessible in Amazon S3 and available in multiple formats, including ORC and PBF (AWS, 2017). These archives encompass both up-to-date snapshots of the current data and historical records, preserving the evolution of the dataset and facilitating various exploration and analysis endeavors.

OSM data offers updated information on utilities, transportation networks, and infrastructure, allowing for efficient resource allocation and infrastructure development (Brandt et al., 2022). This encourages effective resource utilization and minimizes environmental impacts, facilitating informed decision-making by enterprises and governments. By mapping tourist attractions and local businesses, OSM also promotes local economies and tourism, fostering community resilience and prosperity (Steven Moore, 2020). OSM data enables communities and governments to make informed decisions for sustainable land use practices, assisting in the conservation of natural resources and wildlife habitats (Chen et al., 2021). Through the use of geographical data analysis, it helps farmers optimize irrigation schedules and use less water (Singh et al., 2022). According to KLL, (2015), OSM's community-driven methodology promotes a sense of pride, empowerment, and ownership among local communities. OSM fosters social participation and aids in the identification of places with limited access to healthcare and educational resources by incorporating communities in mapping activities (Mooney et al., 2013; Shiau et al., 2018).

To this effect, the primary objective of this chapter is to present a comprehensive overview of how OpenStreetMap contributes to sustainable development. To achieve this objective, a sustainable development model is utilized as the methodology to illustrate OSM's role as a catalyst across the three dimensions of sustainable development: economic, social, and environmental. To achieve the objective, the chapter is organized as follows. Section 2 highlights background concepts. Section 3 describes the methodology. Section 4 designates results and suggestions. The last section serves as the conclusion of the chapter along with future work.

BACKGROUND CONCEPTS

Cloud Computing

In recent times, the field of cloud computing has seen a significant surge in adoption across various sectors, be it public or private. This technology offers users a diverse range of services and resources through a centralized online platform. Despite cloud computing's immense potential, its complete capabilities are yet to be fully harnessed. Nevertheless, its global uptake continues to grow exponentially, with industry giants such as Microsoft and Google spearheading the movement (Alsolami, 2018).

Mikulecky, (2015) has put forward a cloud-based solution to provide intelligent support for diverse human activities in outdoor settings. The application of cloud-based solutions is contributing significantly to the digital transformation of human resource (HR) practices (Al-Rwaidan et al., 2023). A cloud-based solution is proposed by (Mollahoseini Ardakani et al., 2019) for the development and operation of collaborative networked organizations. In the realm of construction projects, cloud-based governance solutions have emerged as innovative tools to foster effective team collaboration (Alreshidi et al., 2018).

Local Cloud-based solutions along with machine learning techniques such as deep learning and artificial neural networks, are being developed to support the healthcare sector in curing various diseases. For example, for Liver cancer detection (Anil et al., 2022), and diabetic retinopathy detection (Pruthviraja et al., 2021). Besides, supporting the remedies for diseases, there is a growing trend of transitioning electronic health records (EHRs) to cloud-based web environments (Johnson et al., 2019). It can support eHealth applications and services (De La Torre-Díez et al., 2015). The concept of smart learning incorporates cutting-edge paradigms to enhance and optimize the teaching and learning process. Alam

& Saiyeda, (2020) introduced a cloud-based solution for smart education, aiming to revolutionize the educational experience.

An advanced farm management system, based on cloud technology, facilitates the seamless integration of internal and external services, leading to the creation of a diverse marketplace offering sophisticated and cutting-edge applications and services for end-users to utilize (Kaloxylos et al., 2014). Cloud-based solutions can be deployed for the development of sustainable solutions for monitoring, control, prediction, and logistics in the context of agriculture (Araújo et al., 2021). Cloud-based Complaints Management and Decision Support System for Sustainable Farming to seek bits of advice on crop diseases, patterns, and adequate prevention actions (K. Munir et al., 2022).

Sustainable Development

Corbett & Rostow, (1960) developed the idea of sustainable growth and development for the first time in 1960. It emphasized the value of sustainability as a key component of real growth, taking into account a variety of dimensions of society, including the environment, well-being, and economy. The Sustainable Development Goals (SDGs) were unveiled by the United Nations (UN) in 2015 as a comprehensive framework to address global issues and promote sustainable development on a global scale. The SDGs, which have 17 objectives and 169 targets, address several interrelated concerns, such as eradicating poverty and ensuring access to education, gender equality, climate change mitigation, and others (UNDP, 2023).

In addressing various questions, including those related to sustainable development goals (SDGs), spatial data becomes indispensable. It plays a crucial role in supporting the study, monitoring, and achievement of SDGs (Scott & Rajabifard, 2017). Spatial data enables the analysis, mapping, and modeling of sustainable development challenges, providing a foundation for worldwide cooperation and informed decision-making (UN-GGIM, 2012).

VGI and OSM

Geospatial information is no longer just produced by public sector organizations in this era of technological advancement; communities and citizens are also contributing to it voluntarily (Ahmad et al., 2022). A good example of this is VGI. Thanks to the contributions of volunteers and humanitarian organizations, VGI acts as a significant resource in support of the SDGs by offering up-to-date geospatial data through public internet platforms. This information is crucial for tracking and reporting SDG progress. For instance, Cramwinckel, (2019) looked at how open geospatial data products from throughout the world can be used to assess SDG indicator 11.7.1: public open spaces.

OpenStreetMap, which was founded in 2004, is a worldwide initiative for developing and curating a freely accessible and editable database and map of the entire world (Grinberger et al., 2022). It is the epitome of VGI because of its enormous volunteer base and the huge amount of spatial data it has produced. OSM's applications have shown that such a project has potential that goes far beyond traditional map-making. These projects range from important initiatives like the Humanitarian OpenStreetMap (HOT) initiative, which serves as a link between the OSM community and humanitarian responders, to cooperative projects like Mapillary, where users capture street-level images to support the system that aims to automate the mapping process (Bertolotto et al., 2020).

METHODOLOGY

The methodology adopted for this research is primarily descriptive and literature-based, focusing on exploring the vast applications of OSM for sustainable development. The methodology employed in this study is grounded on the three essential pillars of sustainable development: economic, environmental, and social (Ranjbari et al., 2019), as illustrated in Figure 1.

The three elements that make up sustainable development are shown in Figure 1. The public's demand for sustainable economic growth is addressed by the first dimension, economic sustainability. Environmental sustainability, the second element, emphasizes the prudent use of natural resources without endangering the ecosystem. The third pillar, referred to as "social sustainability," is concerned with providing for people's necessities, such as social welfare, healthcare, and education.

The objective is to showcase the versatility and applicability of OSM in various domains related to sustainability. Due to the extensive scope of OSM applications and the limitations of a book chapter's length, it is not feasible to cover every individual application area in detail. Therefore, the chapter has focused on presenting a comprehensive overview of the key application domains while referencing relevant literature to support the findings.

RESULTS AND DISCUSSION

This is accomplished by examining how OpenStreetMap contributes to sustainable development. OpenStreetMap plays a crucial role in fostering economic, environmental, and social sustainability. The po-

Figure 1. Interaction of OSM and sustainable development

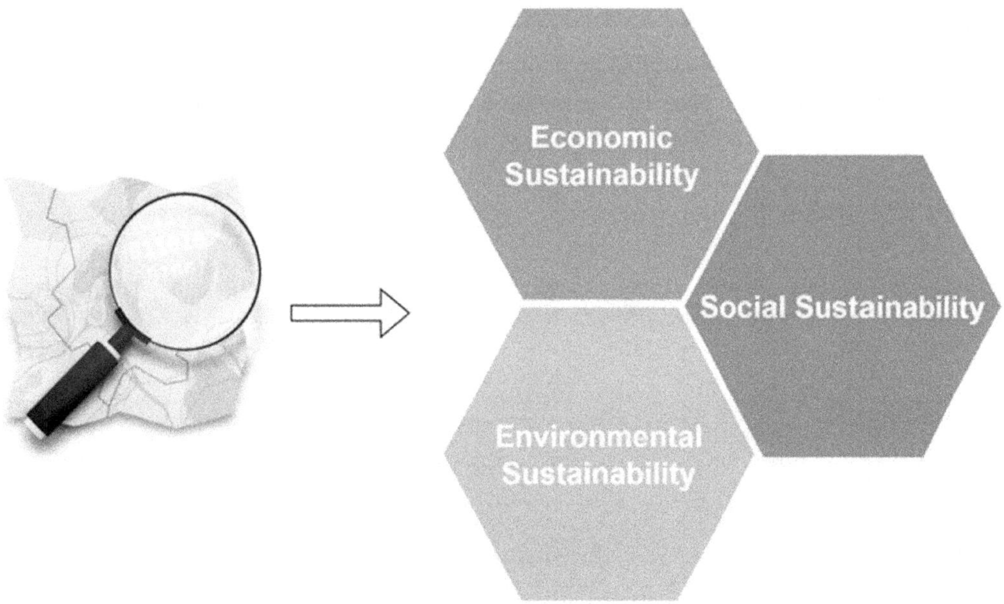

tential contribution of OpenStreetMap to sustainable development in terms of economic, environmental, and social sustainability is summarized in Figure 2 and elaborated as follows:

Economic Sustainability

OpenStreetMap offers a wide range of capabilities that enhance economic sustainability. With an emphasis on economic sustainability, the potential benefits of OpenStreetMap are further explained as follows:

Infrastructure Planning

The planning and development of sustainable infrastructure projects can be facilitated by the comprehensive and current data provided by OpenStreetMap. By providing updated information about utilities, roads, and public transportation, decision-makers can make informed decisions to encourage effective resource use and minimize environmental impact. Governments can derive invaluable insights into regions with inadequate infrastructure development by integrating OSM data with other pertinent factors, including population density, traffic patterns, and transportation networks (Brandt et al., 2022). The mapping of low-income areas—often slums—for efforts aimed at ending poverty can be a crucial support for development initiatives. In particular, monitoring areas with low employment rates and mapping financial service accessibility are essential stages in enabling those living in poverty to achieve more economic stability (Sara Amadi & Jess Beutler, 2019).

Tourism and Local Economies

In every region of the world, tourism is a significant economic engine. The mapping of tourist destinations and neighbourhood businesses by OpenStreetMap (Steven Moore, 2020) encourages travelers to explore other locales and actively engage in local economies, thereby fostering community resilience and prosperity. Through the analysis of spatial data, particular regions can be identified that provide eco-friendly activities, cultural immersion, or other distinctive attractions (S. Liu & Gao, 2022), bringing in money for their companies while also promoting local economic development and cultural heritage preservation. In Larabanga, Ghana, an organization called Leclara has put into place programmes for

Figure 2. OSM for sustainable development

community-run and sustainable tourism to raise money locally, which will help the neighbourhood and encourage economic sustainability (Sara Amadi & Jess Beutler, 2019).

Disaster Response

In order to identify affected areas and create rescue and recovery plans after natural disasters, spatial data is crucial for disaster response. The crowd-sourced data from OpenStreetMap is a priceless asset during emergencies and natural disasters (Latif et al., 2011; Minghini et al., 2017; Tzavella et al., 2022). Real-time maps make it possible for local communities and humanitarian organizations to respond quickly and efficiently, which ultimately reduces economic losses and speeds up the recovery process. OSM has become an essential tool for enabling quick access to current maps and geospatial data for emergency workers during earthquakes in Turkey and Syria as can be seen in Figure 3 (HOTOSM, 2023a; Jessica Pechmann & Claudio de los Reyes, 2023). This useful resource helped search and rescue operations, provided accurate damage assessment, and facilitated the organization of relief and recovery efforts. Responders can strategically allocate resources, make knowledgeable judgments, and act quickly after a disaster by utilizing OSM's capabilities. This helped the impacted communities' economic sustainability through efficient disaster response and recovery.

It is now possible to analyze the precise locations and severity of disasters like floods, wildfires, and earthquakes by integrating real-time spatial data from satellites, weather stations, crowdsourced data, and various sources (X. Hu et al., 2019). This enables quick responses, immediate relief efforts, minimal economic losses, and rapid recovery.

Energy Efficiency

Businesses that provide energy management services can efficiently use spatial data to identify regions with either high or low energy usage (Aldhshan et al., 2021). By combining OSM data with non-spatial data such as energy usage, building density, and infrastructure development, businesses and governments can able to develop well-informed strategies to meet energy demands efficiently. For example, Schiefelbein et al., (2019) presented an urban energy modeling technique by incorporating OSM data to assess the effects of various energy efficiency measures on urban districts. Similarly, Saha et al., (2019) introduced a framework for generating synthetic distribution feeders that align with real geospatial topologies, utilizing existing OpenStreetMap data. These initiatives to improve energy efficiency significantly promote economic efficiency and environmental sustainability on a larger scale (Aldhshan et al., 2021; Alhamwi et al., 2017; Wang et al., 2019).

Environmental Conservation

OpenStreetMap can support conservation activities and promote the adoption of sustainable land use practices by providing comprehensive information about natural resources, wildlife habitats, and ecosystems. For example, Chen et al., (2021) presented an innovative framework aimed at mapping urban land use categories, utilizing a combination of OSM data and various other data sources, including land surface, vertical height, socioeconomic attributes, social media data, demography, and topography. Similarly, Heider et al., (2018) explored the impacts of settlement patterns and major roadways on the peri-urbanization process in Mexico City using OSM data, to support urban planning and sustainable

Figure 3. Mapping in support of Turkey and Syria (HOTOSM, 2023a)

development strategies. Utilizing the data provided by OpenStreetMap, communities, and governments can make informed decisions that promote sustainable practices, the enhancement of ecosystem services, and the preservation of valuable natural resources, ultimately promoting economic resilience and long-term prosperity.

Agricultural Efficiency

Spatial data is essential to improve agricultural productivity and promote economic growth (Singh et al., 2022). Compiling soil profile data (World Soil Information, 2020), spotting areas that need improvement (Jung et al., 2022; Wan et al., 2017), and using precision agriculture methods (Fernández de Sevilla et al., 2021) can be supported by using OSM data. Farmers may strategically optimize irrigation patterns, reduce water use, and eventually increase crop yields by analyzing geographical data on soil composition, moisture levels, and other important characteristics. For example, Nirab Pudasaini, (2016) endorsed the use of OSM data to map crop production areas in Nepal to commercialize agriculture activities. Similarly, Sun et al., (2018) argued the support of OSM to continuous monitoring of farmland area and agricultural production management, which is essential for effective and sustainable farming practices to improve economic sustainability.

Market Analysis

Data from OpenStreetMap is a useful resource that enables local communities, businesses, and entrepreneurs to obtain geographic data that supports economic development activities. It can enable market analysis and assist in locating possible investment regions because of its plethora of location-based information (Wichmann et al., 2023). For example, (Dupre, 2020) conducted research using OSM data to comprehend the features of the housing market for the city of Dublin. By utilizing spatial data for market analysis, it is possible to identify certain geographic areas with a strong demand for particular goods and services. This information can empower business owners to decide whether to launch new ventures or expand existing operations within these borders. This tactical strategy guarantees that companies provide goods or services that perfectly address the requirements of the target market in particular regions (Wieland, 2017).

Supply Chain Management

The identification of the most cost-effective transportation routes and logistical tracks and to maximize efficiency, spatial data is crucial in supply chain management. Businesses can efficiently use geographical data to determine the most efficient routes, transportation modes, and warehouse sites by conducting network analysis of transportation networks (Yusianto et al., 2020). By reducing carbon emissions, this strategic strategy not only lowers the cost of logistics and transportation but also improves the supply chain's overall sustainability. OSM as an open-source spatial data contributor can also support the supply chain sector. For example, Penchev, (2016) used OSM data to estimate the economic consequences of natural disasters on supply chains.

Real Estate Development

In some areas, real estate development is a significant economic force. Real estate developers can better comprehend the dynamics of the market by performing feasibility studies with the help of spatial data. Developers can locate underutilized urban areas with unrealized potential by using spatial data (Rabiei-Dastjerdi et al., 2021). For example, an OSM-based platform (RealHomes, 2023) has been developed to support real estate activities. Entrepreneurs and governments can develop successful plans to improve affordable housing or commercial spaces in the chosen location by integrating geographical data with demographic and infrastructure information, opening up prospects for local populations to experience economic growth.

Environmental Sustainability

OpenStreetMap provides a diverse set of capabilities that contribute to environmental sustainability. The potential benefits of OpenStreetMap, with a focus on environmental aspects, are elaborated below:

Sustainable Resource Management

The effectiveness of management and monitoring of natural resources can be considerably increased by incorporating spatial data (Chaula, 2019). Spatial data is extremely helpful for analyzing the availability of water supplies, forest cover, and wildlife habitats, allowing for data-driven conservation decisions. For example, Yang, (2019) introduced a novel method for creating a forest cover map and quantifying forest fragmentation caused by roads using integration of OpenStreetMap data with remote sensing datasets available in Google Earth Engine. Similarly, Yang et al. (2017) formulated an innovative land-use mapping approach by combining OpenStreetMap (OSM) data with earth observation remote sensing imagery. Moreover, H. Li et al., (2021) suggested an automated workflow for surface water mapping by utilizing a deep learning technique trained on both OSM and RS data.

Sustainable Agriculture

Sustainable agriculture-related businesses can greatly improve their farming practices by utilizing spatial data in conjunction with other crucial elements like topography, soil health, and climate data. This all-encompassing strategy enhances agricultural yields while using less water and toxic chemicals (Singh et al., 2022). Map For Future has undertaken a digitization project to map the farmyards in the Wajaale region of Somaliland (HOTOSM, 2021). To enhance the planning capacity of the Ministry of Agriculture in Somaliland, Map For Future collaborated with local group GIS & Cadastral Survey (GCS) and FORUM-UN.ITALIA to develop a Geoportal for the city of Wajaale, enabling efficient and accessible geographical information for agricultural planning and development in the region. These methods actively support environmental sustainability through such intelligent resource allocation.

Urban Green Spaces

Urban green space studies heavily rely on mapping as a fundamental tool. The comprehensive mapping of parks, greenways, and other urban green spaces by OpenStreetMap, acts as a driving force for sustainable urban planning, promoting a greener and healthier urban environment. For example, (Ju et al., 2022) generated the initial 10 m resolution urban green space (UGS) map covering the main urban clusters across 371 major Latin American cities by employing a supervised classification method utilizing Sentinel-2 satellite images and UGS samples extracted from OSM data. The resulting UGS map provides valuable information for studies involving area measurement, spatial configuration analysis, and assessing human exposure to urban green spaces. Similarly, a public urban green spaces map has been developed through the integration of Sentinel-2 derived Normalized Difference Vegetation Index (NDVI) with OSM data by (Ludwig et al., 2021). Accessible green spaces are essential for fostering both physical and mental health as well as environmental sustainability.

Biodiversity Conservation

The crowd-sourced data from OpenStreetMap is essential for locating and mapping biodiverse zones, including wildlife habitats (Wiki, 2019), protected areas (Levin et al., 2017), and endangered species locations. By assisting with steps aimed at maintaining and safeguarding ecosystems, this important spatial information plays a critical role in assisting conservation efforts. OSM has proven to be a valuable tool in the management of protected areas, as demonstrated in the case of Berchtesgaden National Park (Hennig, 2017). Spatial patterns can be utilized to monitor and document wildlife behavior and migration trends, providing valuable insights that contribute to the protection and conservation of endangered species (Y. Hu et al., 2021).

Climate Change Mitigation and Adaptation

Volunteered geographic information can be instrumental in climate change initiatives (A. Munir et al., 2016a). The information provided by OpenStreetMap data is critical for developing efficient mitigation and adaptation plans. It can help planning and implementation efforts by mapping renewable energy sources, carbon sinks, and climate-vulnerable regions. For example, Mexico is working to create synergies between local and national stakeholders for efficient climate action planning as part of its commitment to decreasing greenhouse gas emissions by 2030 (HOTOSM, 2023b). Tulum, Quintana Roo, is receiving technical assistance from the Humanitarian OpenStreetMap Team (HOT) to develop municipal climate action plans and involve local populations in data creation for monitoring real-time environmental effects. The work of HOT will help identify and rank city-level climate change adaptation and mitigation measures.

Sustainable Land Use Planning

OpenStreetMap is a priceless tool to support the field of sustainable land use planning. It can support the adoption of appropriate land management practices by providing information on land cover, land usage, and land ownership. For example, Fonte et al., (2019) developed two European LULC products using OSM. Similarly, Z. Liu, (2021) demonstrated the effectiveness of OSM data in identifying three distinct land-use social functional clusters through the enhanced DBSCAN algorithm. Utilizing OSM along with RS data, urban land use at the street block level can be effectively mapped as demonstrated by a workflow in the study by (Grippa et al., 2018). This crucial data supported decisions about forestry, agriculture, and urban development, which can ultimately reduce environmental impacts.

Disaster Risk Reduction

OSM data can assist in disaster risk reduction by mapping areas prone to natural hazards, such as floods, landslides, and earthquakes. For example, Michael Wagner et al., (2019) endorsed the use of OSM data to enhance disaster risk management in the Seychelles, enabling better understanding and mapping of risks, especially in coastal areas prone to urban and coastal flooding. Similarly, Scholz et al., (2018) highlighted the capacity of OpenStreetMap and digital volunteers in humanitarian endeavors, specifically emphasizing disaster risk reduction within the Missing Maps project. With the help of this useful information, communities, and authorities may improve disaster preparedness strategies and increase community resilience in susceptible areas. OpenStreetMap has emerged as a valuable resource for researchers and modelers to forecast and comprehend the implications of disasters amid a swiftly evolving climate (Jess Beutler, 2021).

Sustainable Transportation

OSM data can encourage environmentally friendly transportation methods by mapping out walking paths, bicycle lanes, and public transportation routes. For example, Gil, (2015) outlined a methodology for creating a multimodal urban network model by harnessing OSM data to support sustainable accessibility. Similarly, The Trufi Association utilized OSM data as the basis for generating General Transit Feed Specification (GTFS) data, which is an open standard used for public transportation schedules

(TED, 2023). Furthermore, the OpenRouteService also offered routing capabilities using current OSM data across various transportation profiles (Neubauer et al., 2009). This extensive data can help create a more efficient and ecologically friendly transport system by lowering carbon emissions and easing congestion in urban areas.

Environmental Awareness and Education

OpenStreetMap's community-driven approach allows individuals and organizations to engage in environmental mapping initiatives. As a result, more people become educated about the environment and motivated to take part in environmental sustainability initiatives. For example, the collaboration between Perkumpulan OpenStreetMap Indonesia (POI) and Wikimedia Indonesia involved a program called Historical Crowdsources Spatial Data for Sustainable Development and Inclusive Mapping (Zainab Ramadhanis, 2023). The program focused on training participants to contribute to OpenStreetMap and Wikidata by utilizing Galleries, Libraries, Archives, and Museums (GLAM) data. The aim was to promote sustainable development and inclusive mapping through the use of historical spatial data. The initiative also encompassed the GLAM flash competition and edit-a-thon training sessions on OpenStreetMap (OSM) and Wikidata.

Waste Management

Identification of garbage dumps, tracking trends in waste disposal, making it easier to set up recycling centers, and establishing garbage collection sites, spatial data is essential to waste management. These actions support sustainable waste reduction and management techniques and improve waste management policy (Singh, 2019). For example, Schilling & Tränckner, (2020) outlined an approach using OSM data and a dataset from the German official topographic–cartographic Information System (ATKIS) to estimate the volume of wastewater at a building level. Similarly, Dowd et al., (2020) demonstrated an optimization algorithm for waste management collection routes using OpenStreetMap and Google OR-Tools to solve mixed fleet routing problems. Felício et al., (2022) introduced an architecture that utilizes OSM for route planning purposes in the context of solid waste collection.

Social Sustainability

OpenStreetMap offers a versatile range of capabilities that foster social sustainability. The potential benefits of OpenStreetMap, focusing on social aspects, are detailed as follows:

Healthcare Services

Spatial data insights are essential for identifying areas with poor access to healthcare facilities in the context of healthcare services. Businesses and governments can use this data to create and enhance healthcare services for underprivileged communities, making sure that high-quality healthcare is available to everyone (Diggle, 2005). OpenStreetMap greatly increases access to requirements, especially in remote or underserved areas, by mapping important services like healthcare facilities along with other crucial resources. For example, OSM can be used for mapping to monitor HIV hotspot areas (Kihumbe, 2019). Mooney et al., (2013) assessed the potential of OSM for pervasive health computing applications.

Malik & Shaikh, (2019) investigated the potential of OSM data to study spatial distribution of healthcare facilities in Pakistan.

Educational Services

Spatial data analysis enables the identification of regions that lack access to high-quality educational opportunities. With this information, governments or businesses can establish and improve educational services for underrepresented communities. These initiatives support social sustainability by closing educational gaps and empowering communities with education (Muzaffar et al., 2017). For example, (Sezer et al., 2018) analyzed the accessibility of schools by combining OpenStreetMap data with population data. Shiau et al., (2018) explored the factors that influence the diffusion of OSM in Science, Technology, Engineering, and Mathematics (STEM) education to understand the key drivers and barriers to the adoption and integration of OSM in STEM educational settings.

Cultural Preservation

Rich cultural heritage areas can be identified to protect and promote their cultural value with the help of OpenStreetMap (Wiki, 2023). Plans for heritage conservation and tourism can be improved by examining cultural sites (A. Munir et al., 2016b), historical monuments, and travel trends. This can help to preserve cultural identity and foster social sustainability (Zabulis et al., 2020). This also fosters a sense of identity and belonging within communities, promoting social sustainability. The study by Balducci, (2021) investigated the quality of OSM for cultural assets of Italy and found that the cultural information available on OSM is not only abundant in quantity but also demonstrated a high level of positional and semantic accuracy.

Social Inclusion

Governments can discover areas with significant concentrations of poverty using spatial data analysis, creating possibilities to promote social inclusion. By utilizing spatial data, it is possible to meet the needs of poor populations by generating employment possibilities and manufacturing products or services that are specifically catered to their needs. This strategy promotes variety and lessens inequality, which eventually promotes social sustainability (Akinyemi, 2007). OpenStreetMap can play a crucial role in identifying and mapping safe and inclusive public spaces, fostering social interactions, recreation, and a sense of community ownership in both urban and rural environments. The acknowledgment of a person's property rights is greatly helped by geospatial information. With its flexible database design and extensive historical data, OSM makes it possible to precisely trace the origins of data. Since they effectively constitute a social contract between people regarding their land, these qualities are especially advantageous in the context of property rights (Sara Amadi & Jess Beutler, 2019). This guarantees that land ownership is recognized and safeguarded by the law, enabling equal access to resources and minimizing disputes or conflicts over property. As a result, social sustainability is enhanced.

Community Development

By empowering individuals and local groups to actively participate in mapping their neighbourhoods and surroundings, OpenStreetMap's community-driven approach fosters a sense of ownership, pride, and empowerment. Students, teachers, and small farmer cooperatives from the area took an active role in mapping their community and recording the agricultural products in Nepal (KLL, 2015). By working together, they not only improved their skills and expertise in using common technology tools but also put themselves on the map both literally and figuratively. Governments may create better policies and practices for developing socially sustainable infrastructure and empowering communities by looking at housing circumstances, infrastructure availability, and community resources.

Disaster Relief

Spatial data analysis can greatly improve the social sustainability of communities by being incorporated into emergency response plans and programmes for recovering from natural disasters. Spatial analysis provides insightful information that helps to lessen the effects of disasters and eventually save lives by examining past disaster patterns, population density, and vital infrastructure (Jitt-Aer et al., 2022). Real-time mapping features offered by OpenStreetMap are extremely helpful in disaster and emergency situations. This flexible platform makes it easier to coordinate and respond quickly, boosting local resilience and minimizing negative effects. For example, Poiani et al., (2016) presented a novel case study examining the coordination and structure of collaborative mapping activities that emerged in response to the Nepal Earthquake in 2015. The study revealed that a vast network of remote contributors, spanning across the globe, actively participated in coordinated endeavors to support disaster relief activities.

Gender Equality

Spatial data analysis is useful in pinpointing regions with various levels of gender inequality. Governments can offer employment opportunities, skill development programmes, and support services, actively promoting and advancing gender equality in these regions (Jo et al., 2020). The analysis of gender differences and social inequities benefits greatly from the use of OpenStreetMap data. For example, Gardner et al., (2020) found a notable gender gap in the amounts of overall activity, editing, and tagging operations in OSM being predominately dominated by men, offering important new perspectives on gender equality in the context of online communities for digital mapping. Understanding these disparities can be a good place to start to learn more about gender equality and representation in the geospatial data and technology fields.

CONCLUSION

OSM is essential in advancing sustainable development in the fields of economy, the environment, and society. By enabling infrastructure planning, fostering tourist and local economies, enabling effective disaster response, improving energy efficiency, and assisting real estate development and market analysis, OSM's vast capabilities contribute to economic sustainability. OSM offers useful information that enables decision-makers to make well-informed decisions, promoting economic resilience and prosperity.

By promoting sustainable resource management, agricultural efficiency, urban green spaces, biodiversity protection, climate change mitigation and adaptation, sustainable land use planning, and disaster risk reduction, OSM also makes a substantial contribution to environmental sustainability. Governments and organizations can implement efficient conservation policies and practices to safeguard natural resources and ecosystems by utilizing spatial data, promoting environmental resilience and long-term prosperity in the process.

OSM's capabilities in the areas of healthcare, education, cultural preservation, social inclusion, community development, disaster assistance, and gender equality promotion also contribute to social sustainability. OSM enables individuals and local groups through community-driven mapping activities, generating a sense of ownership and pride in their communities. Additionally, OSM's spatial data analysis helps in identifying areas deficient in basic services, allowing for more focused efforts to close social inequalities and advance equality for all.

Even though OSM has shown that it has a huge potential to support sustainable development, there is still room for improvement. Here are some suggestions for future work:

- Continuous efforts should be made to improve data accuracy and completeness in OSM. Utilizing advanced techniques like machine learning and artificial intelligence can assist in automating data validation and improving data quality.
- To further promote social sustainability, initiatives should focus on inclusive mapping that involves marginalized communities and underrepresented groups. Capacity-building programs should be developed to enable their active participation in mapping activities.
- To enhance OSM's disaster response capabilities, efforts can be made to integrate real-time data from various sources like social media, remote sensing, and Internet of Things (IoT) devices. This will enable more timely and accurate updates during emergencies.

REFERENCES

Ahmad, M., Khayal, M. S. H., & Tahir, A. (2022). Analysis of Factors Affecting Adoption of Volunteered Geographic Information in the Context of National Spatial Data Infrastructure. *ISPRS International Journal of Geo-Information, 11*(2), 120. Advance online publication. doi:10.3390/ijgi11020120

Ahmed, I., Zhang, Y., Jeon, G., Lin, W., Khosravi, M. R., & Qi, L. (2022). A blockchain- and artificial intelligence-enabled smart IoT framework for sustainable city. *International Journal of Intelligent Systems, 37*(9), 6493–6507. Advance online publication. doi:10.1002/int.22852

Akinyemi, F. O. (2007). Spatial data needs for poverty management. *Research and Theory in Advancing Spatial Data Infrastructure Concepts, 5*.

Al-Rasheedi, G., & Khan, N. (2021). The Role of Cloud Computing in Education: Saudi Arabian Study. *International Journal of Interactive Mobile Technologies, 15*(24), 191–200. Advance online publication. doi:10.3991/ijim.v15i24.27363

Al-Rwaidan, R. M., Aldossary, N., Eldahamsheh, M. M., Al-Azzam, M. K. A., Al-Quran, A. Z., & Al-Hawary, S. I. S. (2023). The impact of cloud-based solutions on digital transformation of HR practices. *International Journal of Data and Network Science*, *7*(1), 83–90. Advance online publication. doi:10.5267/j.ijdns.2022.12.003

Alam, M. A., & Saiyeda, A. (2020). A Cloud Based Solution for Smart Education. *International Journal of Smart Education and Urban Society*, *11*(2), 28–37. Advance online publication. doi:10.4018/IJSEUS.2020040103

Aldhshan, S. R. S., Abdul Maulud, K. N., Wan Mohd Jaafar, W. S., Karim, O. A., & Pradhan, B. (2021). Energy consumption and spatial assessment of renewable energy penetration and building energy efficiency in Malaysia: A review. *Sustainability (Basel)*, *13*(16), 9244. doi:10.3390u13169244

Alhamwi, A., Medjroubi, W., Vogt, T., & Agert, C. (2017). OpenStreetMap data in modelling the urban energy infrastructure: A first assessment and analysis. *Energy Procedia*, *142*, 1968–1976. Advance online publication. doi:10.1016/j.egypro.2017.12.397

Alreshidi, E., Mourshed, M., & Rezgui, Y. (2018). Requirements for cloud-based BIM governance solutions to facilitate team collaboration in construction projects. *Requirements Engineering*, *23*(1), 1–31. Advance online publication. doi:10.100700766-016-0254-6

Alsolami, E. (2018). Security threats and legal issues related to Cloud based solutions. *International Journal of Computer Science and Network Security, 18*(5).

Amadi, S., & Beutler, J. (2019). *Open Mapping for SDGs*. https://sdgs.hotosm.org/

Anil, B. C., Dayananda, P., Nethravathi, B., & Raisinghani, M. S. (2022). Efficient local cloud-based solution for liver cancer detection using deep learning. *International Journal of Cloud Applications and Computing*, *12*(1), 1–13. Advance online publication. doi:10.4018/IJCAC.2022010109

Araújo, S. O., Peres, R. S., Barata, J., Lidon, F., & Ramalho, J. C. (2021). Characterising the agriculture 4.0 landscape—emerging trends, challenges and opportunities. In Agronomy (Vol. 11, Issue 4). doi:10.3390/agronomy11040667

AWS. (2017). *OpenStreetMap Public Data Set Now Available on AWS*. https://aws.amazon.com/about-aws/whats-new/2017/06/openstreetmap-public-data-set-now-available-on-aws/

Balducci, F. (2021). Is OpenStreetMap a good source of information for cultural statistics? The case of Italian museums. *Environment and Planning. B, Urban Analytics and City Science*, *48*(3), 503–520. Advance online publication. doi:10.1177/2399808319876949

Bertolotto, M., McArdle, G., & Schoen-Phelan, B. (2020). Volunteered and crowdsourced geographic information: The openstreetmap project. *Journal of Spatial Information Science*, *20*(20). Advance online publication. doi:10.5311/JOSIS.2020.20.659

Beutler, J. (2021). *OpenStreetMap for Emergency Preparedness in the US*. https://www.openstreetmap.us/news/2021/06/map-for-emergency

Brandt, D., Alnyme, O., & Heldt, T. (2022). Building a spatial decision support system for tourism and infrastructure planning: Technical solution and data integration challenges. *European Journal of Geography, 13*(1), 94–108. Advance online publication. doi:10.48088/ejg.d.bra.13.1.094.108

Castro, R., Tierra, A., & Luna, M. (2019). Assessing the horizontal positional accuracy in openstreetmap: A big data approach. *Advances in Intelligent Systems and Computing, 931*, 513–523. Advance online publication. doi:10.1007/978-3-030-16184-2_49

Chaula, J. (2019). Opportunities and Challenges for Integrating Statistical and Spatial Data in East African Countries. *American Scientific Research Journal for Engineering, 59*(1).

Chen, B., Tu, Y., Song, Y., Theobald, D. M., Zhang, T., Ren, Z., Li, X., Yang, J., Wang, J., Wang, X., Gong, P., Bai, Y., & Xu, B. (2021). Mapping essential urban land use categories with open big data: Results for five metropolitan areas in the United States of America. *ISPRS Journal of Photogrammetry and Remote Sensing, 178*, 203–218. Advance online publication. doi:10.1016/j.isprsjprs.2021.06.010

Corbett, D. C., & Rostow, W. W. (1960). The Stages of Economic Growth: A Non-Communist Manifesto. *International Journal (Toronto, Ont.), 16*(1), 83. Advance online publication. doi:10.2307/40198523

Cramwinckel, J. (2019). *The role of global open geospatial data in measuring SDG indicator 11.7. 1: Public open spaces*. Wageningen University.

De La Torre-Díez, I., Lopez-Coronado, M., Garcia-Zapirain Soto, B., & Mendez-Zorrilla, A. (2015). Secure cloud-based solutions for different eHealth services in spanish rural health centers. *Journal of Medical Internet Research, 17*(7), e157. Advance online publication. doi:10.2196/jmir.4422 PMID:26215155

Diggle, P. (2005). Applied Spatial Statistics for Public Health Data. *Journal of the American Statistical Association, 100*(470), 702–703. Advance online publication. doi:10.1198/jasa.2005.s15

Dowd, M., Dixon, A., & Kinsella, B. (2020). Optimizing waste management collection routes in Urban Haiti: A collaboration between DataKind and SOIL. *ArXiv, 2012*.

Dupre, D. (2020). Urban and socio-economic correlates of property prices in dublin's area. *Proceedings - 2020 IEEE 7th International Conference on Data Science and Advanced Analytics, DSAA 2020*. 10.1109/DSAA49011.2020.00070

Farooq, M. S., Riaz, S., Abid, A., Abid, K., & Naeem, M. A. (2019). A Survey on the Role of IoT in Agriculture for the Implementation of Smart Farming. In IEEE Access (Vol. 7). doi:10.1109/ACCESS.2019.2949703

Felício, S., Hora, J., Ferreira, M. C., Abrantes, D., Costa, P. D., Dangelo, C., Silva, J., & Galvão, T. (2022). Handling OpenStreetMap georeferenced data for route planning. *Transportation Research Procedia, 62*, 189–196. Advance online publication. doi:10.1016/j.trpro.2022.02.024

Fernández de Sevilla, M., Gutiérrez, Ó., Gómez, J., Tayebi, A., Álvarez, Á., & Sáez de Adana, F. (2021). On the application of radio planning tools in open environments for the improvement of autoguidance systems used in precision agriculture. *Computers and Electronics in Agriculture, 187*, 106258. Advance online publication. doi:10.1016/j.compag.2021.106258

Fonte, C. C., Patriarca, J. A., Minghini, M., Antoniou, V., See, L., & Brovelli, M. A. (2019). Using OpenStreetMap to Create Land Use and Land Cover Maps: Development of an Application. In *Geospatial Intelligence* (Vol. 2). Concepts, Methodologies, Tools, and Applications. doi:10.4018/978-1-5225-8054-6.ch047

Gardner, Z., Mooney, P., De Sabbata, S., & Dowthwaite, L. (2020). Quantifying gendered participation in OpenStreetMap: Responding to theories of female (under) representation in crowdsourced mapping. *GeoJournal*, *85*(6), 1603–1620. Advance online publication. doi:10.100710708-019-10035-z

Gil, J. (2015). Building a multimodal urban network model using OpenStreetMap data for the analysis of sustainable accessibility. *OpenStreetMap in GIScience: Experiences, Research, and Applications*, 229–251.

Grinberger, A. Y., Minghini, M., Juhász, L., Yeboah, G., & Mooney, P. (2022). OSM Science—The Academic Study of the OpenStreetMap Project, Data, Contributors, Community, and Applications. In ISPRS International Journal of Geo-Information (Vol. 11, Issue 4). doi:10.3390/ijgi11040230

Grippa, T., Georganos, S., Zarougui, S., Bognounou, P., Diboulo, E., Forget, Y., Lennert, M., Vanhuysse, S., Mboga, N., & Wolff, E. (2018). Mapping urban land use at street block level using OpenStreetMap, remote sensing data, and spatial metrics. *ISPRS International Journal of Geo-Information*, *7*(7), 246. Advance online publication. doi:10.3390/ijgi7070246

Hamdaqa, M., & Tahvildari, L. (2012). Cloud Computing Uncovered: A Research Landscape. In Advances in Computers (Vol. 86). doi:10.1016/B978-0-12-396535-6.00002-8

Heider, K., Lopez, J. M. R., & Scheffran, J. (2018). The potential of volunteered geographic information to investigate peri-urbanization in the conservation zone of Mexico City. *Environmental Monitoring and Assessment*, *190*(4), 219. Advance online publication. doi:10.100710661-018-6597-3 PMID:29541867

Hennig, S. (2017). OpenStreetMap used in protected area management. The example of the recreational infrastructure in Berchtesgaden National Park. *Eco.Mont*, *9*(2), 30–41. Advance online publication. doi:10.1553/eco.mont-9-2s30

HOTOSM. (2021). *Somaliland Wajaale Farmyards*. https://tasks.hotosm.org/projects/10301

HOTOSM. (2023a). *Hot's Response to the Turkey & Syria Earthquake*. https://www.hotosm.org/projects/join-the-turkey-and-syria-earthquake-response/

HOTOSM. (2023b). *Open Cities: Mexico*. https://www.hotosm.org/open-cities-mexico.html

Hu, X., Gong, J., Renard, E. G., & Parashar, M. (2019). Two-Stage Framework for Big Spatial Data Analytics to Support Disaster Response. *Proceedings - 2019 IEEE International Conference on Big Data, Big Data 2019*. 10.1109/BigData47090.2019.9005613

Hu, Y., Fan, H., Chen, Y., Chang, J., Zhan, X., Wu, H., Zhang, B., Wang, M., Zhang, W., Yang, L., Hou, X., Shen, X., Pan, T., Wu, W., Li, J., Hu, H., & Wei, F. (2021). Spatial patterns and conservation of genetic and phylogenetic diversity of wildlife in China. *Science Advances*, *7*(4), eabd5725. Advance online publication. doi:10.1126ciadv.abd5725 PMID:33523945

Huawei Technologies Co. Ltd. (2023). Introduction to Cloud Computing Computing. In Cloud Computing Technology. doi:10.1007/978-981-19-3026-3_1

Jitt-Aer, K., Wall, G., Jones, D., & Teeuw, R. (2022). Use of GIS and dasymetric mapping for estimating tsunami-affected population to facilitate humanitarian relief logistics: A case study from Phuket, Thailand. *Natural Hazards*, *113*(1), 185–211. Advance online publication. doi:10.100711069-022-05295-x

Jo, A., Lee, S. K., & Kim, J. (2020). Gender gaps in the use of urban space in seoul: Analyzing spatial patterns of temporary populations using mobile phone data. *Sustainability (Basel)*, *12*(16), 6481. Advance online publication. doi:10.3390u12166481

Johnson, L., Callaghan, C., Balasubramanian, M., Haq, H., & Spallek, H. (2019). Cost Comparison of an On-Premise IT Solution with a Cloud-Based Solution for Electronic Health Records in a Dental School Clinic. *Journal of Dental Education*, *83*(8), 895–903. Advance online publication. doi:10.21815/JDE.019.089 PMID:31010892

Ju, Y., Dronova, I., & Delclòs-Alió, X. (2022). A 10 m resolution urban green space map for major Latin American cities from Sentinel-2 remote sensing images and OpenStreetMap. *Scientific Data*, *9*(1), 586. doi:10.103841597-022-01701-y PMID:36153342

Jung, S., Yuan, Y., Del Carpio, M. S., Pawlik, T., Hartmann, S., Estrella, N., Oteros, J., Traidl-Hoffmann, C., Damialis, A., Buters, J., & Menzel, A. (2022). Impact of Local Grasslands on Wild Grass Pollen Emission in Bavaria, Germany. *Land (Basel)*, *11*(2), 306. Advance online publication. doi:10.3390/land11020306

Kaloxylos, A., Groumas, A., Sarris, V., Katsikas, L., Magdalinos, P., Antoniou, E., Politopoulou, Z., Wolfert, S., Brewster, C., Eigenmann, R., & Maestre Terol, C. (2014). A cloud-based farm management system: Architecture and implementation. *Computers and Electronics in Agriculture*, *100*, 168–179. Advance online publication. doi:10.1016/j.compag.2013.11.014

Kalyani, Y., & Collier, R. (2021). A systematic survey on the role of cloud, fog, and edge computing combination in smart agriculture. In Sensors (Vol. 21, Issue 17). doi:10.339021175922

Kamanga, C. T., Bugingo, E., Badibanga, S. N., & Mukendi, E. M. (2023). A multi-criteria decision making heuristic for workflow scheduling in cloud computing environment. *The Journal of Supercomputing*, *79*(1), 243–264. Advance online publication. doi:10.100711227-022-04677-z

Kihumbe. (2019). *Kihumbe- Survey Digitization and Mapping for HIV Monitoring*. Academic Press.

KLL. (2015). *Agriculture Resource Mapping*. http://kathmandulivinglabs.org/our-work/agriculture-resource-mapping

Latif, S., Islam, K. M. R., Khan, M. M. I., & Ahmed, S. I. (2011). OpenStreetMap for the disaster management in Bangladesh. *2011 IEEE Conference on Open Systems, ICOS 2011*, 435–439. 10.1109/ICOS.2011.6079240

Levin, N., Lechner, A. M., & Brown, G. (2017). An evaluation of crowdsourced information for assessing the visitation and perceived importance of protected areas. *Applied Geography (Sevenoaks, England)*, *79*, 115–126. Advance online publication. doi:10.1016/j.apgeog.2016.12.009

Li, H., Zech, J., Ludwig, C., Fendrich, S., Shapiro, A., Schultz, M., & Zipf, A. (2021). Automatic mapping of national surface water with OpenStreetMap and Sentinel-2 MSI data using deep learning. *International Journal of Applied Earth Observation and Geoinformation*, *104*, 102571. Advance online publication. doi:10.1016/j.jag.2021.102571

Liu, S., & Gao, B. (2022). Econometric Analysis of Leisure Agriculture and Rural Tourism Scenery Based on Spatial Data Analysis. *Computational Intelligence and Neuroscience*, *2022*, 1–8. Advance online publication. doi:10.1155/2022/2286803 PMID:35898780

Liu, Z. (2021). Identifying urban land use social functional units: A case study using OSM data. *International Journal of Digital Earth*, *14*(12), 1798–1817. Advance online publication. doi:10.1080/17538947.2021.1988161

Ludwig, C., Hecht, R., Lautenbach, S., Schorcht, M., & Zipf, A. (2021). Mapping public urban green spaces based on OpenStreetMap and Sentinel-2 imagery using belief functions. *ISPRS International Journal of Geo-Information*, *10*(4), 251. doi:10.3390/ijgi10040251

Ma, D., Sandberg, M., & Jiang, B. (2015). Characterizing the heterogeneity of the openstreetmap data and community. *ISPRS International Journal of Geo-Information*, *4*(2), 535–550. Advance online publication. doi:10.3390/ijgi4020535

Malik, N. A., & Shaikh, M. A. (2019). Spatial distribution and accessibility to public sector tertiary care teaching hospitals: Case study from Pakistan. *Eastern Mediterranean Health Journal*, *25*(6), 431–434. Advance online publication. doi:10.26719/emhj.18.049 PMID:31469163

Mikulecky, P. (2015). Cloud – Based solutions for outdoor ambient intelligence support. *Studies in Computational Intelligence*, *598*, 229–238. Advance online publication. doi:10.1007/978-3-319-16211-9_24

Minghini, M., Delucchi, L., Sarretta, A., Lupia, F., Napolitano, M., & Palmas, A. (2017). Collaborative mapping response to disasters through openstreetmap: The case of the 2016 Italian earthquake. In Geoingegneria Ambientale e Mineraria (Vol. 151, Issue 2). Academic Press.

Mollahoseini Ardakani, M. R., Hashemi, S. M., & Razzazi, M. (2019). A Cloud-based solution/reference architecture for establishing collaborative networked organizations. *Journal of Intelligent Manufacturing*, *30*(5), 2273–2289. Advance online publication. doi:10.100710845-017-1387-2

Mooney, P., Corcoran, P., & Ciepluch, B. (2013). The potential for using volunteered geographic information in pervasive health computing applications. *Journal of Ambient Intelligence and Humanized Computing*, *4*(6), 731–745. doi:10.100712652-012-0149-4

Moore, S. (2020). *OpenStreetMap Tourist Attractions for Europe*. https://hub.arcgis.com/datasets/f3dcc26dd868456da288baca83b8a2cd_0/about

Munawar, H. S., Mojtahedi, M., Hammad, A. W. A., Kouzani, A., & Mahmud, M. A. P. (2022). Disruptive technologies as a solution for disaster risk management: A review. In Science of the Total Environment (Vol. 806). doi:10.1016/j.scitotenv.2021.151351

Munir, A., Asmat, A., & Malik Sikander, H. K. (2016a). Potential of Volunteered Geographic Information for Adaptation of Climate Change Effects in Pakistan. *4th International Conference on Energy, Environment and Sustainable Development.*

Munir, A., Asmat, A., & Malik Sikander, H. K. (2016b). Promoting Heritage Tourism in Pakistan using Volunteered Geographic Information: An Investigative Outlook. *24th International Conference of Historians of Asia.*

Munir, K., Ghafoor, M., Khafagy, M., & Ihshaish, H. (2022). AgroSupportAnalytics: A Cloud-based Complaints Management and Decision Support System for Sustainable Farming in Egypt. *Egyptian Informatics Journal, 23*(1), 73–82. Advance online publication. doi:10.1016/j.eij.2021.06.002

Muzaffar, H. M., Tahir, A., Ali, A., Ahmad, M., & McArdle, G. (2017). Quality assessment of volunteered geographic information for educational planning. In *Volunteered Geographic Information and the Future of Geospatial Data*. IGI Global. doi:10.4018/978-1-5225-2446-5.ch005

Neubauer, N., Over, M., Schilling, A., & Zipf, A. (2009). Virtual Cities 2.0: Generating web-based 3D city models and landscapes based on free and user generated data (OpenStreetMap). GeoViz2009.

Pechmann, J., & de los Reyes, C. (2023, March 17). *Using OSM Data in the Turkey and Syria Earthquake Response*. https://www.hotosm.org/updates/using-osm-data-for-the-turkey-and-syria-earthquake-response/

Penchev, G. (2016). Using Space Syntax for Estimation of Potential Disaster Indirect Economic Losses. *Comparative Economic Research, 19*(5), 125–142. Advance online publication. doi:10.1515/cer-2016-0041

Poiani, T. H., Rocha, R. dos S., Degrossi, L. C., & de Albuquerque, J. P. (2016). Potential of Collaborative Mapping for Disaster Relief : A Case Study of OpenStreetMap in the Nepal Earthquake 2015. *Hawaii International Conference on System Sciences*, 1–10. 10.1109/HICSS.2016.31

Pruthviraja, D., Anil, B. C., & Sowmyarani, C. N (2021). Efficient local cloud-based solution for diabetic retinopathy detection. *International Journal of Web-Based Learning and Teaching Technologies, 16*(3), 39–46. Advance online publication. doi:10.4018/IJWLTT.20210501.oa3

Pudasaini, N. (2016). Mapping a sustainable future. *Geospatial World, 7*(1).

Rabiei-Dastjerdi, H., McArdle, G., Matthews, S. A., & Keenan, P. (2021). Gap analysis in decision support systems for real-estate in the era of the digital earth. *International Journal of Digital Earth, 14*(1), 121–138. Advance online publication. doi:10.1080/17538947.2020.1808719

Ramadhanis, Z. (2023). *Promoting Indonesia's historical places through digital mapping with GLAM*. https://openstreetmap.or.id/en/2023/05/22/promoting-indonesias-historical-places-through-digital-mapping-with-glam/

Ranjbari, M., Morales-Alonso, G., Esfandabadi, Z. S., & Carrasco-Gallego, R. (2019). Sustainability and the sharing economy: Modelling the interconnections. *Direccion y Organizacion, 68*, 33–40. Advance online publication. doi:10.37610/dyo.v0i68.549

RealHomes. (2023). https://openstreetmap.realhomes.io/

Saha, S. S., Schweitzer, E., Scaglione, A., & Johnson, N. G. (2019). A Framework for Generating Synthetic Distribution Feeders using OpenStreetMap. *51st North American Power Symposium, NAPS 2019*. 10.1109/NAPS46351.2019.9000187

Schiefelbein, J., Rudnick, J., Scholl, A., Remmen, P., Fuchs, M., & Müller, D. (2019). Automated urban energy system modeling and thermal building simulation based on OpenStreetMap data sets. *Building and Environment*, *149*, 630–639. Advance online publication. doi:10.1016/j.buildenv.2018.12.025

Schilling, J., & Tränckner, J. (2020). Estimation of wastewater discharges by means of OpenStreetMap data. *Water (Basel)*, *12*(3), 628. doi:10.3390/w12030628

Scholz, S., Knight, P., Eckle, M., Marx, S., & Zipf, A. (2018). Volunteered geographic information for disaster risk reduction—The missing maps approach and its potential within the red cross and red crescent movement. *Remote Sensing (Basel)*, *10*(8), 1239. doi:10.3390/rs10081239

Schweizer, J., Poliziani, C., Rupi, F., Morgano, D., & Magi, M. (2021). Building a large-scale microsimulation transport scenario using big data. *ISPRS International Journal of Geo-Information*, *10*(3), 165. Advance online publication. doi:10.3390/ijgi10030165

Scott, G., & Rajabifard, A. (2017). Sustainable development and geospatial information: A strategic framework for integrating a global policy agenda into national geospatial capabilities. *Geo-Spatial Information Science*, *20*(2), 59–76. doi:10.1080/10095020.2017.1325594

Sezer, A., Deniz, M., & Topuz, M. (2018). Analysis of Accessibility of Schools in Usak City via Geographical Information Systems (GIS). *Tarih Kultur Ve Sanat Arastirmalari Dergisi-Journal of History Culture and Art Research, 7*(5).

Shiau, S. J. H., Huang, C. Y., Yang, C. L., & Juang, J. N. (2018). *A derivation of factors influencing the innovation diffusion of the OpenStreetMap in STEM education*. Sustainability. doi:10.3390u10103447

Singh, A. (2019). Remote sensing and GIS applications for municipal waste management. *Journal of Environmental Management*, *243*, 22–29. Advance online publication. doi:10.1016/j.jenvman.2019.05.017 PMID:31077867

Singh, A., Mehrotra, R., Rajput, V. D., Dmitriev, P., Singh, A. K., Kumar, P., Tomar, R. S., Singh, O., & Singh, A. K. (2022). Geoinformatics, artificial intelligence, sensor technology, big data: emerging modern tools for sustainable agriculture. *Sustainable Agriculture Systems and Technologies*, 295–313.

Sun, Z., Wang, D., & Zhong, G. (2018). Extraction of farmland geographic information using OpenStreetMap data. *2018 7th International Conference on Agro-Geoinformatics, Agro-Geoinformatics 2018*. 10.1109/Agro-Geoinformatics.2018.8476088

Tank, B., & Gandhi, V. (2023). A Comparative Study on Cloud Computing, Edge Computing and Fog Computing. *Advances in Transdisciplinary Engineering*, *32*. Advance online publication. doi:10.3233/ATDE221329

TED. (2023). *Mapping the Future: Colombian OpenStreetMappers Impress ITDP Sustainable Transport Expert*. https://www.trufi-association.org/mapping-the-future-colombian-openstreetmappers-impress-itdp-sustainable-transport-expert/

Tzavella, K., Skopeliti, A., & Fekete, A. (2022). Volunteered geographic information use in crisis, emergency and disaster management: A scoping review and a web atlas. *Geo-Spatial Information Science*, 1–32. Advance online publication. doi:10.1080/10095020.2022.2139642

UN-GGIM. (2012). *Monitoring sustainable development: Contribution of geospatial information to the rio+ 20 processes.* United Nations New York.

UNDP. (2023). *Sustainable Development Goals.* https://www.undp.org/sustainable-development-goals

Wagner, M., Doherty, G., & van Zanten, B. (2019). *Leveraging OpenStreetMap to improve disaster risk management in the Seychelles.* https://www.preventionweb.net/news/leveraging-openstreetmap-improve-disaster-risk-management-seychelles

Wan, T., Lu, H., Lu, Q., & Luo, N. (2017). Classification of High-Resolution Remote-Sensing Image Using OpenStreetMap Information. *IEEE Geoscience and Remote Sensing Letters, 14*(12), 2305–2309. Advance online publication. doi:10.1109/LGRS.2017.2762466

Wang, C., Li, Y., & Shi, X. (2019). Information mining for urban building energy models (UBEMs) from two data sources: OpenStreetMap and baidu map. *Building Simulation Conference Proceedings,* 5. 10.26868/25222708.2019.210545

Wichmann, J. R. K., Scholdra, T. P., & Reinartz, W. J. (2023). Propelling International Marketing Research with Geospatial Data. *Journal of International Marketing, 31*(2), 82–102. doi:10.1177/1069031X221149951

Wieland, T. (2017). Market area analysis for retail and service locations with MCI. *The R Journal, 9*(1), 298. Advance online publication. doi:10.32614/RJ-2017-020

Wiki. (2019). *Wildlife sanctuaries of India.* https://wiki.openstreetmap.org/wiki/Wildlife_sanctuaries_of_India

Wiki. (2023). *Niskala Institute: Preservation of Ciamis Regency Cultural Heritage Through OSM Data.* https://wiki.openstreetmap.org/wiki/Niskala_Institute:_Preservation_of_Ciamis_Regency_Cultural_Heritage_Through_OSM_Data

World Soil Information. (2020). *SoilGrids.* SoilGrids250m.

Yang, C., Huang, Q., Li, Z., Liu, K., & Hu, F. (2017). Big Data and cloud computing: innovation opportunities and challenges. In International Journal of Digital Earth (Vol. 10, Issue 1). doi:10.1080/17538947.2016.1239771

Yang, D. (2019). Mapping Regional Landscape by Using OpenStreetMap (OSM): A Case Study to Understand Forest Patterns in Maya Zone, Mexico. In Environmental Information Systems: Concepts, Methodologies, Tools, and Applications (pp. 771–790). IGI Global.

Yang, D., Fu, C. S., Smith, A. C., & Yu, Q. (2017). Open land-use map: A regional land-use mapping strategy for incorporating OpenStreetMap with earth observations. *Geo-Spatial Information Science, 20*(3), 269–281. Advance online publication. doi:10.1080/10095020.2017.1371385

Yusianto, R., Marimin, S., & Hardjomidjojo, H. (2020). Intelligent spatial decision support system concept in the potato agro-industry supply chain. *2020 International Conference on Computer Science and Its Application in Agriculture, ICOSICA 2020*. 10.1109/ICOSICA49951.2020.9243233

Zabulis, X., Meghini, C., Partarakis, N., Beisswenger, C., Dubois, A., Fasoula, M., Nitti, V., Ntoa, S., Adami, I., Chatziantoniou, A., Bartalesi, V., Metilli, D., Stivaktakis, N., Patsiouras, N., Doulgeraki, P., Karuzaki, E., Stefanidi, E., Qammaz, A., Kaplanidi, D., ... Galanakis, G. (2020). Representation and preservation of heritage crafts. *Sustainability (Basel)*, *12*(4), 1461. Advance online publication. doi:10.3390u12041461

Chapter 12
Sustainability in Upstream Oil and Gas:
Integrating Energy and Ecology

Kaustubh Laturkar
https://orcid.org/0000-0002-2643-3942
Michigan State University, USA

Kasturi Laturkar
Validation Associates LLC, USA

ABSTRACT

Oil and gas production plays an important role in meeting global energy needs, but it also has environmental and social implications. In addition to creating ecological degradation and carbon emissions, oil and gas operations also damage the environment. Upstream sustainability requires improvements in several key areas, such as switching to cleaner energy sources including natural gas and renewables, reducing methane emissions, and adopting innovative carbon capture and storage (CCS) technologies. A brief overview of the upstream oil and gas industry is presented in this chapter, along with an assessment of the need to implement sustainable initiatives. An overview of sustainability improvements and challenges associated with implementation follows. The chapter concluded with an exploration of the equitable future of the industry achieved through the execution of these strategies and the promotion of a sustainable culture.

INTRODUCTION

Humanity has been able to explore the planet's natural resources, including fossil fuels like crude oil, largely due to technological advances. Petroleum products such as gasoline, diesel, kerosene, natural gas etc. have transformed everyday life, empowering a wide range of activities. Oil and gas industry has a profound impact on the global economy, the environment, and the lives of millions of people. Due to its high profitability potential, the sector continues to attract investors despite volatility amid energy

DOI: 10.4018/978-1-6684-9979-5.ch012

transitions. The oil and gas market is among the largest in the world, attracting billions of dollars in investment (Kulanov et al., 2020).

The hydrocarbon industry contributes significantly to the global energy supply, which involves the exploration and extraction of hydrocarbon reserves under the Earth's surface. Advanced technologies, complex logistic systems, and multidisciplinary collaboration are all involved in this intricate process (Didenko et al., 2023). In recent years, the global energy landscape has undergone a profound transformation, marked by the growing recognition of sustainability as a fundamental requirement for all sectors of the economy. Among these, the upstream oil and gas industry stands at a critical juncture, facing complex challenges that demand new approaches and responsible practices. Several environmental, health, security, and social issues are associated with upstream oil and gas operations (John O. Robertson, 2017). It faces unprecedented scrutiny as it attempts to reduce its environmental footprint while meeting the world's energy demands. A dynamic array of sustainability measures are necessary to balance economic growth and environmental responsibility. In addition to reducing carbon emissions and minimizing ecosystem impacts, these initiatives also promote local community development and foster circular economies (Foster, 2020). Responses by the industry to these challenges illustrate both its adaptability and the global shift toward sustainable resource management.

In the case of upstream oil and gas exploration and production operations, there can be profound ecological implications, affecting biodiversity and ecosystem health (Albeldawi, 2023). It is possible for marine habitats and terrestrial biomes to be disturbed during the exploration and extraction process (Thakur & Koul, 2022). Drilling platforms and access roads can fragment habitats, disrupt migration patterns, and change the behavior of indigenous species as a result of their construction (Soeder, 2021b). It is important to note that spills and leaks are relatively uncommon, but they are still capable of having devastating effects on aquatic life, including plankton and large marine mammals (Ogunbiyi et al., 2023). A delicate balance must be maintained between the industry's operations and the preservation of Earth's rich biodiversity in order to maintain the delicate balance of ecosystems as well as the industry's operations. Listed below are some of the issues associated with upstream oil and gas operations

1. Issues related to human health and safety

During upstream oil and gas operations, hazardous substances that are released into the air and water can pose health risks (Adebiyi, 2022). The presence of particles, volatile organic compounds, as well as other air pollutants can lead to respiratory and other health problems. It is also important to keep in mind that pipelines and other transportation methods may result in accidental leakages, resulting in immediate harm to residents and biodiversity in the surrounding areas (Weitekamp et al., 2020). It is important for the industry to implement rigorous safety protocols and invest in advanced technologies in order to prevent accidents and minimize the potential health risks.

2. The impact of climate change on the environment

There is no doubt that the upstream oil and gas industry contributes to climate change, but it also releases greenhouse gases, such as carbon dioxide and methane (Wadanambi et al., 2020). It is common for unintentional releases of methane, a potent greenhouse gas, to occur as a result of drilling, production, and transportation. It should be noted that these emissions are aggravating global warming, in addition to causing local air quality problems. For the industry to reduce its carbon footprint, cleaner technologies are being adopted, energy efficiency is being improved, and carbon capture and storage options are being explored (Wilson et al., 2023). The ability to manage energy needs in the face of climate change continues to be a formidable challenge.

This chapter gives a brief overview of the upstream oil and gas industry and examines its environmental footprint, including the activities associated with exploration and production of oil and gas. The following sections offer a more detailed discussion of these ecological, health, and environmental challenges (Wollin et al., 2020) (Walsh & Rose, 2022) The upstream oil and gas industry is employing a multifaceted approach to address these concerns and the chapter will present steps being taken by the industry towards environmental sustainability (Okere & Sheng, 2023). There is no doubt that the journey towards sustainability of the industry is a reflection of a commitment to balancing progress with the preservation of the planet's delicate equilibrium, whether that be by developing innovative technologies that minimize ecological disruption or setting up strict health and safety guidelines.

OVERVIEW OF THE UPSTREAM OIL AND GAS INDUSTRY

In addition to providing a comprehensive picture of the upstream oil and gas sector's contributions to sustainable development, a structural analysis can reveal its obstacles and impacts.

A series of intricate processes leading to the extraction and production of valuable hydrocarbons from the ground are part of the upstream oil and gas process, which is the first stage of the energy production chain. Exploration, drilling, and processing of minerals form a complex and interconnected web of activities that require advanced technologies, multidisciplinary expertise, and strategic decision-making in order to be successful. The following overview provides a detailed overview of the key stages of the upstream sector and highlights its challenges and innovations.

1. **Subsurface exploration**

The first step in the exploration process is the identification of potential oil and gas deposits underground by geologists, geophysicists, and other experts. Through the use of sound waves and the measurement of their reflections, seismic surveys are able to provide information about subsurface structures (Cox et al., 2020). The presence of hydrocarbons is confirmed through drilling exploratory wells after promising areas have been identified. It is important to invest in production infrastructure as soon as exploration is successful, since exploration is a high-risk and high-reward process (Basile et al., 2021).

2. **Exploring underground reserves through drilling**

A well is drilled into the Earth's crust in order to gain access to the underground hydrocarbon reservoirs. There are a variety of drilling methods that are used depending on the depth and location of the reservoir. These methods include rotary drilling, directional drilling, and offshore platform drilling (Islam & Hossain, 2020). The drilling fluids are responsible for carrying the rock cuttings to the surface, which cools the equipment and carries drilling bits that are made of advanced materials and technologies. In order to ensure the integrity of the well and prevent contamination once the casing reaches the hydrocarbon-bearing rock formations, production casing needs to be installed (Yousuf et al., 2021).

3. **Hydrocarbon production: Extraction and handling**

During the production phase, hydrocarbons are extracted from reservoirs after they have been drilled. As a result of the natural pressure within a reservoir, gas and oil are forced to the surface by the pressure on the surface (Burrows et al., 2020). As the reservoir pressure declines, it may be necessary to employ a gas lift system or electric submersible pumps in order to enhance production (K. C. Chen et al., 2019). Following the extraction of hydrocarbons, these substances are separated into crude oil and natural gas, as well as other components. Surface facilities such as wellheads, separators, and pipelines are used for the transportation of the extracted resources to refineries for further processing (Buslaev et al., 2021).

4. **Refining and processing raw resources**

As soon as oil and natural gas are extracted, they are purified by removing any impurities that are present in them, and then they are separated according to their properties. The refineries are responsible for converting crude oil into petrochemical products, such as gasoline, diesel fuel, and diesel fuel, which are used in the production of petrochemicals (Al-Samhan et al., 2022). As a result of the process of processing natural gas, impurities such as water, carbon dioxide, and sulfur compounds are removed from the gas (Ghasem, 2020). With these processes, the quality of the end product as well as its market value will be enhanced, since a variety of energy sources will be created that can be used for consumption.

It is imperative that environmental and safety considerations be taken into account in all of these processes. Regulations and industry standards have been developed to protect the environment, human health, and ecosystems on a global scale (Litvinenko et al., 2022). In regards to reducing emissions, preventing spills, and restoring habitats, there are various initiatives being undertaken. A number of measures are being taken by the industry to reduce its carbon footprint, such as energy-efficient practices, carbon capture and storage, and researching how renewable energy can be integrated into the process (Min et al., 2022).

The upstream sector of the oil and gas industry incorporates scientific knowledge, technological innovation, and environmental responsibility into its operations (Guerrero-Martin et al., 2023). During each phase of the project, from the initial seismic survey to the refinement of valuable resources, meticulous planning, rigorous execution, and continuous adaptation to evolving needs are all necessary. There is no doubt that the upstream energy sector will continue to play an increasingly important role in the global energy landscape as the world strives to achieve a balance between energy needs and sustainability goals.

WHY IS SUSTAINABILITY REQUIRED?

In the upstream oil and gas industry, there are many factors that contribute to the need for sustainable practices and long-term considerations. As a result, it is essential that the industry adopts sustainable solutions for the purpose of minimizing its impacts on the environment, society, and economy. Listed below are some of the most important reasons why it is crucial that upstream oil and gas is sustainable.

1. **The effects of greenhouse gas emissions on climate change**

The majority of carbon dioxide emissions and climate change can be attributed to the burning of fossil fuels, such as oil and gas. As a result of worldwide efforts to mitigate global warming and transition to cleaner energy sources in order to achieve international climate goals, the upstream sector's contribution to reducing emissions is becoming increasingly important (Maniatis et al., 2021). As a result of the combustion of fossil fuels, significant amounts of greenhouse gases are released into the atmosphere, including carbon dioxide (CO_2) and methane (CH_4) (Rafiee et al., 2022). Additionally, these gases trap heat and disrupt climate patterns, thus contributing to global warming.

A major problem with oil and gas infrastructure is that it has a long operational life, which leads to accumulation of emissions as time goes by. It is of the utmost importance that emissions are reduced and renewable energy sources are transitioned to as soon as possible given these cumulative impacts (Ben-Naceur, 2019; Hartmann et al., 2021).

The concept of a global carbon budget has emerged as a result of the Paris Agreement, which aims to limit global warming to well below 2 degrees Celsius (Ekardt et al., 2022). There is a finite amount of CO_2 emissions that can be released so that the temperature threshold limits are not exceeded. As a result, the oil and gas industry needs to carefully manage its contribution to this budget in order to avoid exceeding critical limits.

As a result of climate change, extreme weather events are becoming more frequent and intense, sea levels are rising, and precipitation patterns are changing, which can disrupt the industry's operation as a whole (Sun et al., 2020). The industry can increase its adaptability and resilience if it adopts sustainable practices in light of these changing conditions.

In the event that climate change is left unchecked, it can cause social unrest and economic instability, as well as sea level rise, food scarcity, and displacements (Donaghy et al., 2023). In addition to supporting the well-being of individuals and the economy, sustainable practices contribute to a more stable environment on the global scale.

Investors, as well as oil and gas companies, are facing increasing pressure to adopt responsible environmental practices in the oil and gas industry (X. Chen et al., 2022). As a result, climate-friendly companies are more likely to maintain investor confidence and be perceived as more trustworthy by the public.

2. Exhaustion of resources

In spite of the fact that oil and gas are finite resources, they are being extracted at an alarming rate. In order to ensure sustainability, resources must be managed responsibly, and the industry must consider the long-term availability of these valuable resources as a part of its comprehensive planning.

As a result of the necessity of managing resources responsibly, sustainability has become a vital component of the upstream oil and gas industry. It is essential to keep sustainable practices in mind because of the finite nature of fossil fuel reserves as well as the potential negative effects of unchecked extraction (Soeder, 2021a). For oil and gas to be sustainable, upstream oil and gas resource management is crucial, due to the following reasons

- **Limited Resources:** Non-renewable resources such as oil and gas are depleting over time (Mufutau Opeyemi, 2021). Future energy security is threatened by the industry's continued reliance on these resources. It is imperative to utilize sustainable practices when extracting resources to ensure their long-term availability.

- **Protecting Future Generations:** For future generations to have access to energy, it is crucial to ensure that the current reserves are managed efficiently and effectively. The need for energy can be met both now and in the future if waste is minimized and extraction processes are optimized. An example would be the use of chelating agents such as ethylenediaminetetraacetic acid (EDTA) and glutamic acid diacetic acid (GLDA) to increase the production of wells and reservoirs and improve oil recovery (Hassan et al., 2020).
- **Sustainable Economy:** There can be volatility in global energy markets due to the scarcity of resources. The development of sustainable practices that emphasize efficient resource management is particularly helpful to economies that heavily rely on energy resources as a major component of their growth (Cherepovitsyn et al., 2021; Ionescu, 2021).
- **Displaced Assets:** Displaced assets refer to fossil fuel reserves that are no longer recoverable due to changes in the market, regulatory changes, or climate change concerns (Bos & Gupta, 2019). In the context of sustainable resource management, it's more likely that risk of investments being displaced will reduce with the evolution of the energy landscape.
- **Environmentally Sustainable Production and Consumption:** It is important to note that when circular economy principles are adopted by an industry, consumption and production are promoted in a responsible manner (Foster, 2020). As part of the efforts to minimize the impact of the upstream operations on the environment, waste can be reduced, resource efficiency can be maximized, and materials can be recycled.
- **Introduction of New Technologies:** Technological innovation is often driven by sustainable practices (Jia, 2020; Radnejad et al., 2020). Long-term viability of the industry can be enhanced by pursuing more effective and environmentally friendly extraction methods.
- **Managing the Environment:** Extractive and production activities can be managed in a sustainable manner by minimizing their negative environmental impact (Andersen & Gulbrandsen, 2020). For the industry to operate in harmony with ecosystems, it must minimize the disruption of habitats, pollution, and emissions.
- **Regulatory Compliance:** Many countries are implementing regulations aimed at reducing the impact of resource extraction on the environment, especially in countries with an abundance of natural resources. There is a strong correlation between an industry's social license, as well as the compliance with these regulations. In the event that industries fail to comply with regulations, they can be subjected to significant financial penalties and legal liability (Norouzi, 2021; Yin, 2021). It has been found that the incorporation of sustainability practices reduces the risk of regulatory violations and the costs associated with them.
- **Increasing Expectations From Consumers and Investors:** It is becoming increasingly obvious that consumers and investors expect industries, including the upstream oil and gas industry, to behave responsibly and sustainably in the future (Beck et al., 2020). In order for companies to be trusted by stakeholders, they have to demonstrate a commitment to resource management and sustainability.

It is important to remember that fossil fuel reserves are finite, and that the unchecked extraction of fossil fuels could have negative effects on the environment, the economy, and society. Developing a balance between energy supply, environmental preservation, as well as the well-being of future generations is possible as the industry embraces responsible resource management.

3. **Impact on the Environment:**

In some cases, upstream operations can adversely affect biodiversity, ecosystems, and water bodies in a manner that is detrimental to the environment (Ogolo et al., 2022). As a result of the industry's activities, there is a risk of habitat destruction, contamination of water and soil, and disturbance of marine and terrestrial ecosystems. Taking sustainable measures to mitigate its impact on the environment is imperative for the following reasons.

- **Pollution:** There is a possibility that pollutants can be released into the air and water during upstream operations, affecting the local air quality and aquatic ecosystems (Gonzalez et al., 2022). By adhering to strict environmental standards, extraction and production processes must ensure minimal releases of harmful substances into the environment (X. Chen et al., 2023).
- **Greenhouse Gas Emissions:** In the process of burning oil and gas, carbon dioxide (CO_2) and methane (CH_4) are released into the atmosphere (Allen et al., 2021; Q. Chen et al., 2019). In addition to contributing to global warming, these gases also contribute to climate change. It is essential that the industry adopt sustainable practices in order to reduce its carbon footprint as much as possible.
- **Impacts on ocean life and the environment:** There has been a significant impact on marine life and coastal ecosystems caused by offshore drilling and oil spills. By adopting sustainable practices, such incidents can be prevented and mitigated, in turn protecting the health of the oceans and the livelihoods of coastal communities (Johnston et al., 2019).
- **Managing Oil Spills:** Oil spills and accidents could cause catastrophic environmental damage during the extraction, transportation, and processing of oil. To reduce the risk of spills and accidents, comprehensive spill prevention and response plans are necessary (Kwarto et al., 2022). Sustainable practices, emergency preparedness, and compliance with regulatory standards could help minimize spill incidents.

4. **Public Perception:**

In the upstream oil and gas industry, there has been a lot of scrutiny due to its environmental and social impacts (Green et al., 2022). As discussed below, sustainability can enhance reputations and build trust, as well as demonstrate a commitment to responsible behavior.

- **Environmental Concerns:** There is significant public concern about the environmental impact of oil and gas production, particularly air and water pollution, habitat destruction, and climate change. The oil and gas industry's reputation can be negatively impacted by high-profile environmental disasters, such as oil spills (Kwarto et al., 2022). Companies must take proactive measures to reduce their environmental footprint and ensure their operations are not causing harm to the environment. These measures include investing in renewable energy sources, reducing waste and emissions, and taking steps to ensure the safety of their operations.
- **Skepticism Among the General Public:** Even though the industry is committed to sustainability, there is often skepticism about whether the commitment is genuine or simply greenwashing (Johnston et al., 2019). This is due to the fact that there is often a lack of transparency, and that companies often don't provide clear evidence of their sustainable practices or the results of those practices. This leads to a lack of trust and makes it difficult for consumers to make informed decisions.

- **Differences by Region:** It is possible for public perception to vary significantly from region to region. Some communities may view the industry's contribution to local economies more positively where it is a major employer and economic driver. This variance can be the result of many factors, such as the size of the industry and the local economy, the industry's track record, the level of economic development, and the degree of public engagement. This is why it is important for companies to have local representatives that can connect with the community to ensure that the public understands their mission and values. Companies should also focus on providing positive economic and social benefits to the communities they operate in (Lee et al., 2022). These efforts will help to improve public perception and foster a more positive relationship between the industry and the local population.
- **Social Impact and Community Involvement:** The social and economic impact of the upstream oil and gas industry often causes concern for communities located near these operations. This disrupts local economies and causes noise, pollution, and health hazards. These impacts can be especially concerning for communities that are already economically disadvantaged ("Environmental Accounting and Financial Performance of Oil and Gas Companies in Nigeria," 2019; Kwarto et al., 2022). The construction of oil and gas operations causes disruption to local businesses, can displace residents, and can lower land and property values. Residents within the vicinity may also suffer negative physical and mental health effects as a result of the upstream industry's operations.

STEPS TAKEN TO IMPROVE SUSTAINABILITY

Innovations in technology, operational practices, stakeholder engagement, and environmental stewardship are all factors that play a role in improving sustainability in the upstream oil and gas industry. Several steps have been taken in order to enhance the sustainability of the upstream oil and gas industry which include the following

1. **Use of cleaner energy sources**

Renewable energy technologies should be researched and developed. They are a crucial part of the transition to a low-carbon economy. Renewable energy is clean, abundant, and more cost-effective than traditional energy sources (Abraham-Dukuma, 2021). It is essential for the future of the planet. Renewable energy is critical in reducing pollution and mitigating the impacts of climate change (Pickl, 2019).

Renewable energy sources such as solar and wind have the potential to cut back the reliance on fossil fuels, the main cause of climate change (Diaz et al., 2022; Heidari et al., 2022). Additionally, the use of renewable energy sources can help to create jobs in the renewable energy sector, while also helping to reduce air and water pollution (Garcia-Gonzales et al., 2019). Natural gas is a less polluting alternative to coal and oil and can be used to supplement renewable energy sources to further decrease the dependence on fossil fuels. This combination of renewable energy sources and natural gas can help minimize the overall carbon footprint and make a significant contribution to mitigate global warming.

2. **Strategies to reduce emissions**

Methane emissions can be minimized by implementing advanced leak detection and repair programs (Zimmerle et al., 2020). These help to identify and fix leaks from oil and gas production facilities. It prevents the methane from entering the atmosphere, which can have a significant impact on global warming. These programs can also be used to identify and repair gas flaring, which is the burning of gases and other hydrocarbon effluents that cannot be used or sold (Santibanez-Borda et al., 2021). As flaring releases a large amount of methane into the atmosphere, reducing it can be a significant factor in controlling methane levels.

Fugitive emissions are caused by leaks and other unintentional releases of pollutants. Upgrading equipment and infrastructure can reduce these leaks, thereby reducing emissions (Fox et al., 2019; Hashim et al., 2022). This can be done by retrofitting existing equipment and installing new technologies that are designed to reduce emissions. Additionally, it is important to ensure that equipment is well-maintained and that regular inspections are conducted to identify and repair any leaks. Regular maintenance can help reduce the amount of emissions that are released into the environment, which can help to improve air quality and protect the environment.

3. Wastewater management

Upstream oil and gas Industry needs to ensure that water recycling and reuse systems are implemented in order to minimize freshwater use (Cooper et al., 2022). This is necessary because freshwater is a limited resource and the demand for it is growing rapidly. By reusing and recycling water, the industry can reduce its need for freshwater and increase efficiency, which will assist in reducing operational costs. Additionally, water reuse systems can help protect the environment by reducing the amount of wastewater discharged into rivers and lakes (Cabrera et al., 2021). Finally, by using recycled water in oil and gas production, companies can reduce their total water footprint.

For upstream industries to avoid contamination, produced water must be treated and disposed of properly. The water produced during extraction contains a variety of chemicals, including salts, heavy metals, and hydrocarbons. If not properly treated and disposed of, these chemicals can seep into the environment and contaminate soil and fresh water sources.

Monitoring and managing the use of critical resources such as water and energy during drilling and production can be accomplished with blockchain, similar to downstream oil and gas processes (Laturkar & Laturkar, 2023). Water conservation during drought conditions or energy conservation can be automatically triggered by smart contracts.

4. Use of carbon capture and storage (CCS) technologies

Carbon dioxide emissions from upstream operations can be captured and stored using CCS technologies. By doing this, emissions can be drastically reduced, and can help mitigate the effects of climate change (Ghasem, 2020). Additionally, CCS technologies can be used to capture and reuse the carbon dioxide for other industrial processes, such as enhanced oil recovery, improving the efficiency of the process etc. This is an effective way to reduce the amount of carbon dioxide released into the atmosphere and help combat climate change. In addition, it can also assist in reducing the costs associated with operations and improve their profitability.

5. Achieving energy efficiency

Upstream processes must ensure that drilling, extraction, and transportation processes are energy-efficient. Doing so can reduce the amount of energy required for the processes, resulting in lower energy costs. It can also reduce the amount of emissions released, thus protecting the environment (Wilson et al., 2023). A more efficient process can also help to maximize the use of resources, reducing the need for new materials and resources for the process. This can also lead to an overall reduction in energy and resource use, which can reduce operational costs and environmental impact.

6. A conservation strategy for biodiversity

Protecting local ecosystems and wildlife during exploration and production must be a priority for the upstream oil and gas sector (Delefosse et al., 2020). This is because the upstream oil and gas sector activities have the potential to cause significant environmental damage, including air and water pollution, habitat destruction, and disruption of wildlife species. Additionally, preserving the health of local ecosystems and wildlife can help to ensure the sustainability of the sector in the long term (Dickinson et al., 2020). Before beginning operations, it is important to conduct an environmental impact assessment. It identifies any potential impacts on local ecosystems and wildlife and determines the best course of action to minimize or avoid them.

7. Engaging the community and being socially responsible

Upstream oil and gas operations should engage local communities for their benefit. Engaging local communities can help ensure that operations are conducted in a safe and responsible manner, with potentially positive economic benefits for all involved. Additionally, it can help build trust and facilitate positive relationships between the organizations and the local communities ("Environmental Accounting and Financial Performance of Oil and Gas Companies in Nigeria," 2019; Kwarto et al., 2022). The overall well-being can be improved through educational support and local development projects. Engaging local communities can lead to potential opportunities for employment, training, and other economic benefits. It can also help to ensure that local concerns are heard and addressed.

8. The importance of transparency and reporting

Sustainability reports detailing environmental and social performance can be published by the upstream oil and gas industry (Kwarto et al., 2023). These reports provide investors and the public with information about the company's performance in areas such as greenhouse gas emissions, waste management, and land use. They also provide an opportunity to highlight their efforts to reduce their environmental and social impacts (Adadzi et al., 2022). By doing so, compliance with international sustainability standards and reporting frameworks will be ensured. This helps organizations to demonstrate their commitment to sustainability and helps them to gain trust from the public. Additionally, it provides them with a way to measure their performance against their own goals and against industry standards. This can help them identify areas where they can make improvements and become more sustainable.

9. Regulatory compliance

Environmental regulations and standards must be adhered to by upstream processes. This is to ensure that the extraction and production is performed in a way that does not cause harm to the environment. This includes reducing emissions, limiting hazardous waste, and conserving natural resources. In order to address emerging sustainability issues, industries should work proactively with regulators. When industries are proactive in meeting environmental standards, it can help them save money on compliance costs, reduce their risk of legal action, and potentially gain a competitive advantage by demonstrating their commitment to sustainability (E. Herbert et al., 2020; Yin, 2021). Additionally, it can assist in reducing their environmental footprint and benefit the global community. A blockchain can be used to record safety inspections, equipment maintenance, and safety incident reports. By following safety protocols and complying with regulatory requirements consistently, accidents and environmental damage can be mitigated.

10. **Investing in technology**

Upstream oil and gas operations can be monitored and optimized using advanced technologies such as drones, sensors, and data analytics. By using drones to monitor operations, organizations can get real-time data about equipment, production, and environmental conditions. Sensors can be used to detect leaks and other issues quickly, and data analytics can be used to optimize production and identify opportunities for cost savings (Andersen & Gulbrandsen, 2020; Jia, 2020). Equipment failures and downtime can also be reduced through preventative maintenance. This data can also be used to track energy consumption and emissions, helping to reduce the environmental impact of the company's operations. Additionally, drones can be used to inspect hard-to-reach areas, reducing the need for manual labor which can improve safety and efficiency.

11. **Sustainability in supply chains**

Suppliers and contractors should adopt sustainable practices in upstream processes. This can reduce their environmental impact, improve worker safety, increase user satisfaction, and reduce costs (Li et al., 2019). Sustainable practices can also help organizations become more competitive in the marketplace. Supply chain decisions should be evaluated for their environmental and social impacts. Companies that are seen as socially and environmentally responsible can often attract investors who are looking to support sustainable initiatives. Additionally, implementing sustainable practices can result in significant cost savings, as it can reduce waste and energy usage. Blockchain can be used to record energy consumption data from drilling rigs, processing facilities, and transportation vehicles (Ahmad et al., 2022).

12. **Training and development for workers**

Employees in the upstream oil and gas sector should be trained on sustainability and best safety practices. This is important because of the potential environmental and safety risks associated with the industry, such as oil spills, air and water pollution, and worker safety hazards. Training employees on sustainability and safety practices can help reduce these risks and ensure that the industry is operating in an environmentally friendly and safe manner (Naji et al., 2022). A responsible and sustainable culture should be promoted in all the organizations. This can be accomplished by implementing a comprehensive safety program that includes regular safety and sustainability audits, employee training, and a comprehensive safety and sustainability policy.

13. **Developing exploration and production plans**

Exploration and development of upstream reserves with a low environmental impact should be given priority. This will help reduce the risk of oil spills and other environmental disasters, while also increasing the efficiency of the oil production process (Attia et al., 2019). Furthermore, it will assist in reducing the cost of production, making it more accessible and affordable for consumers. Risk assessments should be conducted to determine how climate change will impact operations (Abdussalam et al., 2021). This will help to ensure that the most suitable areas for exploration and development are chosen and that operations are conducted in a way that minimizes the impacts on the environment and local communities. Additionally, risk assessments are necessary to ensure that operations are not conducted in areas that are prone to weather-related disasters, which can have catastrophic effects if not adequately prepared for.

14. **Innovating and investing in research**

A collaboration between the upstream oil and gas industry and research institutions is needed to develop innovative solutions to ensure sustainable energy production (Jia, 2020). Such collaborations can help to leverage the deep expertise of both industries in order to develop more efficient and sustainable technologies. This could include developing more efficient drilling techniques, advanced seismic imaging, and other technologies. Resource extraction could be made more efficient and responsible by investing in new technologies. For instance, this could be done by utilizing new technologies such as artificial intelligence (AI) and machine learning to identify optimal locations for drilling, or using predictive analytics to estimate the potential of certain oil fields (Elijah et al., 2021). Additionally, these technologies could be used to reduce the environmental impact of projects by monitoring emissions and waste more effectively.

For upstream oil and gas to be sustainable, collaboration between industry stakeholders, governments, and the public is essential. This collaboration will allow the industry to meet energy demands as well as address environmental and social concerns. The collaboration should focus on developing safe and efficient production processes, creating better regulations, and incentivizing innovation. It should also explore ways to reduce the environmental impact of upstream oil and gas production and ensure the responsible use of resources. An overview of some of the steps which can be taken to improve sustainability in the upstream oil and gas are illustrated in Figure 1.

CHALLENGES WITH IMPLEMENTATION

The implementation of sustainability measures within the upstream oil and gas industry is not without its challenges. Sustainability practices are recognized by the industry, but several factors can make it difficult for them to be implemented effectively. The following are some of the challenges that need to be overcome in order to achieve success

1. **The Economic Impact**

Technology, equipment, and infrastructure investments are necessary for sustainability measures. Oil and gas prices can be volatile, making it even more challenging to balance costs with long-term benefits (Foster, 2020; Ionescu, 2021). Organizations need to carefully consider the cost of these investments and weigh it against the potential for long-term gains. Additionally, they need to factor in the potential for fluctuations in oil and gas prices over time, as these can significantly impact the cost-benefit analysis.

Figure 1. Sustainability measures that can be implemented by the upstream oil and gas sector

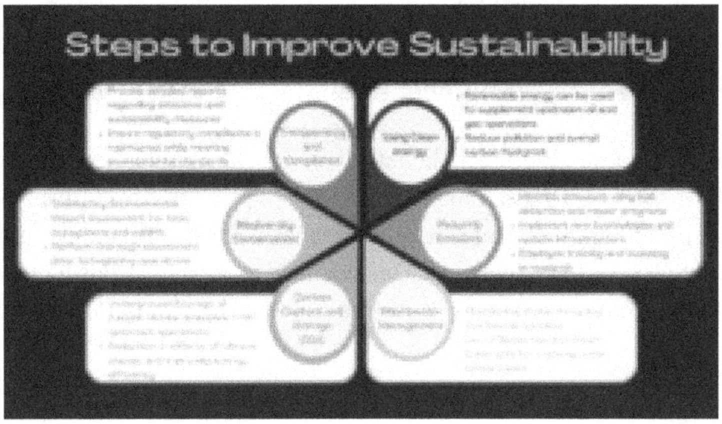

2. Obstacles to Adopting New Technologies

It is often necessary to overcome technical challenges when adopting new technologies and practices. Complex and time-consuming processes are involved in developing and integrating innovative solutions. This is because the existing infrastructure and processes may not be designed to accommodate the new technologies and practices (Fox et al., 2019). Thus, it can be difficult to determine the best way to integrate them into existing systems without compromising their performance.

3. Disruptions in operations

In order to implement sustainability measures, it may be necessary to temporarily stop or modify ongoing operations, which could lead to production downtime and financial losses (Soeder, 2021a). This is because the costs associated with making changes to existing operations may be high, and the process of transitioning to more sustainable practices can take some time. Additionally, the implementation of sustainable practices may require the use of new technologies which could add to the cost of the transition.

4. Environmental regulations and compliance requirements

In the upstream oil and gas industry that is constantly evolving, environmental regulations and compliance requirements can create uncertainty (Norouzi, 2021; Yin, 2021). Planning and decision-making can be affected by changes in rules and standards. This can have a direct impact on the industry's profitability, as well as on the safety and sustainability of its operations. Companies must be aware of and prepared for the ever-changing regulatory landscape, or risk facing hefty fines and other penalties.

5. Infrastructure built on legacy technologies

Due to the design limitations and compatibility issues of older equipment and facilities in the upstream sector, retrofitting existing infrastructure with sustainable technologies can be challenging (Robertson & Chilingar, 2017). These limitations and issues can lead to difficulties in integrating new technologies, such as renewable energy sources, energy-efficient equipment etc. into existing systems. As a result, the cost of retrofitting can be significantly higher than the cost of building new infrastructure from scratch.

6. **Focus on the short-term versus the long-term**

Although sustainability offers long-term benefits, oil and gas companies are often under pressure to demonstrate immediate returns. This can make it difficult to prioritize initiatives that are likely to yield benefits over a longer period of time. Since there is a focus on short-term gains, sustainability initiatives can be seen as too long-term or too expensive. This can lead to organizations focusing on initiatives that can provide more immediate returns, such as cost-cutting measures, rather than long-term sustainability efforts (Desai et al., 2021).

7. **Changes in culture**

Organizations may need to undergo a cultural shift in order to transition to a more sustainable approach. Changing ways of working and thinking can be met with resistance or skepticism by employees and management (Naji et al., 2022). This is because it can be difficult to adjust to new practices and habits, and people may be wary of the unknown. Additionally, it can be difficult to motivate people to make changes when there is no immediate reward or benefit.

8. **Inadequacies in knowledge**

It is often necessary to have specialized knowledge and expertise in order to implement sustainability measures. Developing skilled personnel for emerging technologies and practices may be a challenge for the industry (Amin et al., 2020). Companies need to invest in training staff and providing them with the necessary skills to implement sustainability practices, and many may not have the resources to do so. This can lead to a shortage of workers with the skills and knowledge required to implement sustainability measures.

9. **Reputation and perception of the public**

The oil and gas industry's historical environmental impact can negatively impact public perception and trust, which complicates stakeholder engagement and community relations despite efforts to adopt sustainable practices. The environmental impact of the oil and gas industry includes air and water pollution, land degradation, and the release of hazardous chemicals (Johnston et al., 2019). This damage to the environment can have a long-term impact on local ecosystems and can lead to public mistrust of the industry, even if the company is making an effort to reduce their environmental footprint (Roberts & Flin, 2020).

10. **Variability on a global scale**

Environmental conditions, geographic location and local regulations can significantly affect sustainability challenges. When dealing with diverse operational contexts, it can be difficult to implement a one-size-fits-all approach. Different regions have different environmental conditions and regulations that need to be taken into account when determining the most sustainable approach (Lee et al., 2022). Additionally, some locations may have geographic features that can limit the ways in which sustainability challenges can be addressed. For example, some areas may be prone to natural disasters that can make certain sustainability initiatives difficult or impossible to implement.

11. **Scale and scope of the operations**

In remote or challenging environments, implementing sustainability measures can be logistically challenging due to the sheer scale of oil and gas operations. It requires a large amount of resources, manpower, and capital to properly implement sustainability initiatives in large-scale oil and gas operations (Amin et al., 2020). Moreover, many of these operations take place in areas where access to resources is limited, making it difficult to maintain these measures over time.

12. **Supply chains with complex components**

Multiple stakeholders are involved in the oil and gas industry's operation. It can be challenging to coordinate efforts across these complex networks in order to ensure sustainability (Amin et al., 2020; Naji et al., 2022). The sheer number of parties involved can make it difficult to coordinate efforts and ensure that everyone is on the same page. Additionally, the complexity of the processes can make it difficult to identify areas where sustainability can be improved.

13. **Priorities competing with each other**

Production targets, safety, regulatory compliance, and financial performance are just a few of the priorities faced by oil and gas organizations (Soeder, 2021a). In order to achieve sustainability goals, these priorities must be balanced carefully. Without careful balancing of these priorities, a company may sacrifice safety or compliance in order to increase production or financial performance. This could have long-term negative consequences, such as increased risk of accidents or fines due to non-compliance.

While the upstream oil and gas industry acknowledges the importance of sustainability, various challenges can hinder the implementation of sustainable practices. A sustainable future can be achieved through a long-term outlook that is focused on reducing the environmental impacts of operations, while also providing economic and social benefits. This includes reducing emissions, using renewable energy sources, and conserving resources. In addition, it requires collaboration between all stakeholders, including government, industry, and local communities, to ensure that everyone is working towards the same goal. Figure 2 illustrates some of the challenges faced by the oil and gas industry in implementing sustainability measures.

THE FUTURE OF SUSTAINABILITY

In the upstream oil and gas industry, sustainability is reshaping its trajectory and driving transformative change. It is imperative that the industry envisions a future that balances its economic imperatives, as well as champions responsible resource management and environmental stewardship, as the world transitions toward cleaner energy sources and increased environmental awareness. Creating a sustain-

Figure 2. Challenges faced by upstream oil and gas sector when implementing sustainability measures

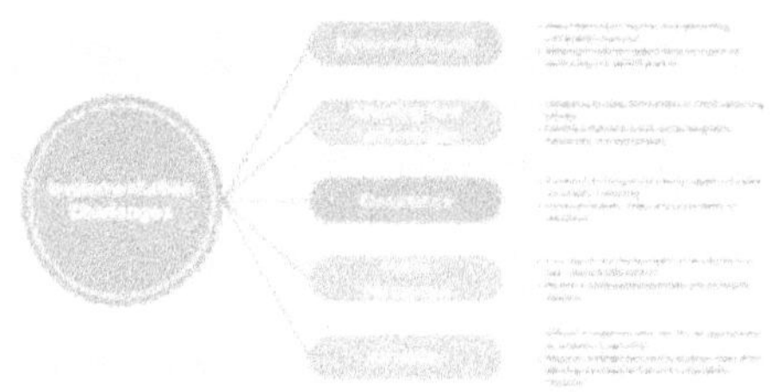

able energy landscape in the upstream sector requires technological advancements, collaboration, and a commitment to innovation as listed below

1. **Innovation and digitalization**

Technology is revolutionizing operations and environmental performance in the industry. Using advanced analytics, machine learning, and artificial intelligence, drilling processes are being optimized, energy efficiency is being improved, and emissions are being reduced (Koroteev & Tekic, 2021; Tontiwachwuthikul et al., 2020). Real-time monitoring and smart sensors minimize environmental risk by detecting leaks and other operational inefficiencies early.

2. **Transition strategies for the energy sector**

Investments in renewable energy sources such as wind, solar, and hydrogen are diversifying the industry's energy portfolio (Andersen & Gulbrandsen, 2020; Hunt et al., 2022). In order to transition towards a low-carbon future, organizations are exploring ways to utilize their expertise and infrastructure. The upstream oil and gas industry is pioneering efforts to align itself with global climate goals through renewable energy solutions, carbon-neutral operations, and research into sustainable aviation fuels.

3. **Principles of the circular economy**

With an eye on reducing waste and maximizing resource efficiency, the circular economy is gaining prominence in the industry (Foster, 2020). Circular approaches are transforming the way the industry perceives and manages its operational footprint, from repurposing drilling materials to repurposing decommissioned platforms for renewable energy.

4. **The importance of collaboration and partnerships**

Collaboration between industries, governments, and stakeholders is essential to solving sustainability's complex challenges. Research institutions, oil and gas businesses, and regulatory bodies are forming partnerships that foster innovation and knowledge sharing (Norouzi, 2021; Yin, 2021). The potential of cross-sector collaboration is demonstrated in projects such as converting oil wells into geothermal energy or carbon storage.

5. **The concept of carbon capture and storage (CCS)**

The use of CCS technologies is becoming a key strategy for reducing emissions from oil and gas operations. The industry can reduce its carbon footprint and contribute to a sustainable energy transition by investing in CCS infrastructure, such as capturing CO_2 emissions at the source and storing them underground (X. Chen et al., 2022; Min et al., 2022). By capturing the CO_2 emissions at the source, the industry can prevent it from entering the atmosphere and contributing to climate change. Storing the captured CO_2 underground helps to reduce the risk of it escaping and further contributing to global warming.

6. **Accountability for the environment**

In the upstream oil and gas industry, sustainability reporting and transparency are becoming increasingly important. Data on emissions, environmental impact assessments, and sustainability goals are being disclosed by corporations (Kwarto et al., 2023). Increasing transparency is a critical element of building public trust and encouraging the industry to meet its sustainability commitments.

7. **Towards a just transition**

As the energy landscape transitions to a sustainable one, it will also affect communities and workers dependent on the oil and gas industry. The goal of just transition initiatives is to ensure that people and regions are supported, retrained, and provided with economic opportunities, ensuring an inclusive and equitable transition to sustainability (Green et al., 2022). These initiatives recognize the need to provide workers with job training, income support, and other measures to ensure that they are not left behind in this transition. They also recognize the importance of engaging communities to ensure that their voices are heard in the process, and that they benefit from the shift to a more energy efficient landscape.

8. **Making data-driven decisions**

The industry's approach to sustainability will be shaped by data-driven decision-making. Informed strategies for continual improvement can be implemented if comprehensive data collection and analysis are in place (Cox et al., 2020; Desai et al., 2021). By gathering information on their energy use, waste production, and other environmental and sustainability factors, organizations can identify which processes are inefficient and which need to be improved. This can enable them to develop better sustainability strategies and make more informed decisions.

The upstream oil and gas industry is challenged to balance economic growth, energy security, and environmental responsibility. With innovation, collaboration, and a forward-looking mindset, the industry can redefine its role in transitioning to a sustainable and resilient future.

CONCLUSION

The pursuit of sustainability stands out in the unforgiving terrain of the upstream oil and gas industry. As tectonic shifts and persistent climate change pressures continue to loom, this chapter emphasized the necessity of sustainable measures. There is a sobering reality that the industry is entangled with challenges that test the limits of responsible resource management, environmental stewardship, and ethical responsibilities.

In a technologically disrupted world, innovation stands as both a sentinel and a savior. Developing cutting-edge technologies reveals hidden reserves, improves extraction efficiency, and refines emissions reduction strategies. Regulatory obstacles and economic uncertainty threaten the industry's future unless these advancements can be leveraged to navigate it successfully. Ultimately, the industry's trajectory will be determined by the interaction between traditional energy giants and the rising tide of renewable integration. Climate change mitigation is an undeniable urgency that encapsulates the relentless push towards cleaner energy sources. In the effort to combat global warming, carbon management is a crucial ally in the mission to arrest emissions. The need to balance energy demands with environmental responsibilities has been recognized by industry experts. Despite this, traditional practices and infrastructure leave a shadow on the industry's commitment.

The commitment to sustainability has reshaped the industry's core operations from the first steps of exploration through production. Technical innovations, including advanced drilling techniques and digital monitoring systems, are poised to revolutionize resource extraction while minimizing environmental disruption. As renewable energy sources are integrated into traditional operations, the industry is committed to a cleaner future through energy diversification and carbon reduction.

Circular economy practices, community engagement, and resilience illuminate the way forward. Sustainability is more than just a concept; it's a roadmap that combines economic growth, environmental stewardship, and social responsibility. It's about building trust, transparency, and shared prosperity by collaborating with stakeholders - both local and global.

In order for the industry to thrive in the future, it must be able to anticipate the evolving energy landscape and be adaptable to it. In response to investor pressure and regulatory frameworks, the industry is becoming more environmentally friendly, accountable, and sustainable. Climate change mitigation, responsible resource management, and ecosystem protection are being addressed by the sector in innovative ways.

REFERENCES

Abdussalam, O., Trochu, J., Fello, N., & Chaabane, A. (2021). Recent advances and opportunities in planning green petroleum supply chains: A model-oriented review. *International Journal of Sustainable Development and World Ecology*, *28*(6), 524–539. Advance online publication. doi:10.1080/13504509.2020.1862935

Abraham-Dukuma, M. C. (2021). Dirty to clean energy: Exploring 'oil and gas majors transitioning''.'. *The Extractive Industries and Society*, *8*(3), 100936. Advance online publication. doi:10.1016/j.exis.2021.100936

Adadzi, F. D., Godson-Amamoo, N. S., & Nunoo, J. (2022). The Ghanaian state and governance of the upstream oil and gas industry. In *Petroleum Resource Management in Africa*. Lessons from Ten Years of Oil and Gas Production in Ghana. doi:10.1007/978-3-030-83051-9_12

Adebiyi, F. M. (2022). Air quality and management in petroleum refining industry: A review. *Environmental Chemistry and Ecotoxicology*, *4*, 89–96. Advance online publication. doi:10.1016/j.enceco.2022.02.001

Ahmad, R. W., Salah, K., Jayaraman, R., Yaqoob, I., & Omar, M. (2022). Blockchain in oil and gas industry: Applications, challenges, and future trends. *Technology in Society*, *68*, 101941. Advance online publication. doi:10.1016/j.techsoc.2022.101941

Al-Samhan, M., Al-Fadhli, J., Al-Otaibi, A. M., Al-Attar, F., Bouresli, R., & Rana, M. S. (2022). Prospects of refinery switching from conventional to integrated: An opportunity for sustainable investment in the petrochemical industry. In Fuel (Vol. 310). doi:10.1016/j.fuel.2021.122161

Albeldawi, M. (2023). Environmental impacts and mitigation measures of offshore oil and gas activities. In Developments in Petroleum Science (Vol. 78). doi:10.1016/B978-0-323-99285-5.00002-8

Allen, D. T., Chen, Q., & Dunn, J. B. (2021). Consistent Metrics Needed for Quantifying Methane Emissions from Upstream Oil and Gas Operations. *Environmental Science & Technology Letters*, *8*(4), 345–349. Advance online publication. doi:10.1021/acs.estlett.0c00907

Amin, I., Zailani, S., & Rahman, M. K. (2020). Predicting employees' engagement in environmental behaviours with supply chain firms. *Management Research Review*, *44*(6), 825–848. Advance online publication. doi:10.1108/MRR-05-2020-0280

Andersen, A. D., & Gulbrandsen, M. (2020). The innovation and industry dynamics of technology phase-out in sustainability transitions: Insights from diversifying petroleum technology suppliers in Norway. *Energy Research & Social Science*, *64*, 101447. Advance online publication. doi:10.1016/j.erss.2020.101447

Attia, A. M., Ghaithan, A. M., & Duffuaa, S. O. (2019). A multi-objective optimization model for tactical planning of upstream oil & gas supply chains. *Computers & Chemical Engineering*, *128*, 216–227. Advance online publication. doi:10.1016/j.compchemeng.2019.06.016

Basile, V., Capobianco, N., & Vona, R. (2021). The usefulness of sustainable business models: Analysis from oil and gas industry. *Corporate Social Responsibility and Environmental Management*, *28*(6), 1801–1821. Advance online publication. doi:10.1002/csr.2153

Beck, C., Rashidbeigi, S., Roelofsen, O., & Speelman, E. (2020). *The future of oil and gas is now: How companies can decarbonize.* McKinsey & Company.

Ben-Naceur, K. (2019). How the Oil and Gas Industry Is Contributing to Sustainability. *Journal of Petroleum Technology*, *71*(03), 38–39. Advance online publication. doi:10.2118/0319-0038-JPT

Bos, K., & Gupta, J. (2019). Stranded assets and stranded resources: Implications for climate change mitigation and global sustainable development. In Energy Research and Social Science (Vol. 56). doi:10.1016/j.erss.2019.05.025

Burrows, L. C., Haeri, F., Cvetic, P., Sanguinito, S., Shi, F., Tapriyal, D., Goodman, A., & Enick, R. M. (2020). A Literature Review of CO2, Natural Gas, and Water-Based Fluids for Enhanced Oil Recovery in Unconventional Reservoirs. *Energy & Fuels*, *34*(5), 5331–5380. Advance online publication. doi:10.1021/acs.energyfuels.9b03658

Buslaev, G., Morenov, V., Konyaev, Y., & Kraslawski, A. (2021). Reduction of carbon footprint of the production and field transport of high-viscosity oils in the Arctic region. *Chemical Engineering and Processing*, *159*, 108189. Advance online publication. doi:10.1016/j.cep.2020.108189

Cabrera, J., Irfan, M., Dai, Y., Zhang, P., Zong, Y., & Liu, X. (2021). Bioelectrochemical system as an innovative technology for treatment of produced water from oil and gas industry: A review. *Chemosphere*, *285*, 131428. Advance online publication. doi:10.1016/j.chemosphere.2021.131428 PMID:34237499

Chen, K. C., Kulkarni, P., McMullen, P., Biddick, D., Nader, L., & Sellers, C. (2019). High reliability protector-less artificial lift technology for electrical submersible pumps/compressors. *Society of Petroleum Engineers - SPE Gulf Coast Section Electric Submersible Pumps Symposium 2019, ESP 2019*. 10.2118/194391-MS

Chen, Q., Dunn, J. B., & Allen, D. T. (2019). Aggregation and Allocation of Greenhouse Gas Emissions in Oil and Gas Production: Implications for Life-Cycle Greenhouse Gas Burdens. *ACS Sustainable Chemistry & Engineering*, *7*(20), 17065–17073. Advance online publication. doi:10.1021/acssuschemeng.9b03136

Chen, X., Shao, N., Zhang, H., Fang, W., Xu, X., Xu, T., Lv, H., & Li, Y. (2022). A Green and Low-Carbon Transformation of Oil and Gas Companies. *Advances in Transdisciplinary Engineering*, *23*. Advance online publication. doi:10.3233/ATDE220317

Chen, X., Wang, M., Wang, B., Hao, H., Shi, H., Wu, Z., Chen, J., Gai, L., Tao, H., Zhu, B., & Wang, B. (2023). Energy Consumption Reduction and Sustainable Development for Oil & Gas Transport and Storage Engineering. In Energies (Vol. 16, Issue 4). doi:10.3390/en16041775

Cherepovitsyn, A., Rutenko, E., & Solovyova, V. (2021). Sustainable development of oil and gas resources: A system of environmental, socio-economic, and innovation indicators. *Journal of Marine Science and Engineering*, *9*(11), 1307. Advance online publication. doi:10.3390/jmse9111307

Cooper, C. M., McCall, J., Stokes, S. C., McKay, C., Bentley, M. J., Rosenblum, J. S., Blewett, T. A., Huang, Z., Miara, A., Talmadge, M., Evans, A., Sitterley, K. A., Kurup, P., Stokes-Draut, J. R., Macknick, J., Borch, T., Cath, T. Y., & Katz, L. E. (2022). Oil and Gas Produced Water Reuse: Opportunities, Treatment Needs, and Challenges. In ACS ES and T Engineering (Vol. 2, Issue 3). doi:10.1021/acsestengg.1c00248

Cox, D. R., Newton, A. M. W., & Huuse, M. (2020). An introduction to seismic reflection data: Acquisition, processing and interpretation. In Regional Geology and Tectonics: Volume 1: Principles of Geologic Analysis. doi:10.1016/B978-0-444-64134-2.00020-1

Delefosse, M., Jacobsen, M. W., Balle, J. D., Hansen, B. K., Middelboe, A. L., Nielsen, E. E., & Teilmann, J. (2020). Marine mammal biodiversity around oil and gas platforms - Challenges and successes of long-term monitoring. *Society of Petroleum Engineers - SPE International Conference and Exhibition on Health, Safety, Environment, and Sustainability 2020, HSE and Sustainability 2020*. 10.2118/199479-MS

Desai, J. N., Pandian, S., & Vij, R. K. (2021). Big data analytics in upstream oil and gas industries for sustainable exploration and development: A review. In Environmental Technology and Innovation (Vol. 21). doi:10.1016/j.eti.2020.101186

Diaz, C. D. P., Hoyos, C. C. M., Páez, E. G. M., & Martin, C. A. G. (2022). Wind Potencial as an Oportunity for Energy Transition in Oil and Gas Industry: Colombian Caribbean Offshore Case of Study. *Proceedings of the Annual Offshore Technology Conference*. 10.4043/31911-MS

Dickinson, S., Plisson-Saune, S., & Mani, P. (2020). Establishing public biodiversity commitments within an oil & gas company. *Society of Petroleum Engineers - SPE International Conference and Exhibition on Health, Safety, Environment, and Sustainability 2020, HSE and Sustainability 2020*. 10.2118/199508-MS

Didenko, N., Skripnuk, D., Merkulov, V., Kikkas, K. N., & Skripniuk, K. (2023). Methodology for the Formation of a Digital Model of the Life Cycle of an Offshore Oil and Gas Platform. *Resources*, *12*(8), 86. Advance online publication. doi:10.3390/resources12080086

Donaghy, T. Q., Healy, N., Jiang, C. Y., & Battle, C. P. (2023). Fossil fuel racism in the United States: How phasing out coal, oil, and gas can protect communities. In Energy Research and Social Science (Vol. 100). doi:10.1016/j.erss.2023.103104

Ekardt, F., Bärenwaldt, M., & Heyl, K. (2022). The Paris Target, Human Rights, and IPCC Weaknesses: Legal Arguments in Favour of Smaller Carbon Budgets. *Environments, 9*(9). doi:10.3390/environments9090112

Elijah, O., Ling, P. A., Rahim, S. K. A., Geok, T. K., Arsad, A., Kadir, E. A., Abdurrahman, M., Junin, R., Agi, A., & Abdulfatah, M. Y. (2021). *A Survey on Industry 4.0 for the Oil and Gas Industry: Upstream Sector*. IEEE. doi:10.1109/ACCESS.2021.3121302

Environmental Accounting and Financial Performance of Oil and Gas Companies in Nigeria. (2019). *Research Journal of Finance and Accounting*. doi:10.7176/RJFA/10-10-21

Foster, G. (2020). Circular economy strategies for adaptive reuse of cultural heritage buildings to reduce environmental impacts. *Resources, Conservation and Recycling*, *152*, 104507. Advance online publication. doi:10.1016/j.resconrec.2019.104507

Fox, T. A., Barchyn, T. E., Risk, D., Ravikumar, A. P., & Hugenholtz, C. H. (2019). A review of close-range and screening technologies for mitigating fugitive methane emissions in upstream oil and gas. In Environmental Research Letters (Vol. 14, Issue 5). doi:10.1088/1748-9326/ab0cc3

Garcia-Gonzales, D. A., Shonkoff, S. B. C., Hays, J., & Jerrett, M. (2019). Hazardous Air Pollutants Associated with Upstream Oil and Natural Gas Development: A Critical Synthesis of Current Peer-Reviewed Literature. In Annual Review of Public Health (Vol. 40). doi:10.1146/annurev-publhealth-040218-043715

Ghasem, N. (2020). CO2 removal from natural gas. In *Advances in Carbon Capture*. Methods, Technologies and Applications. doi:10.1016/B978-0-12-819657-1.00021-9

Gonzalez, D. J. X., Francis, C. K., Shaw, G. M., Cullen, M. R., Baiocchi, M., & Burke, M. (2022). Upstream oil and gas production and ambient air pollution in California. *The Science of the Total Environment*, *806*, 150298. Advance online publication. doi:10.1016/j.scitotenv.2021.150298 PMID:34844318

Green, J., Hadden, J., Hale, T., & Mahdavi, P. (2022). Transition, hedge, or resist? Understanding political and economic behavior toward decarbonization in the oil and gas industry. *Review of International Political Economy*, *29*(6), 2036–2063. Advance online publication. doi:10.1080/09692290.2021.1946708

Guerrero-Martin, C. A., Ortega-Ramírez, A. T., Rodríguez, P. A. P., López, S. J. R., Guerrero-Martin, L. E., Salinas-Silva, R., & Camacho-Galindo, S. (2023). Analysis of Environmental Sustainability through a Weighting Matrix in the Oil and Gas Industry. *Sustainability (Basel)*, *15*(11), 9063. Advance online publication. doi:10.3390u15119063

Hartmann, J., Inkpen, A. C., & Ramaswamy, K. (2021). Different shades of green: Global oil and gas companies and renewable energy. *Journal of International Business Studies*, *52*(5), 879–903. Advance online publication. doi:10.105741267-020-00326-w

Hashim, H. H., Abdul Talip, N. A., Muhamad Pikri, M. H., Abu Bakar, N. A., & Wan Zakaria, A. (2022). Sustainable Methane Emissions Management. *IOP Conference Series. Materials Science and Engineering*, *1257*(1), 012013. Advance online publication. doi:10.1088/1757-899X/1257/1/012013

Hassan, A., Mahmoud, M., Bageri, B. S., Aljawad, M. S., Kamal, M. S., Barri, A. A., & Hussein, I. A. (2020). Applications of chelating agents in the upstream oil and gas industry: A review. In Energy and Fuels (Vol. 34, Issue 12). doi:10.1021/acs.energyfuels.0c03279

Heidari, H., Akbari, M., Souhankar, A., & Hafezi, R. (2022). Review of global energy trends towards 2040 and recommendations for Iran oil and gas sector. *International Journal of Environmental Science and Technology*, *19*(8), 8007–8018. Advance online publication. doi:10.100713762-022-03963-w

Herbert, W., Nwaorgu, I., Onyilo, F., & Iormbagah, J. (2020). Sustainability Reporting and Performance of Listed Upstream Oil and Gas Firms in Nigeria: A Content Evaluation Approach. *International Journal of Applied Economics. Finance and Accounting*, *8*(1). Advance online publication. doi:10.33094/8.2017.2020.81.46.61

Hunt, J. D., Nascimento, A., Nascimento, N., Vieira, L. W., & Romero, O. J. (2022). Possible pathways for oil and gas companies in a sustainable future: From the perspective of a hydrogen economy. *Renewable & Sustainable Energy Reviews*, *160*, 112291. Advance online publication. doi:10.1016/j.rser.2022.112291

Ionescu, L. (2021). Transitioning to a low-carbon economy: Green financial behavior, climate change mitigation, and environmental energy sustainability. *Geopolitics, History, and International Relations*, *13*(1), 86. Advance online publication. doi:10.22381/GHIR13120218

Islam, M. R., & Hossain, M. E. (2020). Drilling Engineering: Towards achieving total sustainability. In Drilling Engineering: Towards Achieving Total Sustainability. doi:10.1016/C2019-0-00943-0

Jia, C. (2020). Development challenges and future scientific and technological researches in China's petroleum industry upstream. *Shiyou Xuebao. Acta Petrolei Sinica*, *41*(12). Advance online publication. doi:10.7623yxb202012001

Johnston, J. E., Lim, E., & Roh, H. (2019). Impact of upstream oil extraction and environmental public health: A review of the evidence. In Science of the Total Environment (Vol. 657). doi:10.1016/j.scitotenv.2018.11.483

Koroteev, D., & Tekic, Z. (2021). Artificial intelligence in oil and gas upstream: Trends, challenges, and scenarios for the future. *Energy and AI*, *3*, 100041. Advance online publication. doi:10.1016/j.egyai.2020.100041

Kulanov, A., Issakhova, A., Koshkina, O., Issakhova, P., & Karshalova, A. (2020). Venture financing and the fuel and energy complex: Investing in alternative energy. *International Journal of Energy Economics and Policy*, *10*(5), 531–538. Advance online publication. doi:10.32479/ijeep.9963

Kwarto, F., Nurafiah, N., Suharman, H., & Dahlan, M. (2022). *The potential bias for sustainability reporting of global upstream oil and gas companies: a systematic literature review of the evidence*. Management Review Quarterly. doi:10.100711301-022-00292-7

Kwarto, F., Nurafiah, N., Suharman, H., & Dahlan, M. (2023). Sustainability Reporting Reliability: An Industry Worker's Perspective In Indonesia. *International Journal of Business*, *28*(3). Advance online publication. doi:10.55802/IJB.028(3).002

Laturkar, K., & Laturkar, K. (2023). *Environmental Sustainability Measures in the Oil and Gas Industry*. doi:10.4018/978-1-6684-8223-0.ch008

Lee, J., Sorensen, C., Lemery, J., Workman, C. F., Linstadt, H., & Bazilian, M. D. (2022). Managing upstream oil and gas emissions: A public health oriented approach. In Journal of Environmental Management (Vol. 310). doi:10.1016/j.jenvman.2022.114766

Li, M., Wiedmann, T., & Hadjikakou, M. (2019). Enabling Full Supply Chain Corporate Responsibility: Scope 3 Emissions Targets for Ambitious Climate Change Mitigation. *Environmental Science & Technology*. Advance online publication. doi:10.1021/acs.est.9b05245 PMID:31766851

Litvinenko, V., Bowbrick, I., Naumov, I., & Zaitseva, Z. (2022). Global guidelines and requirements for professional competencies of natural resource extraction engineers: Implications for ESG principles and sustainable development goals. In Journal of Cleaner Production (Vol. 338). doi:10.1016/j.jclepro.2022.130530

Maniatis, K., Chiaramonti, D., & van den Heuvel, E. (2021). Post covid-19 recovery and 2050 climate change targets: Changing the emphasis from promotion of renewables to mandated curtailment of fossil fuels in the eu policies. *Energies*, *14*(5), 1347. Advance online publication. doi:10.3390/en14051347

Min, J., Yan, G., Abed, A. M., Elattar, S., Amine Khadimallah, M., Jan, A., & Elhosiny Ali, H. (2022). The effect of carbon dioxide emissions on the building energy efficiency. *Fuel*, *326*, 124842. Advance online publication. doi:10.1016/j.fuel.2022.124842

Mufutau Opeyemi, B. (2021). Path to sustainable energy consumption: The possibility of substituting renewable energy for non-renewable energy. *Energy*, *228*, 120519. Advance online publication. doi:10.1016/j.energy.2021.120519

Naji, G. M. A., Isha, A. S. N., Alazzani, A., Brough, P., Saleem, M. S., Mohyaldinn, M. E., & Alzoraiki, M. (2022). Do Leadership, Organizational Communication, and Work Environment Impact Employees' Psychosocial Hazards in the Oil and Gas Industry? *International Journal of Environmental Research and Public Health*, *19*(8), 4432. Advance online publication. doi:10.3390/ijerph19084432 PMID:35457300

Norouzi, N. (2021). Post-COVID-19 and globalization of oil and natural gas trade: Challenges, opportunities, lessons, regulations, and strategies. In International Journal of Energy Research (Vol. 45, Issue 10). doi:10.1002/er.6762

Ogolo, N. A., Anih, O. C., & Onyekonwu, M. O. (2022). Sources and effects of environmental pollution from oil and gas industrial operations. *Arabian Journal of Chemical and Environmental Research, 9*(1).

Ogunbiyi, O., Al-Rewaily, R., Saththasivam, J., Lawler, J., & Liu, Z. (2023). Oil spill management to prevent desalination plant shutdown from the perspectives of offshore cleanup, seawater intake and onshore pretreatment. In Desalination (Vol. 564). doi:10.1016/j.desal.2023.116780

Okere, C. J., & Sheng, J. J. (2023). Review on clean hydrogen generation from petroleum reservoirs: Fundamentals, mechanisms, and field applications. *International Journal of Hydrogen Energy*. Advance online publication. doi:10.1016/j.ijhydene.2023.06.135

Pickl, M. J. (2019). The renewable energy strategies of oil majors – From oil to energy? *Energy Strategy Reviews*, *26*, 100370. Advance online publication. doi:10.1016/j.esr.2019.100370

Radnejad, A. B., Osiyevskyy, O., & Vredenburg, H. (2020). Barriers to radical process innovation: A case of environmental technology in the oil industry. *Journal of Strategy and Management*, *13*(4), 453–476. Advance online publication. doi:10.1108/JSMA-11-2019-0206

Rafiee, J., Sarma, P., Gutierrez, F., Hilliard, R., Calad, C. M., Angulo, O., & Boyer, B. (2022). Energy Transition Meets Digital Transformation: Design and Implementation of a Comprehensive Carbon Emissions Estimation and Forecasting Platform. *Proceedings of the Annual Offshore Technology Conference*. 10.4043/31747-MS

Roberts, R., & Flin, R. (2020). Unlocking the Potential: Understanding the Psychological Factors That Influence Technology Adoption in the Upstream Oil and Gas Industry. *SPE Journal*, *25*(1), 515–528. Advance online publication. doi:10.2118/198903-PA

Robertson, J. O., & Chilingar, G. V. (2017). Environmental Aspects of Oil and Gas Production. In Environmental Aspects of Oil and Gas Production. doi:10.1002/9781119117421

Santibanez-Borda, E., Korre, A., Nie, Z., & Durucan, S. (2021). A multi-objective optimisation model to reduce greenhouse gas emissions and costs in offshore natural gas upstream chains. *Journal of Cleaner Production*, *297*, 126625. Advance online publication. doi:10.1016/j.jclepro.2021.126625

Soeder, D. J. (2021a). Balancing Energy, Environment, and Economics. In Fracking and the Environment. doi:10.1007/978-3-030-59121-2_11

Soeder, D. J. (2021b). Fracking and the Environment. In *Fracking and the Environment*. Springer International Publishing. doi:10.1007/978-3-030-59121-2_6

Sun, Y., Yang, Y., Huang, N., & Zou, X. (2020). The impacts of climate change risks on financial performance of mining industry: Evidence from listed companies in China. *Resources Policy*, *69*, 101828. Advance online publication. doi:10.1016/j.resourpol.2020.101828

Thakur, A., & Koul, B. (2022). Impact of oil exploration and spillage on marine environments. In Advances in Oil-Water Separation. doi:10.1016/B978-0-323-89978-9.00018-5

Tontiwachwuthikul, P., Chan, C. W., Zeng, F. B., Liang, Z. H., Sema, T., & Min, C. (2020). Recent progress and new developments of applications of artificial intelligence (AI), knowledge-based systems (KBS), and Machine Learning (ML) in the petroleum industry. *Petroleum*, *6*(4), 319–320. Advance online publication. doi:10.1016/j.petlm.2020.08.001

Wadanambi, R. T., Wandana, L. S., Chathumini, K. K. G. L., Dassanayake, N. P., Preethika, D. D. P., & Arachchige, U. S. P. R. (2020). The effects of industrialization on climate change. *Journal of Research Technology and Engineering*, *1*(4).

Walsh, K. B., & Rose, J. (2022). A Review of Restoration Techniques and Outcomes for Rangelands Affected by Oil and Gas Production in North America. *Ecological Restoration*, *40*(4), 259–269. Advance online publication. doi:10.3368/er.40.4.259

Weitekamp, C. A., Stevens, T., Stewart, M. J., Bhave, P., & Gilmour, M. I. (2020). Health effects from freshly emitted versus oxidatively or photochemically aged air pollutants. In Science of the Total Environment (Vol. 704). doi:10.1016/j.scitotenv.2019.135772

Wilson, I., Saini, S., Sreenivasan, H., Sahu, C., Krishna, S., & Gupta, P. (2023). Review and Perspectives of Energy-Efficient Methane Production from Natural Gas Hydrate Reservoirs Using Carbon Dioxide Exchange Technology. In Energy and Fuels. doi:10.1021/acs.energyfuels.3c00715

Wollin, K. M., Damm, G., Foth, H., Freyberger, A., Gebel, T., Mangerich, A., Gundert-Remy, U., Partosch, F., Röhl, C., Schupp, T., & Hengstler, J. G. (2020). Critical evaluation of human health risks due to hydraulic fracturing in natural gas and petroleum production. In Archives of Toxicology (Vol. 94, Issue 4). doi:10.100700204-020-02758-7

Yin, C. (2021). International law regulation of offshore oil and gas exploitation. *Environmental Impact Assessment Review*, *88*, 106551. Advance online publication. doi:10.1016/j.eiar.2021.106551

Yousuf, N., Olayiwola, O., Guo, B., & Liu, N. (2021). A comprehensive review on the loss of wellbore integrity due to cement failure and available remedial methods. In Journal of Petroleum Science and Engineering (Vol. 207). doi:10.1016/j.petrol.2021.109123

Zimmerle, D., Vaughn, T., Bell, C., Bennett, K., Deshmukh, P., & Thoma, E. (2020). Detection Limits of Optical Gas Imaging for Natural Gas Leak Detection in Realistic Controlled Conditions. *Environmental Science & Technology*, *54*(18), 11506–11514. Advance online publication. doi:10.1021/acs.est.0c01285 PMID:32786569

KEY TERMS AND DEFINITIONS

Carbon Capture and Storage (CCS): It is the process of capturing CO_2 emissions from industrial processes, such as oil and gas production, and then storing them. CCS is a potential solution to reduce the amount of CO_2 in the atmosphere, thus helping to reduce the effects of climate change. It can be used in conjunction with other climate change mitigation strategies, such as renewable energy, to help reduce the overall emissions of greenhouse gases.

Energy Efficiency: By using less energy, the same tasks can be performed and results achieved, reducing energy waste. This reduces emissions, air pollution, and energy costs. It also helps to conserve natural resources and protect the environment. Energy efficiency also contributes to improved economic performance, resulting in increased competitiveness in the global marketplace.

Greenhouse Gas (GHG) Emissions: Upstream Oil and gas operations contribute to the release of greenhouse gases such as carbon dioxide (CO_2). By releasing large amounts of CO_2 and other GHGs into the atmosphere, these activities contribute to global warming, which can have a range of impacts on the environment. This includes melting glaciers, rising sea levels, extreme weather patterns, and droughts.

Regulatory Compliance: Environmental, safety, and social practices governing the upstream oil and gas industry are adhered to by following laws, regulations, and standards. Companies must ensure compliance with local laws and regulations in order to protect their reputation and brand. Additionally, it helps to ensure that operations are conducted safely and responsibly.

Renewable Energy: Solar, wind, and hydropower are examples of renewable energy sources. Adopting renewable energy is often viewed as a sustainable alternative to fossil fuels. Renewable energy sources have a lower environmental impact than fossil fuels, and can reduce the dependence on fossil fuels in the long-term. The use of renewable energy can also help reduce air pollution and greenhouse gas emissions.

Sustainability: In order to meet current energy needs without compromising future energy demands, steps must be taken to ensure future generations can meet theirs as well. Oil and gas sustainability aims to minimize environmental and social impacts as well as ensure long-term success. This includes minimizing waste and emissions, protecting water resources, and investing in renewable sources of energy.

Upstream Oil and Gas Industry: Initial stage in the oil and gas industry that involves exploration, drilling, and production. Exploration includes seismic surveying and exploratory drilling. Drilling involves creating wells and installing production infrastructure. Production is the process of retrieving the oil and gas from the reservoir and transporting it to refineries or processing facilities.

Chapter 13
Eco-Friendly Colors:
A Review of Recent Advances in Natural Dyes

Rupayan Roy
Department of Fashion Technology, National Institute of Fashion Technology, Kannur, India

ABSTRACT

This chapter reviews recent developments in natural dyes, highlighting the advancements made in natural dye research, benefits of using modern natural dyes, examples of recent natural dyes and the colors they produce, and the methods used to extract and process these modern natural dyes. Additionally, the sustainability and environmental impact of using natural dyes are discussed, including a comparison and contrast to synthetic dyes, and the importance of using eco-friendly dyes in the fashion industry is emphasized. The chapter concludes with potential future developments in natural dye research, the benefits of using natural dyes, and their importance for sustainability. Overall, the chapter emphasizes the significance of recent developments in natural dyes and their potential impact on the fashion industry.

1.0 INTRODUCTION

One of the most environmentally damaging sectors of the global economy is the textile sector. Given that they are frequently made of toxic compounds that can be damaging to the environment as well as human health, synthetic dyes used in the production of textiles play a significant role in this impact (Asia InCH, 2023; Roy Choudhury, 2014; Selvaraj et al., 2021). The use of natural dyes has drawn more attention as a more environmentally responsible solution in response to these worries. The usage of natural dyes, which are produced from plants, animals, and minerals, dates back thousands of years and is used in several cultures throughout (Kadolph, 2008; Shahid & Mohammad, 2013). Natural dyes offer local communities involved in their manufacturing economic opportunities in addition to reducing the textile industry's negative environmental effects. A little sample of natural dye is shown in Figure 1.

Natural dyes have been used to colour fabrics for thousands of years and have a lengthy history of use in many different cultural traditions all around the world. Prior to the development of synthetic dyes

DOI: 10.4018/978-1-6684-9979-5.ch013

Figure 1. Various natural dye sources (Asia InCH, 2023)

in the 19th century, natural dyes were the only option for dying fabrics (Ferreira et al., 2004; Mussak & Bechtold, 2009; Tamburini et al., 2020). However, there has been a resurgence in interest in natural dyes in recent years due to growing environmental and health concerns around synthetic dyes. Natural dyes provide a number of advantages over synthetic dyes, such as less of an impact on the environment, the use of renewable resources, and potential health advantages (Saxena & Raja, 2014; Samanta & Konar, 2011). In a number of ways, using natural colours can lessen the environmental effect of textile production. The method of extracting natural dyes is frequently less resource-intensive than the process of creating synthetic dyes, in part because natural dyes are frequently generated from renewable resources such as plants, which may be grown without the use of hazardous chemicals. Last but not least, because natural colours are frequently biodegradable, they do not add to the buildup of dangerous chemicals in the environment (Egan & Salmon, 2022).

Natural dyes have a long history of traditional use, and different civilizations have developed distinctive techniques for extracting and utilising natural colours. For instance, indigo has been used for centuries to dye blue clothes in India and other regions of Asia (Balfour-Paul, 1999). Cochineal insects are used to create a vivid red dye in South America. Many tribes have used natural dyes as a way to show their cultural identity and maintain long-standing customs (Endensor, 2004; Mungmachon, 2012). New and improved natural dyes have been created as a result of recent developments in natural dye research. One illustration is the use of nanotechnology to produce natural dyes with a larger variety of hues and greater durability (Scotter, 2011). New techniques for extracting natural dyes have also been developed by researchers, such as employing enzymes to break down plant fibres and liberate pigment molecules. As a practical substitute for synthetic dyes, these advancements have rekindled interest in natural dyes among manufacturers and designers (Kristensen et al., 2016; Mariotti et al., 2020).

Depending on the raw material, several extraction and processing techniques are employed to create current natural colours. For instance, water, alcohol, or other solvents are frequently used to extract plant-based colours (Prakash et al., 2022). Different extraction techniques, such as boiling the insects that create the pigment, may be necessary for animal-based dyes. After the dye has been removed, it may undergo additional processing to get rid of impurities and intensify the colour (Chu, 2001). When contrasting natural and synthetic colours, it's crucial to take sustainability and environmental impact into account. Although natural dyes are typically seen as more environmentally benign and sustainable than synthetic dyes, there are still some negative effects on the environment related to their manufacture and use (Arora et al., 2017; Handayaniet al., 2018).

This article seeks to offer a comprehensive summary of recent developments in natural dye research and their importance to the apparel sector. The advantages of employing natural dyes over synthetic dyes, the historical use of natural dyes, recent advances in natural dye research, and the procedures used to extract and refine contemporary natural dyes will all be covered in this article. The use of natural colours and synthetic dyes will also be contrasted and compared for sustainability and environmental impact, emphasising the significance of eco-friendly methods in the fashion sector. The overall goal of this article is to give readers a thorough overview of the most recent advancements in natural dyes. Readers will have a greater understanding of the significance of natural dyes and their potential influence on the fashion industry by looking at traditional natural dyes, contemporary developments, sustainability and environmental impact, and future directions.

2.0 TRADITIONAL NATURAL DYES

Natural dyes have cultural value and have been used for thousands of years. They have been utilised for a number of different things, including food, cosmetics, and textile colouring (Ekin, 2005; Dikshit & Tallapragada, 2018). This article will cover the background and conventional applications of natural dyes, as well as the benefits and drawbacks of doing so. It will also give samples of conventional natural dyes and the colours they produce. The Indus Valley Civilization's use of madder, indigo, and turmeric to colour cotton garments provides the earliest documentation of natural dyeing, which dates back to 2600 BCE (Kulkarni et al., 2011). Natural dyes were employed to colour textiles and create cosmetics in ancient Egypt, and in ancient Rome, they were used to tint the emperors' robes. Natural dyes were employed to dye garments, leather, and even feathers in other cultures, such as those in Asia, Africa, and the Americas (Abel, 2012; Linh, 2009).

2.1 Advantages of Traditional Natural Dyes

The fact that traditional natural dyes have cultural and traditional importance is one benefit of employing them. Many civilizations have carried down the art of natural dyeing from generation to generation, and the use of natural dyes is frequently connected to traditional attire and ceremonies. Traditional natural dyes can also produce distinctive hues and tones that are frequently challenging to duplicate with synthetic dyes. (Maxwell, 2012; Widiawati et al., 2012)

2.2 Disadvantages of Traditional Natural Dyes

Nevertheless, there are several drawbacks to utilising conventional natural colours. The fact that they can be challenging to extract and process might make them expensive and time-consuming to use, which is one of the major issues. Additionally, the use of natural dyes may be restricted in some locations due to regional and seasonal availability issues (Linchant et al., 2015; Kasiri & Safapour, 2014).

2.3 Examples of Traditional Natural Dyes

Traditional natural dyes come in a wide variety and have been used for centuries. Indigo is a rich blue dye made from the leaves of the indigo plant that is renowned for its ability to produce a variety of blue colours and resilience to fading (Stasiak et al., 2014). The madder plant, which was well-liked in Europe throughout the Middle Ages, yields the red dye known as madder from its roots. Henna is a reddish-orange dye made from the leaves of the henna plant that has been used for hundreds of years to colour hair, skin, and nails throughout India, the Middle East, and North Africa. The Aztecs and other American cultures utilised cochineal, a crimson dye made from the dried corpses of the cochineal bug, to colour fabrics and other items (Adeel et al., 2018). Since ancient times, India and other regions of Asia have used turmeric, a yellow dye made from the turmeric plant's root, to colour clothing and create cosmetics.

2.4 Colors Produced by Traditional Natural Dyes

Reds, blues, greens, yellows, and browns are just a few of the colours that may be created by using these natural dyes. The type of material being dyed, the dye's concentration, and the processing technique utilised can all affect the colours that natural dyes generate (Gürses et al., 2016; Sigurdson et al., 2017; Wrolstad & Culver, 2012). The conventional natural dyes have a lengthy history and are still useful today. They produce distinctive hues and are significant culturally. However, there are drawbacks to their use, including cost and a lack of supply. Traditional natural dyes offer an overall intriguing window into history and the variety of human cultures.

3.0 RECENT DEVELOPMENTS IN NATURAL DYES

Because they are sustainable and eco-friendly, natural dyes have experienced a recent spike in popularity. Modern natural dyes have been created as a result of considerable research developments in the field of natural dyes (Samanta, 2020). In addition to highlighting the advantages of using contemporary natural dyes, this article will go over recent developments in natural dye research, give examples of contemporary natural dyes and the colours they produce, and go over the techniques used to extract and refine these contemporary natural dyes (Cesaratto et al., 2019). Some recent advancements in natural dyes are shown in Figure 2.

3.1 Advancements in Natural Dye Research

Natural dye research has recently made strides that have improved the colorfastness, stability, and accessibility of natural dyes. To increase yields and lower prices, researchers have found new sources of natural dyes and enhanced extraction techniques (Bhuiyan et al., 2017). They have also experimented with blending natural dyes with other substances to produce new hues and improve colorfastness.

3.2 Benefits of Using Modern Natural Dyes

The fact that current natural dyes are sustainable and kind to the environment is one of their key advantages. Natural colours don't create harmful waste or add to pollution like manufactured dyes do (Křížová, 2015). Utilizing natural dyes also helps the communities and local farmers who create them. Modern natural dyes are more adaptable and appropriate for contemporary fashion and design since they provide a greater variety of colours and tones than traditional natural dyes (Behera et al., 2019).

Figure 2. A few newly investigated dyes turmeric, the rind of a pomegranate, marigolds, dragonfruit

3.3 Examples of Recent Natural Dyes

Some examples of recent natural dyes include:

Avocado: Avocado pits and skins can be used to produce a range of pink and peach shades (Tesfaye et al., 2022).

Onion: Onion skins can produce warm yellow and orange hues (Zubairu & Mshelia, 2015).

Blackcurrant: Blackcurrant leaves and stems can produce a range of greens, from light to dark (Yasukawa et al., 2017).

Safflower: Safflower petals can produce bright shades of pink and orange (Zarkogianni et al., 2011).

Logwood: Logwood can produce deep shades of purple and blue (Zarkogianni et al., 2011).

3.4 Commercial Advantages

Natural dyes such as Avocado, Onion, Blackcurrant, Safflower, and Logwood offer several commercial advantages. They provide unique and distinctive color palettes that can set products apart in the market. Moreover, these dyes align with the increasing demand for eco-friendly and sustainable products, making them appealing to environmentally conscious consumers. Brands can also leverage the safety and health aspects of natural dyes, emphasizing their non-toxic nature. Additionally, the cultural significance and artisanal appeal of these dyes can resonate with consumers interested in traditional craftsmanship. Brands can position themselves in premium and niche markets, catering to consumers willing to pay a premium for sustainable and distinct items. Furthermore, the transparency and traceability of natural dyes can build trust among consumers who value ethical and sustainable practices. Finally, compliance with stringent environmental regulations and alignment with prevailing market trends make natural dyes a compelling choice for businesses seeking commercial success with a conscience (Behara et al., 2019; Křížová, 2015).

3.5 Methods for Extracting and Processing Modern Natural Dyes

Modern natural dyes are extracted and processed using a variety of techniques depending on the colour and the material being dyed. Natural dyes are often extracted with the aid of water or solvents from plant, animal, or mineral sources (Samanta & Konar, 2011). The dye is then painted or immersed onto the material, fixed with heat or other chemicals, and left to dry. Mordants, which are compounds that aid in the dye's adhesion to the cloth and enhance colorfastness, are frequently mixed with modern natural dyes (Ado et al., 2014).

It's interesting to note that recent advances in natural dye research have led to the creation of contemporary natural dyes that are sustainable, eco-friendly, and adaptable (Venil et al., 2020). These dyes, which may be extracted and processed using a number of techniques, provide a larger spectrum of colours and shades than conventional natural dyes (Bechtold et al., 2003). It is expected that demand for contemporary natural dyes will continue to rise as awareness of the negative social and environmental effects of synthetic dyes rises, spurring additional study and innovation in this area.

4.0 SUSTAINABILITY AND ENVIRONMENTAL IMPACT

Environmental impact and sustainability are important considerations in the textile and apparel industries. Natural dyes are a sustainable and environmentally friendly alternative to synthetic dyes, and the environment is a crucial factor to take into account (Muthu, 2014). This article compares and contrasts the environmental effects of natural dyes and synthetic dyes, discusses the significance of utilising eco-friendly dyes in the fashion business, and discusses the sustainability and environmental impact of using natural dyes.

4.1 Sustainability and Environmental Impact of Using Natural Dyes

Because natural dyes are made from renewable resources including plants, animals, and minerals, their use is environmentally friendly. Natural dyes don't include hazardous chemicals or heavy metals, thus their manufacture has less of an impact on the environment than synthetic dye production (Yusuf, 2019). Due to their frequent local production and lack of energy-intensive processing, natural dyes also have a minimal carbon footprint. Natural dyes are produced by regional farmers and communities, thus using them benefits both the economy and society.

4.2 Comparison of Environmental Impact of Natural Dyes and Synthetic Dyes

Due to the hazardous substances and heavy metals employed in their manufacture, synthetic dyes have a substantial negative influence on the environment. These substances pose substantial health risks to both workers and people living close to dyeing operations since they can pollute the air, water, and land. Additionally energy- and water-intensive to create, synthetic dyes need a lot of water for washing and processing. Natural dyes, on the other hand, have a lower environmental impact and are biodegradable, making them a more environmentally responsible choice (Chavan, 2013; Mukherjee, 2015).

4.3 Importance of Using Eco-Friendly Dyes in the Fashion Industry

The fashion sector is one of the biggest global pollutants, and it has a big effect on the environment. This environmental impact is a result of the usage of synthetic dyes in the textile industry (Sarkar et al., 2017). The fashion industry has a duty to implement sustainable techniques, such as using eco-friendly colours, to lessen its influence on the environment. The fashion sector can lessen its carbon impact, assist local economies, and advance sustainable agriculture by adopting natural colours. Additionally, utilising natural dyes can produce exquisite hues that are impossible to achieve with synthetic dyes, enhancing the worth and originality of garments (Rukhaya et al., 2021).

A sustainable and environmentally beneficial substitute for synthetic dyes in the manufacture of textiles is the use of natural dyes. Natural dyes produce distinctive and lovely hues, have little effect on the environment, and help local communities. Synthetic dyes, on the other hand, are unhealthy for communities and employees, as well as the environment (Parvin et al., 2020). Utilizing eco-friendly colours, such as natural dyes, is a critical first step in reducing the environmental effect of the fashion sector. It is anticipated that as environmental effect and sustainability become more widely understood, demand for eco-friendly dyes will rise as well, spurring research and development in this area (Bradu et al., 2022; Cai & Choi, 2020).

5.0 FUTURE DIRECTIONS

Due to their eco-friendliness, safety, and potential health benefits, natural dyes are becoming more popular. As customers become more conscious of the effects of synthetic colours on the environment and their health, the natural dye business has experienced tremendous growth in recent years (Rymbai et al., 2011; Olas et al., 2021). This article will summarise the advantages of using natural dyes and their significance for sustainability, discuss potential future developments in natural dye research, and offer some closing remarks on the significance of recent advancements in natural dyes and their potential effects on the fashion industry.

5.1 Potential Future Developments in Natural Dye Research

Research on natural dyes is still in its infancy, although major strides have been made recently. There are a number of possible areas for advancement in the current research into the use of natural dyes in textiles (Eadie & Ghosh, 2011; O'Connor et al., 2009). The improvement of natural dye extraction and processing techniques to boost production and enhance colour fastness is one field of research. In order to provide a more plentiful and sustainable source of natural dyes, researchers are also exploring new natural dye sources, such as microbes and algae (Kanchana et al., 2013; Kasiri & Safapour, 2014). A greater degree of control over the colour and design of textiles may also be made possible by the development of new dyeing techniques, such as digital printing and 3D printing.

5.2 Benefits of Using Natural Dyes and their Importance for Sustainability

Natural dyes have a variety of advantages, and people are beginning to see them as a viable and environmentally responsible alternative to synthetic dyes (Ebrahim et al., 2022). Natural colours are healthy for the environment and human health because they degrade naturally and don't contain harmful chemicals or heavy metals. In addition, they are produced locally and help farmers and local communities, leaving a less carbon footprint than synthetic dyes (Elgarahy et al., 2021; Ojuederie & babalola, 2017; Vardhan et al., 2019). Additionally, natural dyes offer distinctive and brilliant colours that synthetic dyes are unable to match. The use of natural dyes is a critical step towards the fashion industry's sustainability and has the potential to lessen the industry's environmental impact (Adrosko, 2012; Salma et al 2021; Samsami et al., 2020).

6.0 CONCLUSION

In summary, natural dyes offer a viable and environmentally friendly substitute for synthetic colours. There is ongoing study into the improvement of natural dye extraction and processing techniques as a result of the industry's recent tremendous growth. There may be new business opportunities for the industry as a result of the development of new natural dye supplies and dyeing methods. Natural dyes have a variety of advantages, including their eco-friendliness, safety, and distinctive colours. For the garment business to be sustainable and lessen its negative environmental effects, natural colours must be used. It is anticipated that as environmental effect and sustainability become more widely understood, demand for natural dyes will rise as well, spurring further research and development in this area.

REFERENCES

Abel, A. (2012). The history of dyes and pigments: From natural dyes to high performance pigments. In *Colour design* (pp. 557–587). Woodhead Publishing. doi:10.1016/B978-0-08-101270-3.00024-2

Adeel, S., Abrar, S., Kiran, S., Farooq, T., Gulzar, T., & Jamal, M. (2018). *Sustainable application of natural dyes in cosmetic industry. In Handbook of Renewable Materials for Coloration and Finishing*. John Wiley & Sons, Inc.

Ado, A., Yahaya, H., Kwalli, A. A., & Abdulkadir, R. S. (2014). Dyeing of textiles with eco-friendly natural dyes: A review. *International Journal of Environmental Monitoring and Protection*, *1*(5), 76–81.

Adrosko, R. J. (2012). *Natural dyes and home dyeing* (Vol. 281). Courier Corporation.

Arora, J., Agarwal, P., & Gupta, G. (2017). Rainbow of natural dyes on textiles using plants extracts: Sustainable and eco-friendly processes. *Green and Sustainable Chemistry*, *7*(1), 35–47. doi:10.4236/gsc.2017.71003

Asia InCH. (2023). *Natural Dye Sources*. Retrieved from https://asiainch.org/craft/natural-dye-sources/

Balfour-Paul, J. (1999). Indigo in South and South-East Asia. *Textile History*, *30*(1), 98–112. doi:10.1179/004049699793710688 PMID:22523776

Bechtold, T., Turcanu, A., Ganglberger, E., & Geissler, S. (2003). Natural dyes in modern textile dye-houses—How to combine experiences of two centuries to meet the demands of the future? *Journal of Cleaner Production*, *11*(5), 499–509. doi:10.1016/S0959-6526(02)00077-X

Behera, S., Khandual, A., & Luximon, Y. (2019). An Insight In To the Ikat Technology in India: Ancient To Modern Era. *IOSR Journal of Polymer and Textile Engineering*.

Bhuiyan, M. R., Islam, A., Ali, A., & Islam, M. N. (2017). Color and chemical constitution of natural dye henna (Lawsonia inermis L) and its application in the coloration of textiles. *Journal of Cleaner Production*, *167*, 14–22. doi:10.1016/j.jclepro.2017.08.142

Bradu, P., Biswas, A., Nair, C., Sreevalsakumar, S., Patil, M., Kannampuzha, S., Mukherjee, A. G., Wanjari, U. R., Renu, K., Vellingiri, B., & Gopalakrishnan, A. V. (2022). Recent advances in green technology and Industrial Revolution 4.0 for a sustainable future. *Environmental Science and Pollution Research International*, 1–32. doi:10.100711356-022-20024-4 PMID:35397034

Cai, Y. J., & Choi, T. M. (2020). A United Nations' Sustainable Development Goals perspective for sustainable textile and apparel supply chain management. *Transportation Research Part E, Logistics and Transportation Review*, *141*, 102010. doi:10.1016/j.tre.2020.102010 PMID:32834741

Cesaratto, A., Leona, M., & Pozzi, F. (2019). Recent advances on the analysis of polychrome works of art: SERS of synthetic colorants and their mixtures with natural dyes. *Frontiers in Chemistry*, *7*, 105. doi:10.3389/fchem.2019.00105 PMID:30886843

Chavan, R. B. (2013). *Health and environmental hazards of synthetic dyes*. Garment and Fashion Design at the University of Bahir Dar.

Chu, W. (2001). Dye removal from textile dye wastewater using recycled alum sludge. *Water Research*, *35*(13), 3147–3152. doi:10.1016/S0043-1354(01)00015-X PMID:11487111

Dikshit, R., & Tallapragada, P. (2018). Comparative study of natural and artificial flavoring agents and dyes. In *Natural and artificial flavoring agents and food dyes* (pp. 83–111). Academic Press. doi:10.1016/B978-0-12-811518-3.00003-X

Eadie, L., & Ghosh, T. K. (2011). Biomimicry in textiles: Past, present and potential. An overview. *Journal of the Royal Society, Interface*, *8*(59), 761–775. doi:10.1098/rsif.2010.0487 PMID:21325320

Ebrahim, S. A., Mosaad, M. M., Othman, H., & Hassabo, A. G. (2022). A valuable observation of eco-friendly natural dyes for valuable utilisation in the textile industry. *Journal of Textiles. Coloration and Polymer Science*, *19*(1), 25–37.

Edensor, T. (2004). Automobility and national identity: Representation, geography and driving practice. *Theory, Culture & Society*, *21*(4-5), 101–120. doi:10.1177/0263276404046063

Egan, J., & Salmon, S. (2022). Strategies and progress in synthetic textile fiber biodegradability. *SN Applied Sciences*, *4*(1), 1–36. doi:10.100742452-021-04851-7

Ekin, Z. (2005). Resurgence of safflower (Carthamus tinctorius L.) utilization: A global view. *Journal of Agronomy*, *4*(2), 83–87. doi:10.3923/ja.2005.83.87

Elgarahy, A. M., Elwakeel, K. Z., Mohammad, S. H., & Elshoubaky, G. A. (2021). A critical review of biosorption of dyes, heavy metals and metalloids from wastewater as an efficient and green process. *Cleaner Engineering and Technology*, *4*, 100209. doi:10.1016/j.clet.2021.100209

Ferreira, E. S., Hulme, A. N., McNab, H., & Quye, A. (2004). The natural constituents of historical textile dyes. *Chemical Society Reviews*, *33*(6), 329–336. doi:10.1039/b305697j PMID:15280965

Gürses, A., Açıkyıldız, M., Güneş, K., Gürses, M. S., Gürses, A., Açıkyıldız, M., ... Gürses, M. S. (2016). Historical development of colorants. *Dyes and Pigments*, 1–12.

Handayani, W., Kristijanto, A. I., & Hunga, A. I. R. (2018). Are natural dyes eco-friendly? A case study on water usage and wastewater characteristics of batik production by natural dyes application. *Sustainable Water Resources Management*, *4*(4), 1011–1021. doi:10.100740899-018-0217-9

Kadolph, S. (2008). Natural dyes: A traditional craft experiencing new attention. *Delta Kappa Gamma Bulletin*, *75*(1), 14.

Kanchana, R., Fernandes, A., Bhat, B., Budkule, S., Dessai, S., & Mohan, R. (2013). Dyeing of textiles with natural dyes-an eco-friendly approach. *International Journal of Chemtech Research*, *5*(5), 2102–2109.

Kasiri, M. B., & Safapour, S. (2014). Natural dyes and antimicrobials for green treatment of textiles. *Environmental Chemistry Letters*, *12*(1), 1–13. doi:10.100710311-013-0426-2

Kristensen, A., Yang, J. K., Bozhevolnyi, S. I., Link, S., Nordlander, P., Halas, N. J., & Mortensen, N. A. (2016). Plasmonic colour generation. *Nature Reviews. Materials*, *2*(1), 1–14. doi:10.1038/natrevmats.2016.88

Křížová, H. (2015). *Natural dyes: their past, present, future and sustainability. In Recent Developments in Fibrous Material Science*. Kosmas Publishing.

Kulkarni, S. S., Bodake, U. M., & Pathade, G. R. (2011). Extraction of Natural Dye from Chili (Capsicum Annum) for Textile Coloration. *Universal Journal of Environmental Research and Technology*, *1*(1).

Linchant, J., Lisein, J., Semeki, J., Lejeune, P., & Vermeulen, C. (2015). Are unmanned aircraft systems (UAS s) the future of wildlife monitoring? A review of accomplishments and challenges. *Mammal Review*, *45*(4), 239–252. doi:10.1111/mam.12046

Linh, H. T. (2009). Natural dyes in Eastern Asia (Vietnam and neighbouring countries). Handbook of Natural Colorants, 65-72.

Mariotti, N., Bonomo, M., Fagiolari, L., Barbero, N., Gerbaldi, C., Bella, F., & Barolo, C. (2020). Recent advances in eco-friendly and cost-effective materials towards sustainable dye-sensitized solar cells. *Green Chemistry*, *22*(21), 7168–7218. doi:10.1039/D0GC01148G

Maxwell, R. (2012). *Textiles of Southeast Asia: Trade, tradition and transformation*. Tuttle Publishing.

Mukherjee, S. (2015). Environmental and social impact of fashion: Towards an eco-friendly, ethical fashion. *International Journal of Interdisciplinary and Multidisciplinary Studies*, *2*(3), 22–35.

Mungmachon, M. R. (2012). Knowledge and local wisdom: Community treasure. *International Journal of Humanities and Social Science*, *2*(13), 174–181.

Mussak, R. A., & Bechtold, T. (2009). Natural colorants in textile dyeing. Handbook of natural colorants, 315-335.

Muthu, S. S. (2014). *Roadmap to sustainable textiles and clothing: Eco-friendly raw materials, technologies, and processing methods* (1st ed.). Springer.

O'Connor, A. E., Gallagher, W. M., & Byrne, A. T. (2009). Porphyrin and nonporphyrin photosensitizers in oncology: Preclinical and clinical advances in photodynamic therapy. *Photochemistry and Photobiology*, *85*(5), 1053–1074. doi:10.1111/j.1751-1097.2009.00585.x PMID:19682322

Ojuederie, O. B., & Babalola, O. O. (2017). Microbial and plant-assisted bioremediation of heavy metal polluted environments: A review. *International Journal of Environmental Research and Public Health*, *14*(12), 1504. doi:10.3390/ijerph14121504 PMID:29207531

Olas, B., Białecki, J., Urbańska, K., & Bryś, M. (2021). The effects of natural and synthetic blue dyes on human health: A review of current knowledge and therapeutic perspectives. *Advances in Nutrition*, *12*(6), 2301–2311. doi:10.1093/advances/nmab081 PMID:34245145

Parvin, F., Islam, S., Akm, S. I., Urmy, Z., & Ahmed, S. (2020). A study on the solutions of environment pollutions and worker's health problems caused by textile manufacturing operations. *Biomedical Journal of Scientific & Technical Research*, *28*, 21831–21844.

Prakash, P., Balasundaram, J., Al-Enizi, A. M., Ubaidullah, M., & Pandit, B. (2022). Effect of photovoltaic performance of plant-based cocktail DSSCs and adsorption of nano TiO2 onto the solvent-influenced dye sensitizers. *Optical Materials*, *133*, 113031. doi:10.1016/j.optmat.2022.113031

Roy Choudhury, A. K. (2014). Environmental impacts of the textile industry and its assessment through life cycle assessment. *Roadmap to sustainable textiles and clothing: environmental and social aspects of textiles and clothing supply chain*, 1-39.

Rukhaya, S., Yadav, S., Rose, N. M., Grover, A., & Bisht, D. (2021). Sustainable approach to counter the environmental impact of fast fashion. *The Pharma Innovation Journal*, *10*(8S), 517–523.

Rymbai, H., Sharma, R. R., & Srivastav, M. (2011). Bio-colorants and its implications in health and food industry–a review. *International Journal of Pharmacological Research*, *3*(4), 2228–2244.

Samanta, A. K., & Konar, A. (2011). Dyeing of textiles with natural dyes. *Natural Dyes, 3*, 30-56.

Samanta, P. (2020). A review on application of natural dyes on textile fabrics and its revival strategy. *Chemistry and technology of natural and synthetic dyes and pigments*, 1-25.

Samsami, S., Mohamadizaniani, M., Sarrafzadeh, M. H., Rene, E. R., & Firoozbahr, M. (2020). Recent advances in the treatment of dye-containing wastewater from textile industries: Overview and perspectives. *Process Safety and Environmental Protection*, *143*, 138–163. doi:10.1016/j.psep.2020.05.034

Sarkar, S., Banerjee, A., Halder, U., Biswas, R., & Bandopadhyay, R. (2017). Degradation of synthetic azo dyes of textile industry: A sustainable approach using microbial enzymes. *Water Conservation Science and Engineering*, *2*(4), 121–131. doi:10.100741101-017-0031-5

Saxena, S., & Raja, A. S. M. (2014). Natural dyes: sources, chemistry, application and sustainability issues. *Roadmap to sustainable textiles and clothing: eco-friendly raw materials, technologies, and processing methods*, 37-80.

Scotter, M. J. (2011). Emerging and persistent issues with artificial food colours: Natural colour additives as alternatives to synthetic colours in food and drink. *Quality Assurance and Safety of Crops & Foods*, *3*(1), 28–39. doi:10.1111/j.1757-837X.2010.00087.x

Selvaraj, V., Karthika, T. S., Mansiya, C., & Alagar, M. (2021). An over review on recently developed techniques, mechanisms and intermediate involved in the advanced azo dye degradation for industrial applications. *Journal of Molecular Structure*, *1224*, 129195. doi:10.1016/j.molstruc.2020.129195

Shahid, M., & Mohammad, F. (2013). Recent advancements in natural dye applications: A review. *Journal of Cleaner Production*, *53*, 310–331. doi:10.1016/j.jclepro.2013.03.031

Sigurdson, G. T., Tang, P., & Giusti, M. M. (2017). Natural colorants: Food colorants from natural sources. *Annual Review of Food Science and Technology*, *8*(1), 261–280. doi:10.1146/annurev-food-030216-025923 PMID:28125346

Slama, H. B., Chenari Bouket, A., Pourhassan, Z., Alenezi, F. N., Silini, A., Cherif-Silini, H., Oszako, T., Luptakova, L., Golińska, P., & Belbahri, L. (2021). Diversity of synthetic dyes from textile industries, discharge impacts and treatment methods. *Applied Sciences (Basel, Switzerland)*, *11*(14), 6255. doi:10.3390/app11146255

Stasiak, N., Kukula-Koch, W., & Glowniak, K. (2014). Modern industrial and pharmacological applications of indigo dye and its derivatives-a review. *Acta Poloniae Pharmaceutica*, *71*(71), 215–221. PMID:25272640

Tamburini, D., Breitung, E., Mori, C., Kotajima, T., Clarke, M. L., & McCarthy, B. (2020). Exploring the transition from natural to synthetic dyes in the production of 19th-century Central Asian ikat textiles. *Heritage Science*, *8*(1), 1–27. doi:10.118640494-020-00441-9

Tesfaye, T., Ayele, M., Gibril, M., Ferede, E., Limeneh, D. Y., & Kong, F. (2022). Beneficiation of avocado processing industry by-product: A review on future prospect. *Current Research in Green and Sustainable Chemistry*, *5*, 100253. doi:10.1016/j.crgsc.2021.100253

Vardhan, K. H., Kumar, P. S., & Panda, R. C. (2019). A review on heavy metal pollution, toxicity and remedial measures: Current trends and future perspectives. *Journal of Molecular Liquids*, *290*, 111197. doi:10.1016/j.molliq.2019.111197

Venil, C. K., Velmurugan, P., Dufossé, L., Renuka Devi, P., & Veera Ravi, A. (2020). Fungal pigments: Potential coloring compounds for wide ranging applications in textile dyeing. *Journal of Fungi (Basel, Switzerland)*, *6*(2), 68. doi:10.3390/jof6020068 PMID:32443916

Widiawati, D., Sn, S., Rosandini, M., & Ds, S. (2012). Natural Dyes on Indonesian Traditional Textiles-A Case Study: Geringsing Woven Fabric, In Tenganan Pegeringsingan Village Bali. *The Research Journal of the Costume Culture*, *20*(1), 111–120. doi:10.7741/rjcc.2012.20.1.111

Wrolstad, R. E., & Culver, C. A. (2012). Alternatives to those artificial FD&C food colorants. *Annual Review of Food Science and Technology*, *3*(1), 59–77. doi:10.1146/annurev-food-022811-101118 PMID:22385164

Yasukawa, A., Chida, A., Kato, Y., & Kasai, M. (2017). Dyeing silk and cotton fabrics using natural blackcurrants. *Textile Research Journal*, *87*(19), 2379–2387. doi:10.1177/0040517516671125

Yusuf, M. (2019). Synthetic dyes: A threat to the environment and water ecosystem. *Textiles and clothing*, 11-26.

Zarkogianni, M., Mikropoulou, E., Varella, E., & Tsatsaroni, E. (2011). Colour and fastness of natural dyes: Revival of traditional dyeing techniques. *Coloration Technology*, *127*(1), 18–27. doi:10.1111/j.1478-4408.2010.00273.x

Zubairu, A., & Mshelia, Y. M. (2015). Effects of selected mordants on the application of natural dye from onion skin (Allium cepa). *Science and Technology*, *5*(2), 26–32.

Chapter 14
Optimization of Parameters and Modelling for Breakthrough Curve for Chromium Removal in Fixed Bed

Sunil Jayant Kulkarni
https://orcid.org/0000-0002-5988-3448
Gharda Institute of Technology, India

Saima Munawar Parkar
Gharda Institute of Technology, India

Rutuja Santosh Mulukh
Gharda Institute of Technology, India

Gautami Jagdish Narhari
Gharda Institute of Technology, India

ABSTRACT

Wastewater released from industrial activities is loaded with heavy metals such as nickel, arsenic, lead, zinc, copper, cadmium, chromium. The current investigation is conducted to remove the hexavalent chromium (Cr VI) from the simulated wastewater by adsorption process using activated carbon. Experimental studies were carried out in batch and continuous mode. For studying the fixed bed, namely bed height, concentration and flow rate at the inlet were varied. In batch studies, as the initial concentration of the sample, contact time of adsorption and adsorbent dosage for the given initial concentration increases, adsorption also increases. The optimal pH value is in the acidic range. Isotherms were plotted and studied to understand the adsorption. For fixed bed, increasing concentration decreases the breakthrough time. Experiments show that the time required for exhaustion reduces when there is increase in flow rate and concentration. Fixed bed adsorption results were fitted to the models like Yoon-Nelson and Thomas.

DOI: 10.4018/978-1-6684-9979-5.ch014

1. INTRODUCTION

Rapid industrialization and technological advancements result in heavy waste water pollution which is a global concern due to disposal of elements like heavy harmful metals such as copper, zinc, nickel, chromium, lead and iron into the water streams (Kulkarni et al., 2018). Pollution of lakes, rivers and other water streams is caused by the release of heavy harmful metals in them which is a huge threat to not only to human beings but also to the flora and fauna of various water bodies. In high dosage, chromium and lead are carcinogenic in nature which creates a risk for human health and flora and fauna of water bodies. Lead, which is a heavy metal, is classified under toxic category while other heavy metal i.e., chromium is classified under micronutrient category and thus, intake of even small concentration of chromium is lethal to the people's health. Heavy metals are introduced in the human body via various ways like drinking water, animal feed, pharmaceutical medicines, green fodder, etc. These heavy metals can be accumulated in the human body directly or indirectly via the food chain. These toxic metals can enter drinking supplies as the industrial waste water is introduced in the water streams. According to WHO, chromium concentration in industrial wastewater should be 1 mg/l maximum. Excessive uptake of chromium in the human body results in capillary damage, irritation to central nervous system, gastrointestinal irritation and hepatic and renal damage. These metals can get accumulated in the soil causing the agricultural damage. In drinking water, concentration of lead and chromium must be 0.05 mg/l at maximum which is recommended by the World Health Organization (WHO). Because of all these harmful effects, it is extremely necessary to remove these toxic heavy metals before releasing them into the environment. For removal of heavy metals, various methods like reverse osmosis, electro floatation, ion exchange, chemical precipitation, electro dialysis, membrane separation, biological separation, adsorption, etc. are employed. From these different methods, we will use the adsorption method.

During the recent years, the usage of adsorption for treatment of waste water has become more prevalent due to its effectiveness and cost-effectiveness for removing pollutants (heavy metals) from waste water. Chromium has broad applications in different industrial processes. The inappropriate disposal of chromium from various industrial sources leads to the pollution and it has adverse effects. The adverse effects of chromium on the plants, humans, environment and animals include mutagenic, carcinogenic effects, tissue damage and teratogenic effects (2). Cr(III) and Cr(VI) are the two states of chromium ions known as trivalent and hexavalent state respectively. Cr(III)can be used as an important trace element. On the other hand, Cr(VI) is non-essential and toxic in nature which causes various diseases like dermatitis, gastric and kidney damage, lung cancer, respiratory tract and eye irritation (Mitra et al., 2017). Therefore, it is necessary to reduce its concentration in waste water before getting released into the aquatic system and getting included in the food chain. The conventional processes used for chromium removal include Cr(VI) immobilisation or reduction of toxic Cr(VI) to less toxic Cr(III). Some methods for removal of hexavalent chromium involve methods like solvent extraction, membrane filtration, electrokinetic procedures and leaching (Mitra et al., 2017). The mentioned processes are of high cost. Thus, adsorption of chromium on the various adsorbent materials like activated carbon, titanium dioxide, goethite, zeolites, etc. is a low-cost method which has design flexibility and also easy to operate.

Adsorption is a surface phenomenon. In this mass transfer process, molecules get accumulated on the surface of another slid surface. It can be also stated as a physical or chemical bonding of molecules on the surface(active sites) of material through Van der Wall forces or by chemical bonding on the surface with the active sites (Amal, 2018). In adsorption, adsorbate is a material which gets attached to the surface of adsorbent. Adsorbate is the solid particle or molecule which is going to adsorb on another surface i.e., effluent particle while adsorbent is the solid surface on which molecules are going to adsorb; e.g., carbon, zeolite, porous clays, etc. For the treatment of waste water, the commonly used adsorbent is activated carbon as it has ability of adsorbing various organic compounds.

Adsorption is of two types; physisorption and chemisorption. Adsorption process where adsorbate get attached to surface of solid using weak intermolecular forces of attraction between the adsorbate and adsorbent is called physisorption (Shanmugasundaram and Sudalaimani, 2012). Physical adsorption is reverse phenomenon. Thus, it is possible to remove contaminant from the adsorbent surface. Adsorption where adsorbate get attached to solid surface of adsorbent through chemical bond formation between them is called as chemisorption. Chemisorption process cannot be reversed and thus is hard to separate contaminants and adsorbent (Shanmugasundaram and Sudalaimani, 2012). Adsorption process is exothermic in nature. Adsorption process and condensation process are similar. On the contrary, the reverse process is known as desorption and it is endothermic process. At low temperature, adsorption process is favoured. Thus, increase in temperature will reduce adsorption and as the temperature is reduced, the adsorption will increase.

The data obtained from batch experiments is difficult to apply to column experiments for fixed bed. This is because the isotherms do not provide precise data for scaling up because of the reasons like non-equilibrium condition of flow column, uneven flow patterns in the fixed beds and recycling and regeneration problem which cannot be studied (Samarghandi et al., 2014). Thus, it is important to perform column flow tests before obtaining design models. The problem with column design is the difficulty in anticipating how long the column will last before it needs to be regenerated or replaced (Samarghandi et al., 2014). Fixed-bed column studies are important for anticipating the column breakthrough or the wavefront shape. This helps in determining the operational lifespan of bed, and studying the behaviour of the continuous flow systems is important (Samarghandi et al., 2014).

2. ADSORPTION

Adsorption is a process where one or more components of a mixture (gas or liquid) selectively concentrate or retain on the surface of a solid. In majority of natural, physical, biological and chemical systems, adsorption process occurs. During adsorption, a material is removed from one phase and then accumulates or concentrates on another surface. Adsorption is the process of retaining the sorbed molecules or atoms concentrated or accumulated near or at the surface. Adsorbent refers to the solid that adsorbs a component, and adsorbate refers to the component that is adsorbed. The contact between the molecules of the adsorbate and the surface of the adsorbent leads to the adsorption process (Dutta, 2009). Adsorption of a material on a solid surface occurs due to the surface's affinity for the specific substance. Van der Waals forces and electrostatic interactions between adsorbate molecules and atoms on the adsorbent surface are involved in physical adsorption. Adsorption capacity is affected by adsorbent properties, chemical properties of adsorbate, pH, temperature, etc. Several adsorbents have been employed to treat industrial wastewater, including activated carbon, silica gel, and activated alumina.

2.1. Types of Adsorptions

Adsorption processes are classified in different ways. One of the ways of classification of adsorption is based on force of attraction among the adsorbent and adsorbate. Based on these, adsorption can either be physical or chemical.

2.1.1. Physical Adsorption

When the forces of interaction are relatively weak, the adsorption is known as physical adsorption. The physical adsorption forces are weak Van der Waal's forces (or dispersion force) and electrostatic forces. The heat of adsorption is just a few times (1-4 times) the heat of vaporization of the component in consideration (Dutta,2009). Nature of physical adsorption is reversible.

2.1.2. Chemical Adsorption

The adsorption where chemical bond is formed between solute and adsorbent when the interaction between them is strong is called chemisorption or chemical adsorption. The heat of chemical adsorption significantly more than the heat of physisorption (Dutta,2009). Chemical adsorption occurs as a necessary step in a heterogenous catalytic reaction. Nature of chemical adsorption is irreversible.

2.2. Factors Affecting Adsorption

- Initial concentration of chromium in the sample
- Contact time for the adsorption
- Quantity of the adsorbent in the sample
- pH of the sample

2.3. Adsorption Isotherms

Adsorption is described by isotherms because of the fact that temperature has significant impact on whole process. Adsorption isotherm is very crucial. Adsorption isotherm is defined at constant temperature as the relationship between the adsorbate in the liquid phase and adsorbate adsorbed on adsorbent surface at equilibrium condition (Batool et al., 2020).

Adsorption isotherms are extremely important in environmental protection and adsorption research. The adsorption capacity of a given adsorbent material can be predicted by two simple isotherms known as Freundlich isotherm and Langmuir isotherm. Other isotherm models include the BET isotherm model, Dubinin-Radushkevich isotherm model, the Hill isotherm model, the Temkin isotherm model, the Flory-Huggins isotherm model, etc. (Batool et al., 2020).

Adsorption equilibrium information is very crucial which is required for understanding adsorption process in detail. Understanding and well interpreting adsorption isotherms is much needed for overall adsorption process improvement and effective adsorption system design (Sircar et al., 1996).

2.3.1. Freundlich Isotherm

In recent years, the Freundlich isotherm has primarily been used in heterogeneous systems, particularly for organic compounds or highly interacting species on activated carbon.

The Freundlich isotherm is given by following relation,

$$ln\frac{x}{m} = \frac{1}{n}\ln Ce + \ln k \tag{1}$$

where,
x - the amount of gas adsorbed,
m -mass of adsorbent (gram),
k and n - adsorption constant,
C_e - concentration after adsorption

Adsorption intensity or surface homogeneity is measured by the slope of the isotherm, which varies from 0 to 1. As the slope value approaches 0, it gets increasingly heterogeneous. If value of slope is less than 1, then it is chemisorption process. If the slope is greater than 1, then it is cooperative adsorption (Liakos et al., 2021).

2.3.1.1. Limitations of Freundlich Isotherm

- It is completely empirical, with no theoretical foundation(based on experiments and practical experience, not on ideas).
- The equation is only valid up to a specific pressure and becomes invalid at higher pressures.
- The constants of this isotherm equation of K & n vary with temperature.
- It fails when the adsorbate concentration is high.
- It could not be explained for multilayer adsorption process.

2.3.2. Langmuir Isotherm

It was initially developed to describe gas-solid phase adsorption on activated carbon. Langmuir isotherm is given by the following relation,

$$\frac{1}{qe} = \frac{k}{qo.Ce} + \frac{1}{qo} \tag{2}$$

where, q_e - concentration of solid phase (kg adsorbate / kg adsorbent),

q_0 - kg adsorbate/kg solid
C_e- concentration after adsorption
k - adsorption constant

Every adsorption isotherm gives an important information, e.g., q_e from Langmuir isotherm is important when we design our batch system for waste water treatment (Liakos et al., 2021).

2.3.2.1. Limitations of Langmuir Isotherm

- Multilayer adsorption is possible in this isotherm.
- Ignores the interactions between adsorbate-adsorbate interaction.
- Fails for surface roughness of adsorbent

When the adsorbent and adsorbate remain in contact with each other for an extended period of time, an equilibrium is formed between the amount of adsorbate adsorbed and the amount of adsorbate in solution. Study of adsorption can be done through graphs called as isotherms. Isotherms show the amount of adsorbate (X) adsorbed on the surface of the adsorbent (m) and concentration (C) at constant temperature. After concentration reaches to its saturation point then the adsorption does not occur anymore (Kulkarni, 2020).

2.3.3. Difference Between Freundlich and Langmuir Isotherm

Freundlich and Langmuir isotherms differ in different ways which are listed in Table 1.

2.4 Breakthrough Curve

The breakthrough curve from the adsorption data collected from the tests can be used to study fixed bed adsorption or continuous adsorption in the packed column. Figure 3 depicts the adsorbate distribution, with horizontal lines indicating the relative concentration of adsorbate. As the influent solution flows through the column, the adsorption zone moves downward at a rate that is typically much slower than the velocity of the fluid flowing through the bed (Treybal, 1981). After some time, as shown in the part (b) of the Figure 1, when bed is roughly half saturated, outlet concentration is still zero. When lower portion just reaches the bottom of the bed, breakpoint is achieved, after getting appreciable value of effluent concentration. Now effluent concentration increases rapidly and reaches initial value C_0. Portion between breakpoint and saturation point is termed as breakthrough curve. The curves are often S-shaped; however, they can be sharp, flat, or twisted in other situations. In general, the breakpoint time reduces with decreasing bed height, increasing adsorbent particle size, increasing fluid flowrate across the bed and increasing initial solute content of the feed. There is a critical minimum bed height below which the solute concentration in the effluent will rapidly rise from the time the effluent first appears.

Table 1. Difference between Freundlich and Langmuir isotherm

Freundlich Isotherm	Langmuir Isotherm
It describes the quality of effluent adsorbed by per unit mass of solid adsorbent with concentration.	It specifies the number of active sites undergoing adsorption and the concentration.
Adsorption site is heterogeneous.	Adsorption site is homogeneous
No any informative data about adsorption and desorption.	It is based on adsorption and desorption equilibrium.

Figure 1. Breakthrough curve

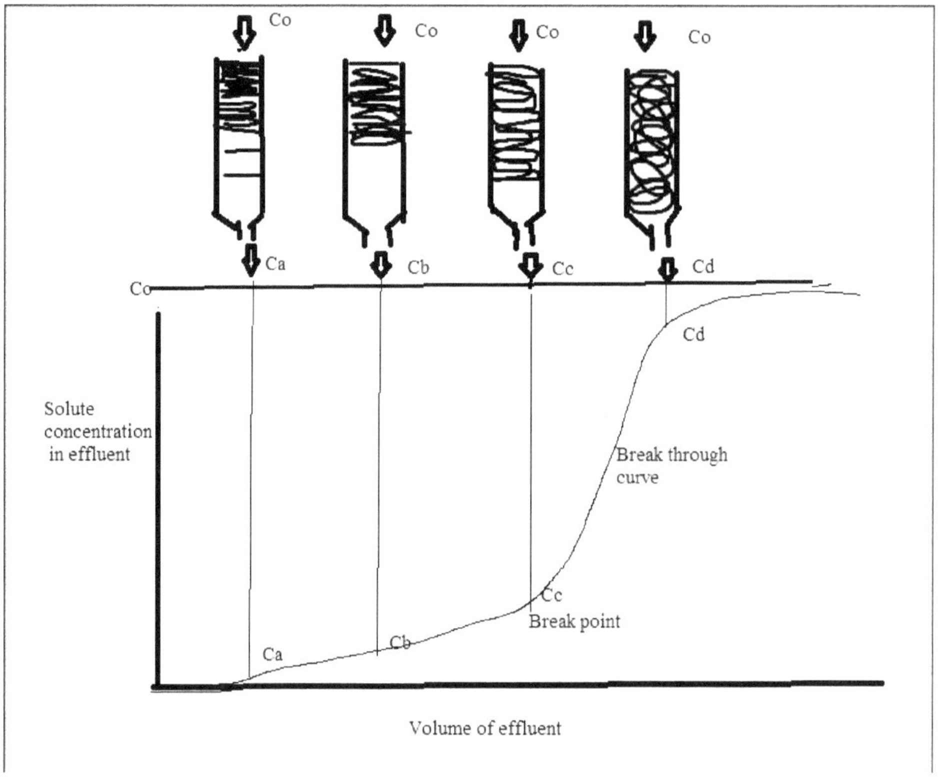

2.4. Factors Affecting Breakthrough Curve

The factors affecting breakthrough curve in the continuous adsorption column i.e., fixed bed adsorption are listed below:

- Bed length: as bed length increases breakpoint shifts towards right along with the time scale. Smaller bed length of have less amount of adsorbent & bed gets saturated in less time.
- Velocity: as the velocity increases breakpoint shifts towards the left and also curve becomes much steeper, because of the residence time i.e., time spent by the solute in the packed column. Before the equilibrium is reached, the adsorbate solution leaves the packed column when the velocity is higher.
- Mass transfer coefficient: with decrease in value of coefficient C/C_0 approaches to unity but when coefficient increases C/C_0 adsorbate removal value less than unity initially and then approaches to constant value (Babu and Gupta, 2004).
- Diffusivity
- particle size and isotherm properties from the breakthrough curve.

The factors considered during the study which affect the breakthrough curve are as follows:

- Initial concentration at the inlet of fixed bed column

- Height of the adsorbent bed inside column
- Initial flow rate of adsorbate at the column inlet

2.5. Mathematical Models for Fixed Bed Adsorption

Mathematical modelling in adsorption process plays vital role in waste water treatment. They are used in order to ensure that the proper adsorbent material is used to reduce the hazardous waste material in water by removing it effectively, at the proper reaction rate and under the right conditions. These mathematical models contain vital data that can be used to manufacture a large-scale adsorption reactor (Unuabonah et al., 2019). Fix packed bed dynamics models include Yoon-Nelson model, Thomas model, Bohart-Adams Model and others. Simple mathematical models used to describe packed bed adsorption process & governing equations of model are used to solved numerically to obtain single component monolayer adsorption. To analyze the breakthrough curve in our study, we employed the Thomas model and the Yoon-Nelson model. Because their model equations can be linearized and their unknown parameters can be estimated using linear regression analysis, these two packed bed mathematical models are very popular. We did not use the Bohart-Adams model because it does not fit the breakthrough data well. Additionally, the Bohart-Adams model equation is an exponential function that predicts an exponentially increasing breakthrough value over time and is unable to fit typical breakthrough curves that are S-shaped and sigmoidal. Fitting abilities of these models are studied by comparing the coefficient of determination R^2 (Chu., 2020).

2.5.1. Thomas Model

The Thomas Model's mathematical expression for a fixed bed adsorption column is given by the following equation:

$$\frac{C}{Co} = \frac{1}{1 + exp\left[\frac{Kt}{Q(qoM - CoV)}\right]} \tag{3}$$

This equation is expressed in linear form as:

$$\ln\left(\frac{Co}{C} - 1\right) = \frac{KtqoM}{Q} - \text{ktCot} \tag{4}$$

where, C (mg/l) - outlet concentration of effluent
C_0 (mg/l) - Inlet concentration of effluent
K_t (1/min mg) - Thomas rate constant
t - Time in hour
Q (ml/min) - volumetric flowrate
M (g) - quantity of adsorbent in the column
q_0 (mg/g) - equilibrium chromium uptake per gm of adsorbent

2.5.2. Yoon and Nelson Model

This is a simple model that is based on the assumption that the likelihood of sorbate sorption and sorbate breakthrough on the sorbent are proportional to the rate at which the chance of sorption for each sorbate molecule decreases (El-Naas, 2013).

Yoon and Nelson Model is expressed by the following equation:

$$\frac{C}{Co} = \frac{1}{1 + exp\left(K\left(T - t\right)\right)} \tag{5}$$

This equation in the linear form is given by:

$$\ln\frac{C}{Co - C} = kt - Tk \tag{6}$$

where, K_t (min^{-1}) - rate constant
T (min) - time needed for 50% adsorbate breakthrough

3. EXPERIMENTAL SETUP AND METHODOLOGY

3.1. Preparation of Stock Solution

By dissolving 0.1414 gm of Potassium Dichromate ($K_2Cr_2O_7$) in 100 ml of distilled water, the simulated stock solution was prepared. This was 500 ppm solution (Figure 2). 1 ml of this 500 ppm stock solution was equivalent to 500 μg Cr.

Then, 1 ml sample was taken from 500 ppm solution and it was diluted to 100 ml. This was 5 ppm stock solution (Figure 3). 1 ml of this 5 ppm stock solution was equivalent to 5 μg Cr.

3.2. Diphenyl Carbazide Reagent Preparation

Diphenyl Carbazide reagent was made by mixing 0.25 gm of 1,5-Diphenyl Carbazide powder with 50 ml of acetone (Figure 4). This reagent solution was stored in amber coloured reagent bottle and was further used during the analysis process.

3.3. Batch Experiments

To remove hexavalent chromium; Cr(VI), from aqueous solution, various initial samples of varying concentrations were prepared from the stock solution of 5 ppm and were diluted using distilled water (Figure 5). Twenty-five ml distilled water was added in the samples and they were shaken well. To reduce the pH of the samples, 5 drops of ortho-phosphoric acid were added in the samples and they were shaken well. After that, pH of samples was checked (must be in the acidic range 1.5 to 2.5).

Figure 2. Stock solution (500 ppm)

Figure 3. Stoch solution (5 ppm)

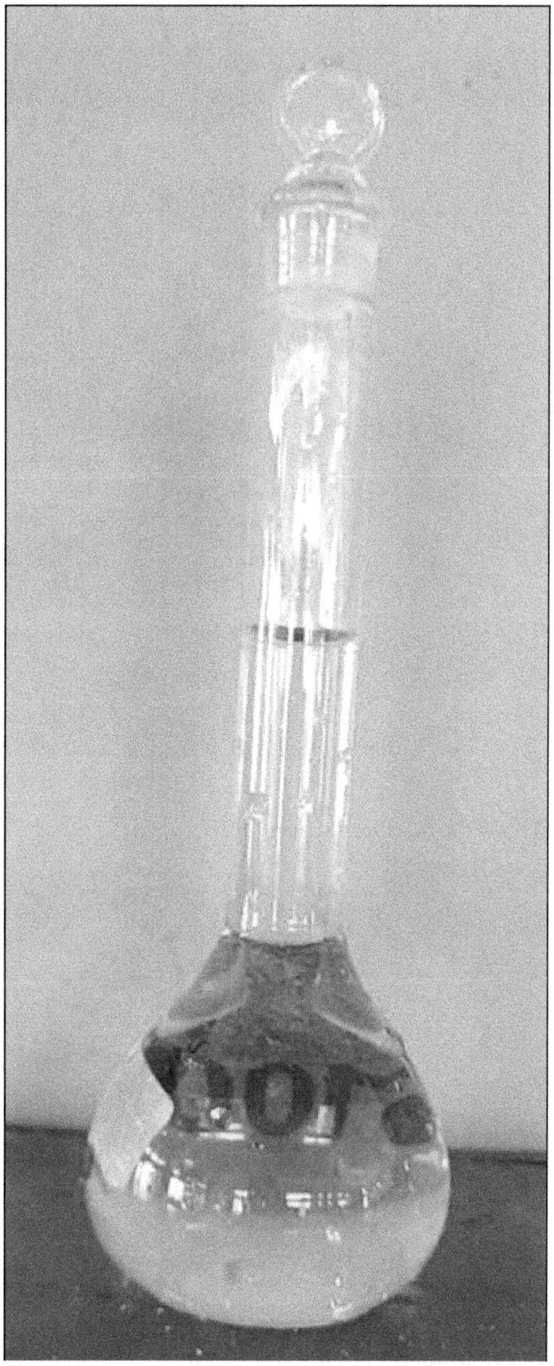

The samples were then diluted up to 100 ml. For analysing the amount of chromium in these aqueous solutions before adsorption, 2 ml of diphenyl carbazide reagent was added. After adding the reagent, samples were kept aside for 8-10 minutes for full colour development (Figure 6). The initial absorbance of

Figure 4. Diphenyl carbazide reagent solution

Figure 5. Samples before adsorption

these samples was checked by using UV spectrophotometer (Model: UV-1800; Serial No.: A11635581333) and were noted down. Calibration curve (Figure 14) was plotted using these values.

The initial samples were prepared again. The parameters (adsorbent dosage, initial concentration, pH of the sample and contact time) were varied in order to optimize them. The samples were shaken on the rotary flask shaker then the samples were filtered after adsorption (Figures 7-8). Two ml of diphenyl

Figure 6. Colour development in sample before adsorption

Figure 7. Samples on rotary flask shaker

Figure 8. Samples after adsorption

Figure 9. Colour development in samples after adsorption

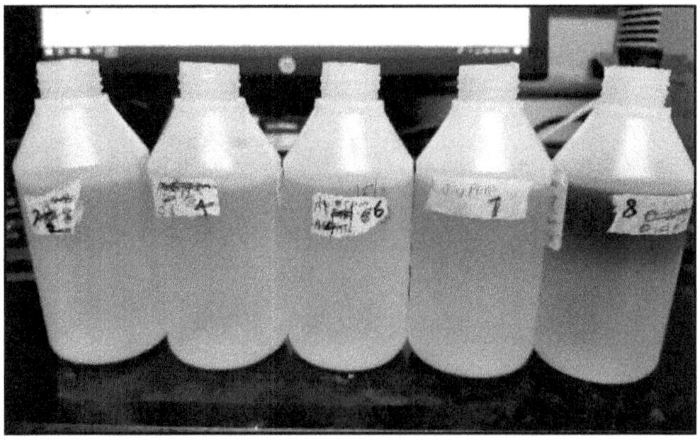

carbazide reagent were added in each sample after adsorption and were kept aside for 8 to 10 minutes for full colour development (Figure 9).

Absorbance of sample after adsorption was checked using UV spectrophotometer (Model: UV-1800; Serial No.: A11635581333).Unknown concentration of chromium after adsorption was determined from calibration curve (Figure 12) (initial absorbance vs initial known concentration).

The optimised parameters i.e., adsorbent dosage, initial concentration, pH of the sample and contact time were obtained. The experiment was repeated with all optimised parameters.

3.4. Column Experiments

A continuous fixed bed experiment was carried out to remove harmful Cr (VI) from waste water, in which activated carbon was used as an adsorbent. The experimental setup for the column experiments consists of a separating funnel used as storage tank at the top followed by fixed glass columns with internal diameter of 1.8 cm and the total height of glass column of30 cm. At the bottom the effluent is collected in the conical flask as shown in Figure 10. Between the separating funnel (storage tank) and the glass

Figure 10. Experimental setup for column experiments

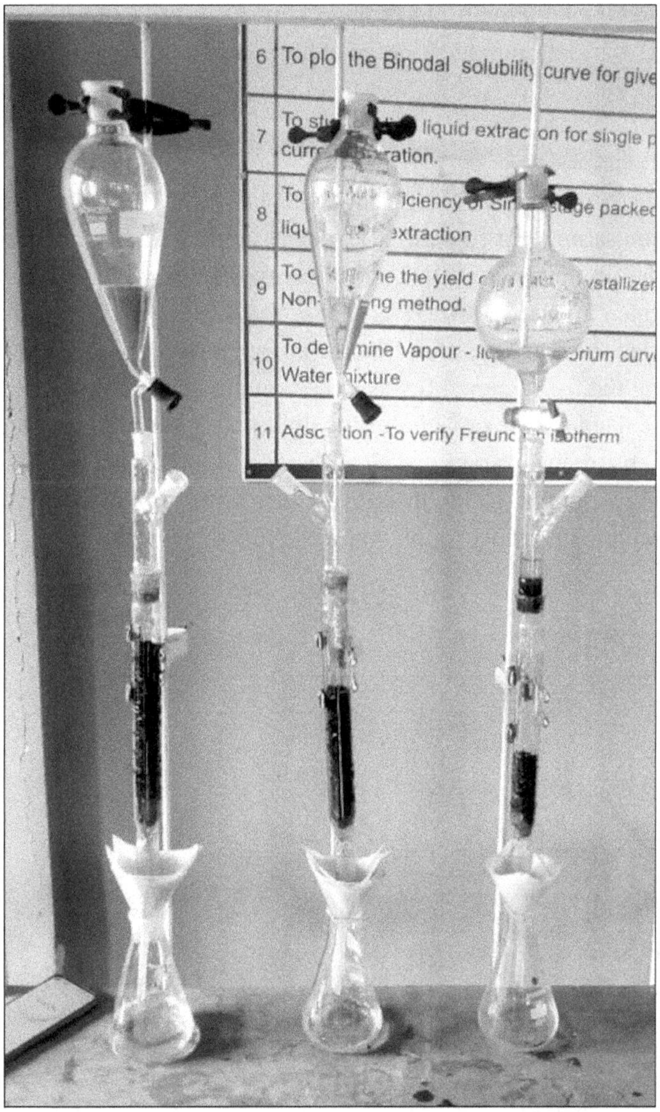

column (fixed bed column), the Y shaped glass column was used as a distributor. The simulated stock solution of Cr(VI) was charged from the separating funnel and made to pass over and about the glass column in a down flow mode. The valve of the separating funnel is used to control the inlet flow rate.

During this project, parameters such as inlet column flow rate, bed height in the column and inlet chromium concentration were varied. The effluent was collected after every 15 minutes at the end of column until the adsorbent bed gets saturated (Figure 10). Diphenyl carbazide was used to analyse the amount the chromium in the effluent solution. This method has been frequently used in numerous studies at low pH (Malkoc and Nuhoglu, 2007). Diphenyl carbazide reagent was added to the effluent, which resulted in the development of a purple-violet coloration to the effluent i.e., chromium sample collected from the fixed column experiment. Further UV-Spectrophotometer (Model: UV-1800; Serial No.: A11635581333) was used to determine the amount of chromium present in the effluent (Figure

11). In this study, the diphenyl carbazide reagent serves as a complexing agent. At 540 nm, the absorbance of the effluent after the addition of the diphenyl carbazide reagent was determined. The unknown concentration of Cr(VI) was determined from the calibration curve (Figure 12).

3.5. Analysis

A UV spectrophotometer (Model: UV-1800; Serial No.: A11635581333) (Figure 11) was used for analysis of chromium in the given sample. At first the UV spectroscopy was calibrated using the distilled water. Distilled water was filled in both of cuvettes and baseline was drawn and then it was set to auto zero. After the calibration, one cuvette was filled as it is with distilled water and the other was filled with the sample to be analysed. It was made sure that the cuvettes were filled up to two third part and were dried from outside using the tissue paper. It was done to make sure that no small particles were present on the smooth transparent side of the cuvettes as the light passes through the smooth transparent side of cuvette. The graphs of absorbance versus wavelength were plotted. The absorbance of the samples was noted down at 540 nm wavelength. Using this absorbance and calibration curve (Figure 12), unknown concentrations of the sample after adsorption were calculated.

Figure 11. UV spectrophotometer

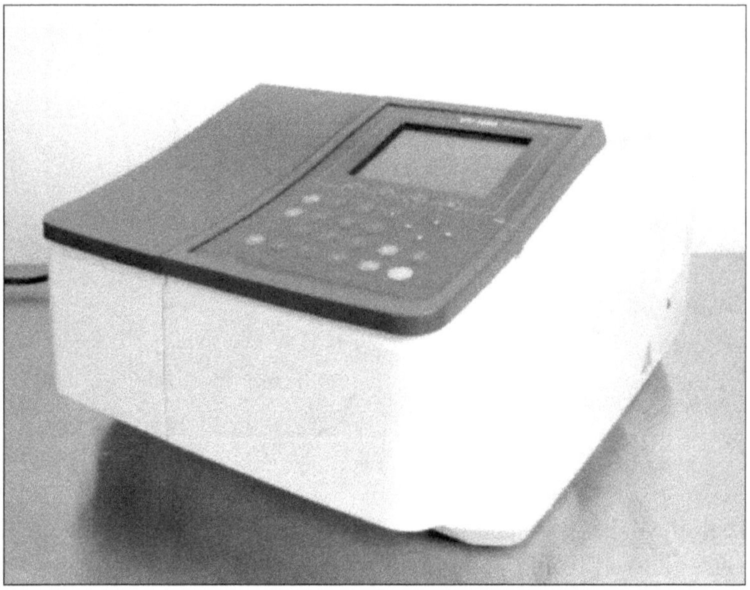

A UV spectrophotometer monitors the intensity of light as it travels through a sample solution in a cuvette and compares it to the intensity of light before the sample. The main components of a UV spectrophotometer are a light source, a sampling holder, a dispersive device for separating distinct wavelengths of light, and an appropriate detector. The use of a UV spectrophotometer is based on the relationship between sample absorbance and sample concentration. UV spectroscopy is used as a quantitative analytical technique in segments such as a water testing, food and beverages, pharmaceutical and biotech industries (Foo and Hameed, 2010).

Principle: The amount of light absorbed by the substance is calculated by spectrophotometer. The spectrophotometer measures the sample's light intensity by passing a light beam through it.

The absorption of light rays by a sample with unknown concentration is the basis for UV spectroscopy. Light is absorbed to the limited extent depending on the various substances and transmitted light is recorded as a function of wavelength by a detector that provides UV spectrum of a sample (Kaur, 2017). UV spectrophotometer have many advantages such as:

- Workflow with fast
- Easy to use
- Trustworthy analytical instruments
- Gives accurate and precise results
- Method does not require large amount of sample for analysis purpose
- It does not require any additional reagent or incubation time

4. RESULTS AND DISCUSSION

4.1. Calibration Curve

The plot of absorbance vs concentration is the calibration curve; shown in Figure 12, which is used to determine the unknown concentration of any sample by measuring its absorbance. The main goal of calibration curve is to minimize the measurement uncertainty by ensuring the accuracy of test equipment. We got the linear nature of calibration curve which is desirable because it result in the best accuracy and precision. Absorbance is the dependent variable and hence placed on y-axis. As the concentration goes on increasing, more UV light adsorbed and absorbance also increases. Therefore, absorbance is directly proportional to the concentration.

Figure 12. Calibration curve

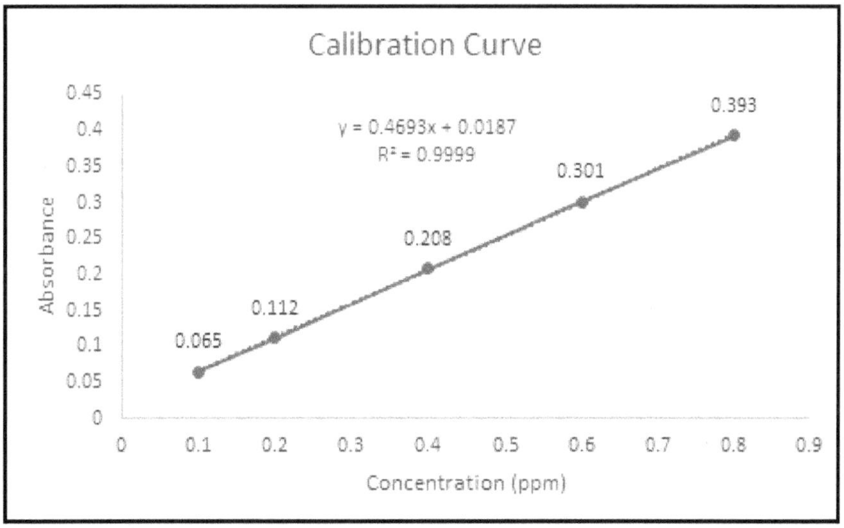

4.2. Effects of Various Parameters on Adsorption in Batch Experiments

4.2.1. Initial Concentration of Sample

Figure 13 shows the effect of initial Cr concentration on the process of adsorption. At higher initial chromium concentration, the adsorbent's surface area is surrounded by a greater number of Cr ions in the solution, increasing the adsorption capacity of activated carbon. Hence, adsorption capacity increases with increasing number of Cr ions; i.e., increase in initial chromium concentration. The adsorption increases until the saturation point of adsorbent reaches where the adsorption is maximum. If the initial concentration is increased in excess, then the back leaching process takes place. It occurs because the chromium ions could not reach the adsorption sites of adsorbent due to large number of chromium ions. The optimum initial concentration obtained is of 0.4 ppm.

Figure 13. Effect of initial concentration

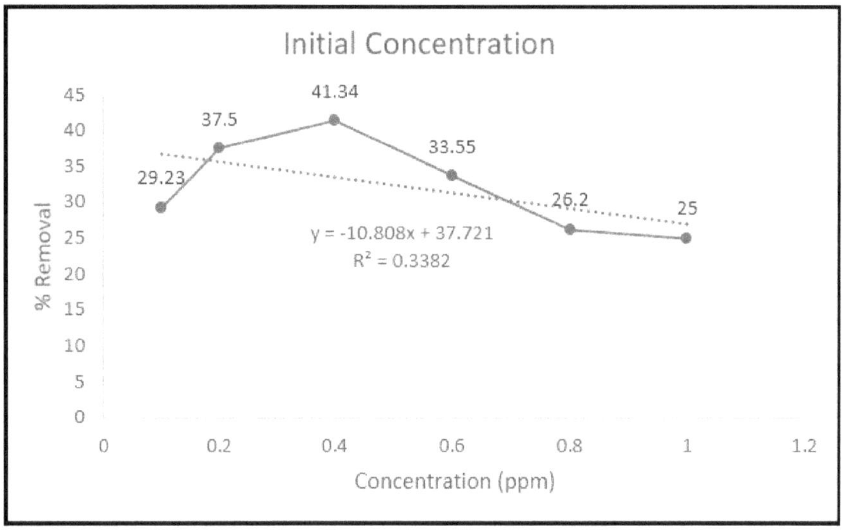

4.2.2. Contact Time for Adsorption

The contact time between the adsorbent and the adsorbate is an important parameter that indicates the extent of adsorption process. This is due to the fact that it provides information on the sorption kinetics of adsorbate for a given initial adsorbent dosage. Figure 14 shows the percentage removal of chromium with time. The optimum contact time for adsorption was determined to be one hour. To a certain extent, adsorption increases with increasing contact time.

Figure 14. Effect of contact time

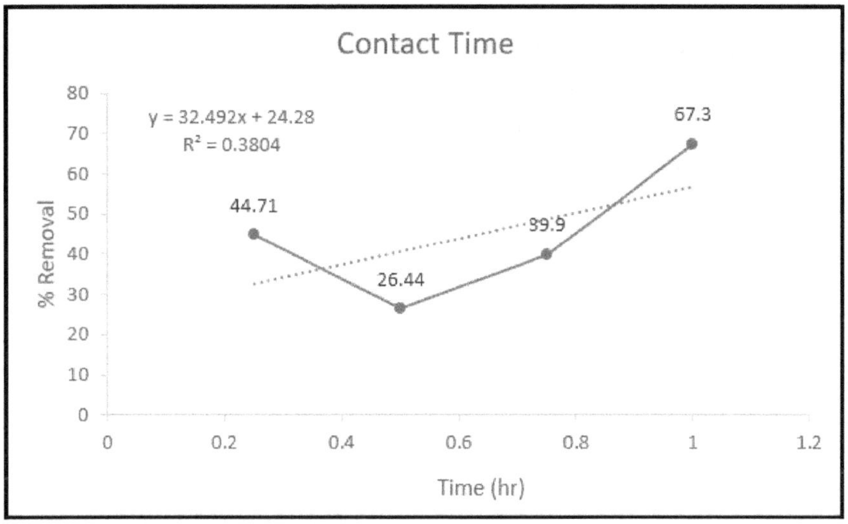

4.2.3. Quantity of Adsorbent Dosage

Adsorbent dose is an essential parameter to consider while studying adsorption since it is used to determine the adsorbent's capacity for initial chromium concentration in solution. For a given initial chromium concentration, concentration of chromium decreases as the adsorbent dosage is increased. For given initial concentration, chromium removal is increased as the adsorbent dosage is increased because it gives more availability of surface area of adsorbent and this is shown in the Figure 15. The optimum quantity of adsorbent dosage is 0.3 gm.

Figure 15. Effect of adsorbent dosage

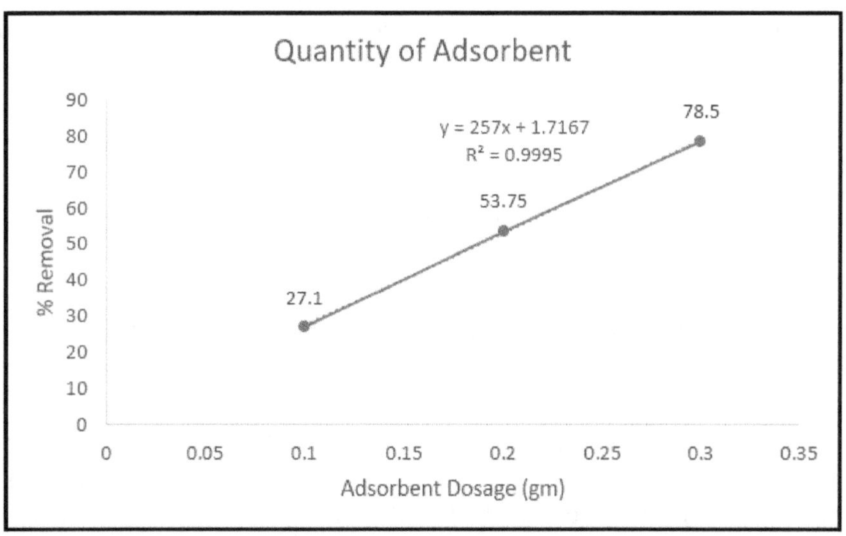

4.2.4. pH of Sample

The pH of aqueous solution gives the adsorption of chromium ion. Results from the Figure 16 indicated that the acidic pH strongly favoured the adsorption of chromium and increase in pH reduces the adsorption. Optimum pH for adsorption of chromium ranges from 1 to 3. Removal efficiency of hexavalent chromium is higher at low pH. The results obtained showed that the optimum pH is 2.

Figure 16. Effect of pH of sample

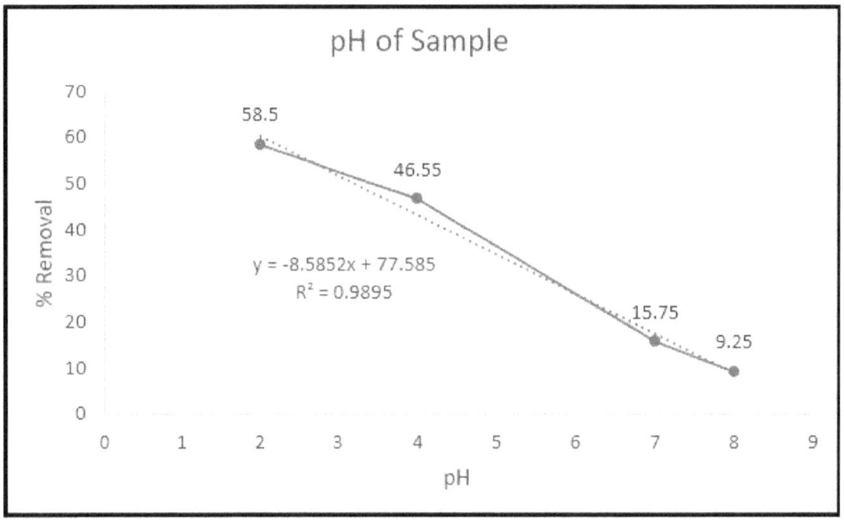

4.3. Effects of Various Parameters on Adsorption in Column Experiments

4.3.1. Initial Concentration of the Sample at the Inlet of the Column

In order to determine the adsorption capacity of activated carbon to remove hexavalent chromium from the simulated solution, studies were done on the change of adsorption capacities for different initial concentration with respect to time. Two different initial concentration of 6 ppm and 8 ppm were used to carry out experiments in order to remove the remove Cr from the simulated solution. Inlet flowrate and bed height in the columns were fixed at 10 ml/min and 13 cm respectively. For the given bed height and flowrate, it was observed that the maximum removal was 87.28% for the 6 ppm concentration. The maximum removal was obtained after 2.25 hr as shown in Figure 17. The removal at this time for 8 ppm was found to be 76.08%. The initial concentration of 6 ppm was found to be optimum.

To study the concentration of Cr at various time intervals, a number of experiments were conducted by varying the initial concentration of Cr (6 ppm and 8 ppm). Breakthrough curves were plotted for these two different concentrations as shown in Figure 18. It is evident that the breakthrough time gets shorter as concentration increases. When the concentration of chromium ions at the column's inlet increases, the adsorption capacity of the adsorbents also increases. Breakthrough point is obtained earlier as the concentration goes on increasing. This is because increasing inlet concentration increases the adsorption capacity and more chromium ions gets adsorbed on the adsorbate in less time. The breakthrough time

Figure 17. Effect of concentration on Cr adsorption

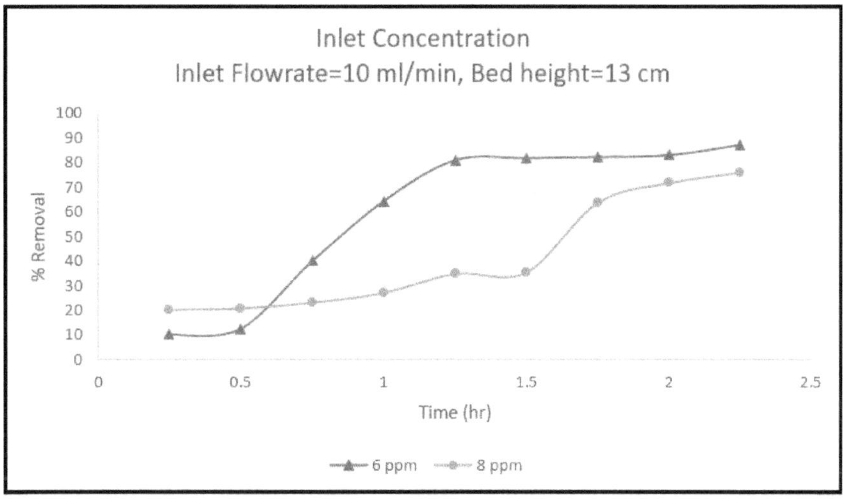

Figure 18. Effect of concentration (C/C_0)

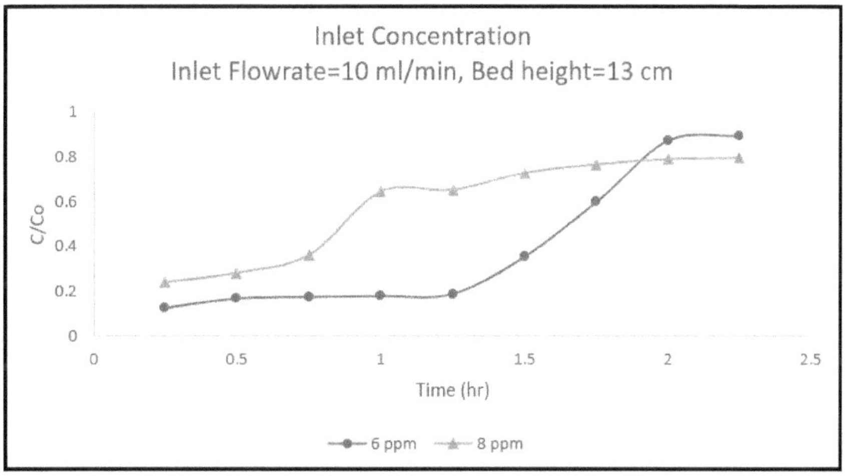

for 6 ppm initial concentration was obtained between 1.25 to 1.5 hr while it was around 0.75 hr for the 8 ppm initial Cr concentration.

4.3.2. Bed Height in the Column

For the column experiments, the bed height in the column influences the adsorption capacity. Numerous experiments were conducted by varying bed heights as 9 cm, 13 cm and 18 cm by using 30 gm, 55 gm and 70 gm of activated carbon in the column respectively. Inlet concentration and inlet flowrate in the columns were fixed at 6 ppm and 10 ml/min respectively. Figure 19 shows the removal of influent i.e., Cr with continuous adsorption for different fixed bed heights in the column. It has been observed that as bed height increases, adsorption increases, i.e., chromium removal increases till the saturation of the

Figure 19. Effect of bed height on Cr adsorption

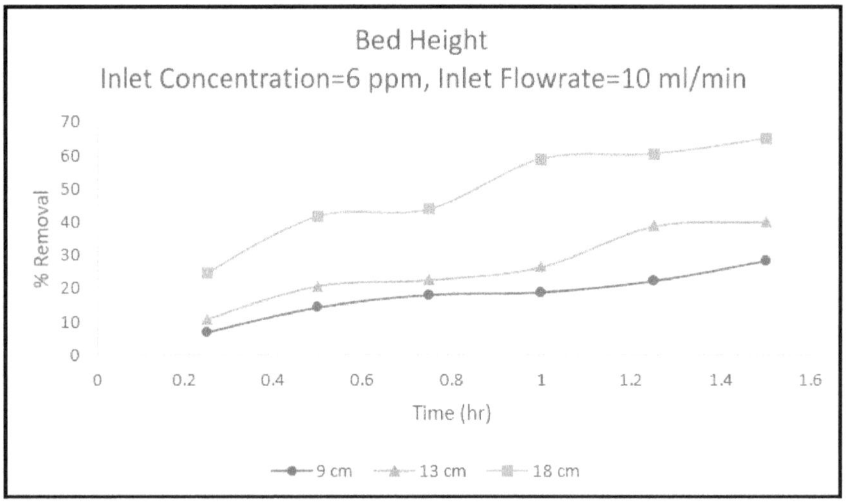

bed is achieved.% removal of chromium in sample increases with increasing bed height due to increase in not only area for mass transfer but also the contact time. Removal of Cr was maximum i.e., 65.08% for the bed height of 18 cm at 1.5 hr. At the same time, the removal was 40.27% and 28.4% for the 13 and 9 cm height respectively.

The influence of bed height in Cr removal was studied by conducting a series of experiments of continuous adsorption in the column with varying bed heights of 9 cm, 13 cm and 18 cm and constant inlet concentration and inlet flowrate of 6 ppm and 10 ml/min respectively. The capacity of Cr removal with increasing bed height was more and rapid initially and it slowed down as the bed got saturated. Figure 20 shows the influence of flow rate on adsorption.

Figure 20. Effect of bed height (C/C_0)

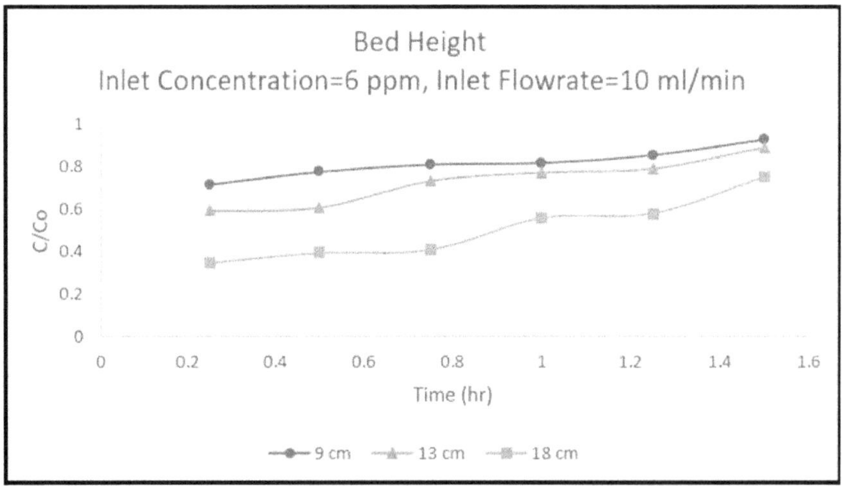

4.3.3. Flowrate of the Sample at Column Inlet

The influence of flowrate on Cr adsorption was investigated by varying the flow rate, with the same inlet concentration of 6 ppm and adsorbent bed height of 18 cm. A number of experiments were carried out by continuous adsorption in the column with different inlet flow rates of simulated solution (10 ml/min, 21 ml/min and 27 ml/min).

Figure 21. Effect of flowrate on Cr adsorption

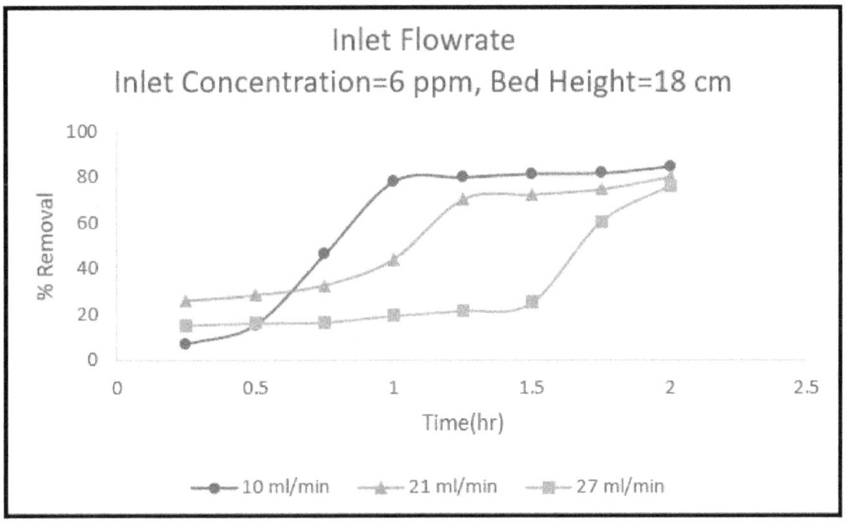

It has been discovered that when the flow rate increases, the adsorption capacity of adsorbent also increases. The % removal of chromium with respect to time for varying flow rate is shown in Figure 21. For the given column conditions, the optimum flow rate was found out to be 10 ml/min. The % removal of Cr at this flow rate was found to be maximum i.e., 84.61% at around 2 hr. % removal of Cr at this time for 21 ml/min and 27 ml/min of flow rate were 80.06% and 75.98% respectively. In order to determine the % removal of Cr, the residence time of simulated solution plays a vital role. The residence time increases in the column with increase in the flow rate. Thus, activated carbon can be used in the industries for treatment of waste water with lower value of flow rates.

The results showed that increase in flow rate of chromium sample decreases the breakthrough time. Also, as the flow rate increased, the total amount of Cr adsorbed on the adsorbent decreased. It is because the solution with higher flow rates passes the column quickly and the contact time of adsorbate and adsorbent was reduced and bed start to saturate early. When the flow rate is low, the adsorbate has much more time to contact the adsorbent, resulting in slow bed saturation and thus a delay in breakthrough time. These results are shown in the Figure 22.

Figure 22. Effect of flowrate (C/C_0)

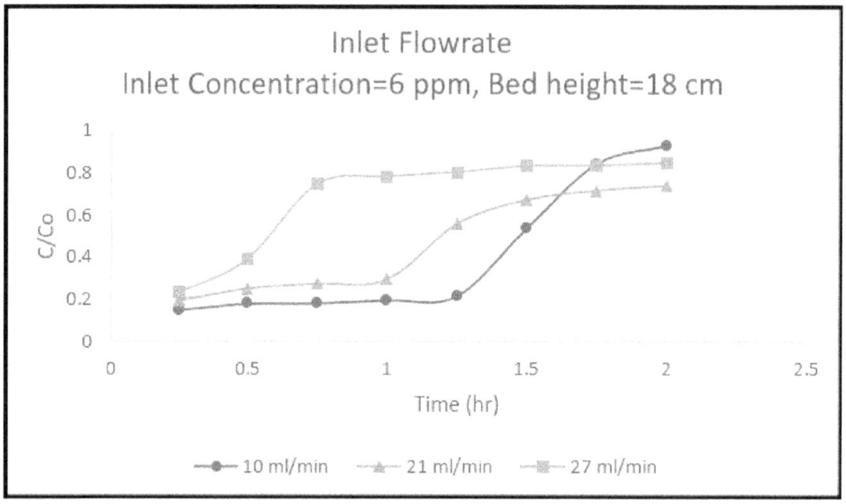

5. MATHEMATICAL MODELLING

5.1. Adsorption Isotherms

5.1.1. Freundlich Adsorption Isotherm

Freundlich isotherm is applicable only for physical adsorption and thus, it is applicable for multi-layer adsorption. Figure 23 shows the Freundlich isotherm nature for chromium adsorption. The slope (1/n) of Freundlich isotherm is 0.3146 which lies between 0 to 1 which is favorable for the Freundlich isotherm.

Figure 23. Freundlich isotherm

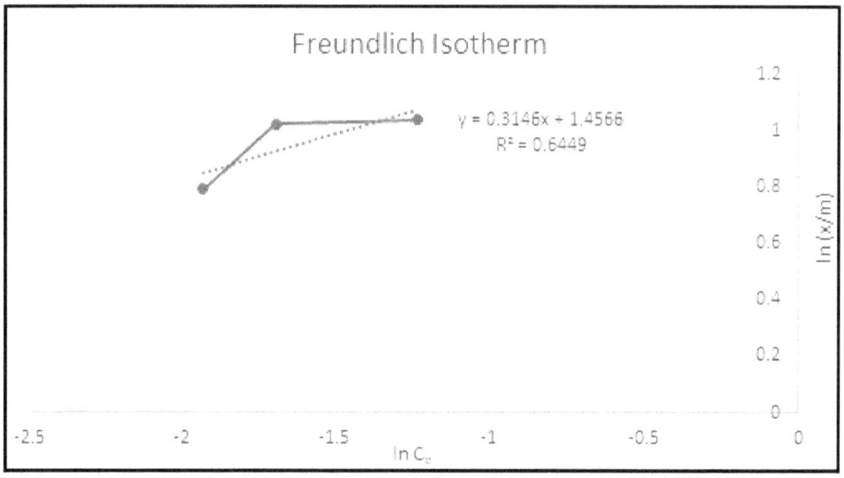

5.1.2. Langmuir Adsorption Isotherm

Langmuir isotherm is applicable for physical as well as chemical adsorption. It is applicable only for monolayer adsorption. Figure 24 shows the nature of Langmuir isotherm for the adsorption of chromium.

Figure 24. Langmuir Isotherm

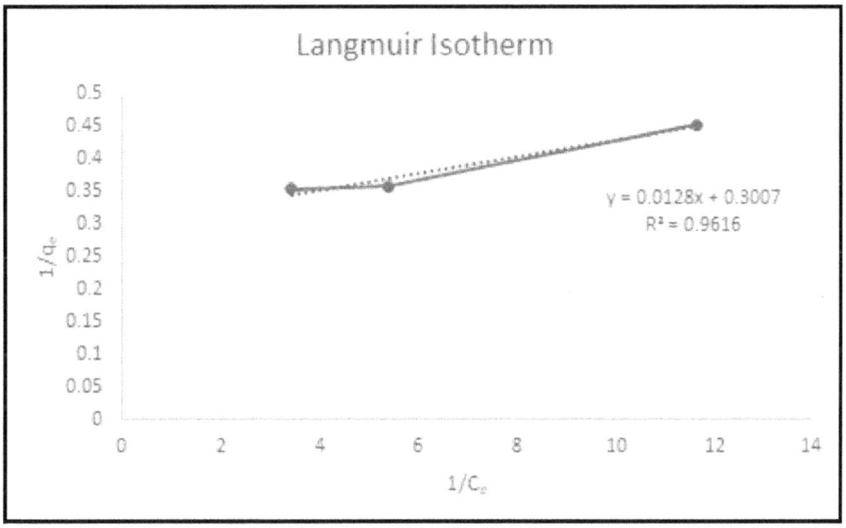

5.2. Thomas Model for Inlet Concentration

The graphs (Figures 25-26) showed that the value of R^2 was increased from 0.7674 to 0.9011 when the concentration was increased from 6 ppm to 8 ppm respectively. Thomasmodel for 6 ppm inlet concentration fits the data of adsorption well than the Yoon and Nelson Model.

Table 2. Effect of inlet concentration for Thomas model

Parameters	Concentrations at Fixed Bed Height and Flowrate		Units
	6 ppm	8 ppm	
K_t	-0.317	-0.2312	ml/min/mg
q_o	0.8558	1.831	mg/g
R^2	0.7674	0.9011	

The values of the Thomas model constant (K_t) and the maximum adsorption capacity (q_0) for different inlet concentrations were calculated from graphs and were tabulated in the Table 2. The maximum adsorption capacity is directly proportional to the inlet chromium concentration.

Figure 25. Thomas model for inlet concentration of 6 ppm

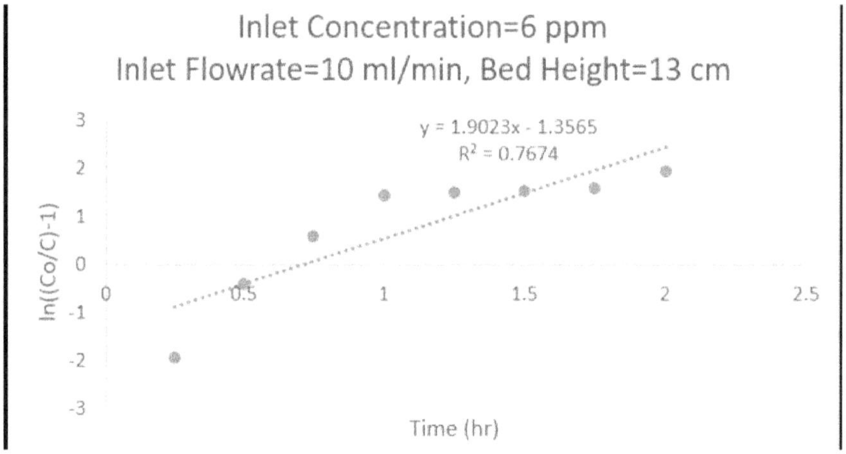

Figure 26. Thomas model for inlet concentration of 8 ppm

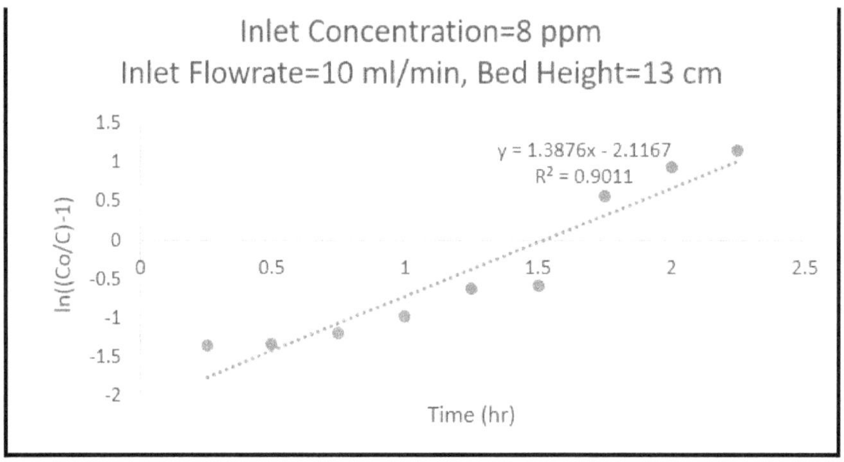

5.3. Yoon and Nelson Model for Inlet Concentration

The rate constant (k) and the time required for 50% adsorbate breakthrough (T) were calculated from the graphs (Figures 27-28) and are given in the table 3. It was noticed from the graphs that value of R^2 increases from 0.7673 to 0.9011 with increases in the concentration of Cr from 6 ppm to 8 ppm respectively at the inlet. The value of T increases as the inlet concentration increases.

5.4. Thomas Model for Bed Height

The values of the Thomas model constant (K_t) and the maximum adsorption capacity (q_o) for different bed heights were calculated from graphs (Figures 29-31). The value of R^2 changes from 0.8747 to 0.929 to 0.9103 for bed height of 9 cm, 13 cm and 18 cm respectively (Table 4). These values are tabulated in the Table 4. Adsorption capacity decreases as the bed height increases.

Table 3. Effect of inlet concentration for Yoon and Nelson model

Parameters	Concentrations at Fixed Bed Height and Flowrate		Units
	6 ppm	8 ppm	
k	-1.9025	-1.3876	min^{-1}
T	0.7129	1.5254	min
R^2	0.7673	0.9011	

Figure 27. Yoon and Nelson model for inlet concentration of 6 ppm

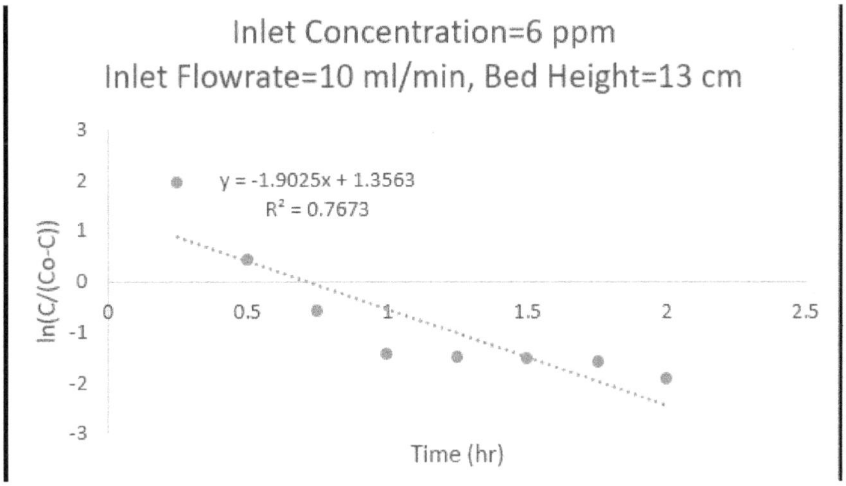

Figure 28. Yoon and Nelson model for inlet concentration of 8 ppm

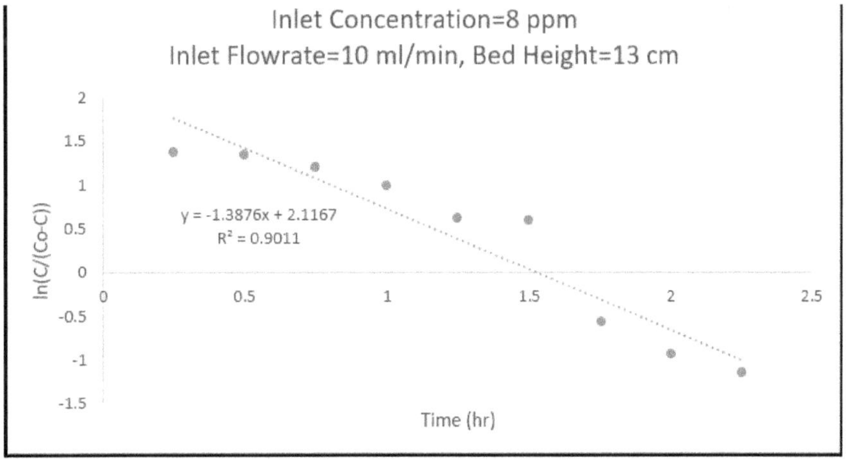

Table 4. Effect of bed height for Thomas model

Parameters	Bed Heights at Fixed Concentration and Flowrate			Units
	9 cm	13 cm	18 cm	
K_t	-0.1939	-0.2154	-0.2195	ml/min/mg
q_o	4.4881	1.8692	0.779	mg/g
R^2	0.8747	0.929	0.9103	

Figure 29. Thomas model for bed height of 9 cm

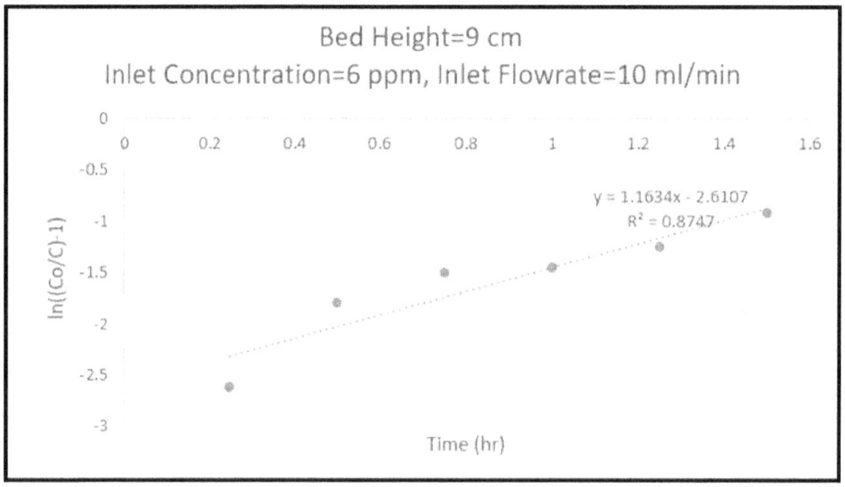

Figure 30. Thomas model for bed height of 13 cm

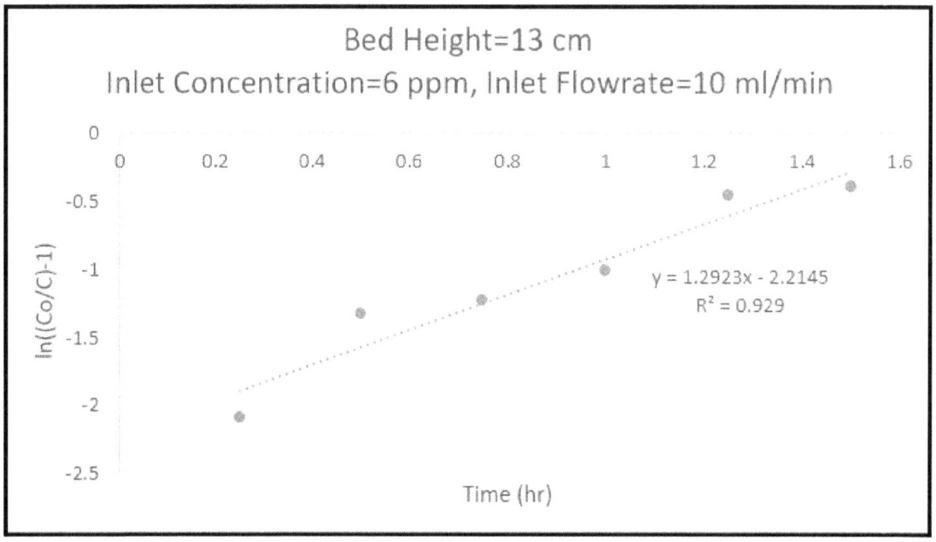

Figure 31. Thomas model for bed height of 18 cm

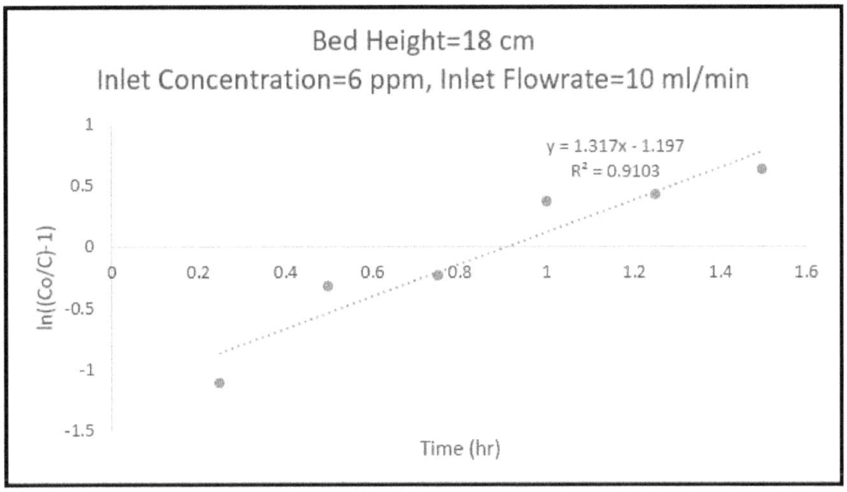

5.5. Yoon and Nelson Model for Bed Height:

The rate constant (k) and the time required for 50% adsorbate breakthrough (T) were calculated from the graphs (Figures 32-34) and are given in the Table 5. It was found from the graphs that value of R^2 changes from 0.8747 to 0.929 to 0.9103 with increasing bed height from 9 cm to 13 cm to 18 cm respectively (Table 5). Flow rate and the time required for 50% adsorbate breakthrough (T) are inversely proportional.

Table 5. Effect of bed height for Yoon and Nelson model

Parameters	Bed Heights at Fixed Concentration and Flowrate			Units
	9 cm	13 cm	18 cm	
k	-1.1634	-1.2923	-1.317	min^{-1}
T	2.244	1.7136	0.9089	min
R^2	0.8747	0.929	0.9103	

5.6. Thomas Model for Inlet Flowrate

The values of the Thomas model constant (K_t) and the maximum adsorption capacity (q_0) for different inlet flow rates were calculated from graphs and were tabulated in the Table 6. Results from graphs (Figures 35-37) shows that the value of R^2 changes from 0.7984 to 0.9302 to 0.7472 when the flow rate changes from 10 ml/min to 21 ml/min to 27 ml/min respectively. K_t decreases with increasing flow rate (Table 6).

5.7. Yoon and Nelson Model for Inlet Flowrate

With increase in flow rate time T decreases due to rapid availability of adsorbate. Table 7 and Figure 38 depicts effect of flow rate on Yoon Nelson parameters.

Figure 32. Yoon and Nelson model for bed height of 9 cm

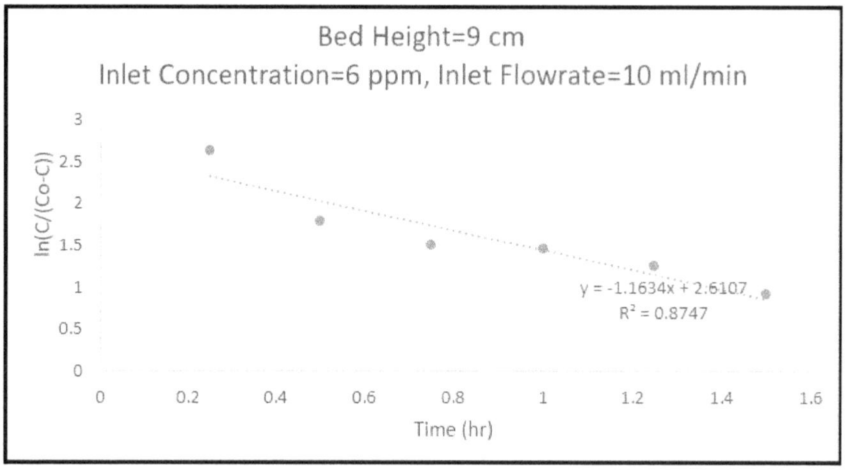

Figure 33. Yoon and Nelson model for bed height of 13 cm

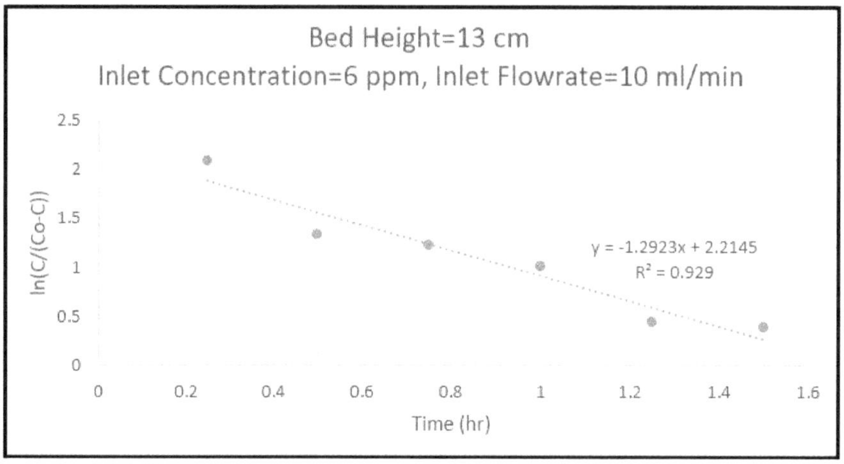

Figure 34. Yoon and Nelson model for bed height of 18 cm

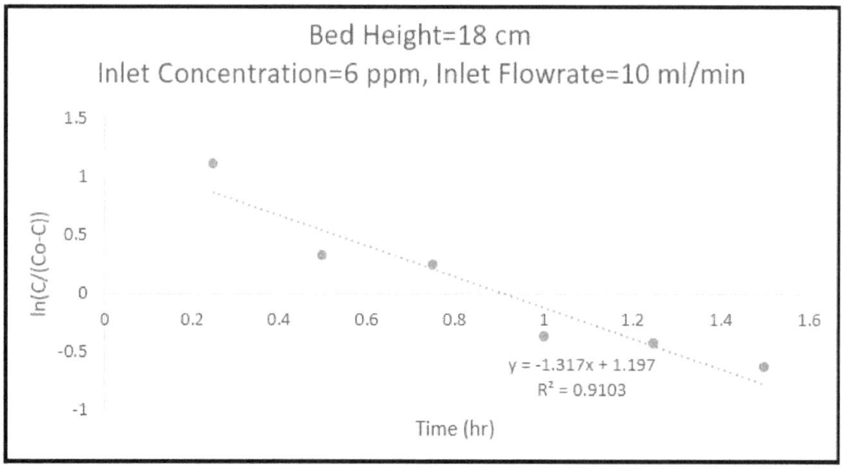

Table 6. Effect of inlet flowrate for Thomas model

Parameters	Flow Rates at Fixed Concentration and Bed Height			Units
	10 ml/min	21 ml/min	27 ml/min	
K_t	0.4053	0.2644	0.2443	ml/min/mg
q_o	1.0914	2.2135	1.1325	mg/g
R^2	0.7984	0.9302	0.7472	

Figure 35. Thomas model for inlet flowrate of 10 ml/min

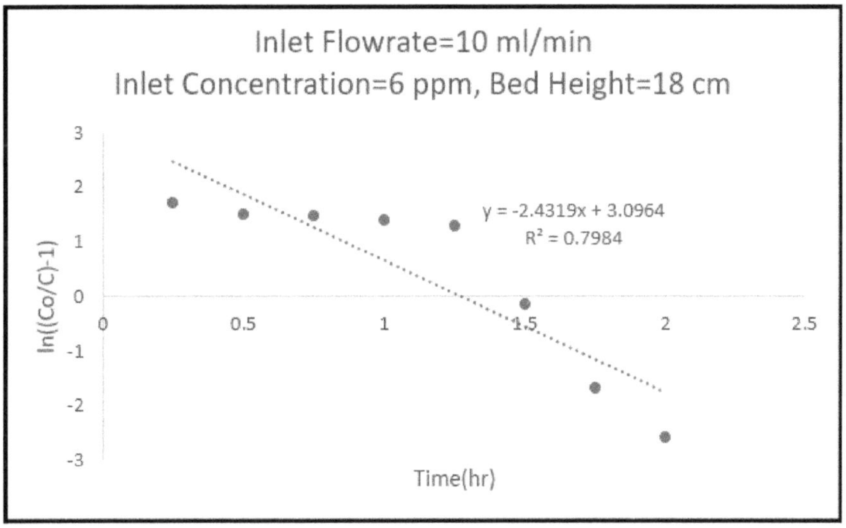

Figure 36. Thomas model for inlet flowrate of 21 ml/min

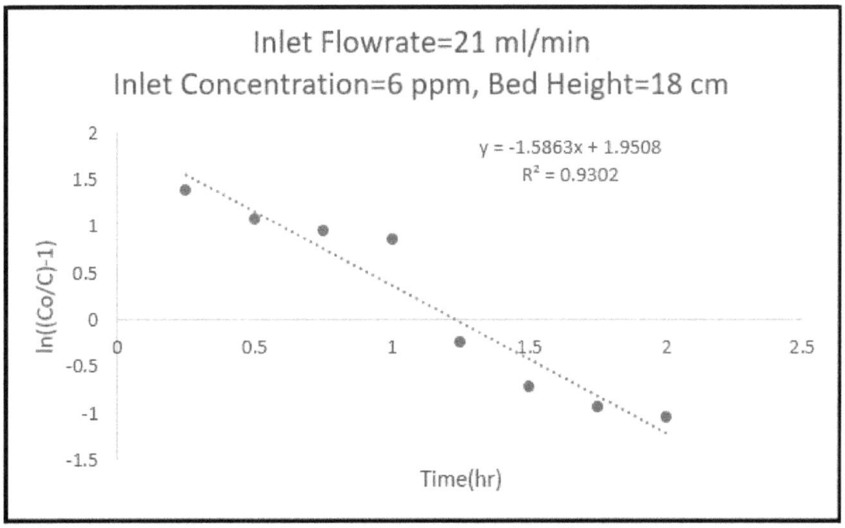

Figure 37. Thomas model for inlet flowrate of 27 ml/min

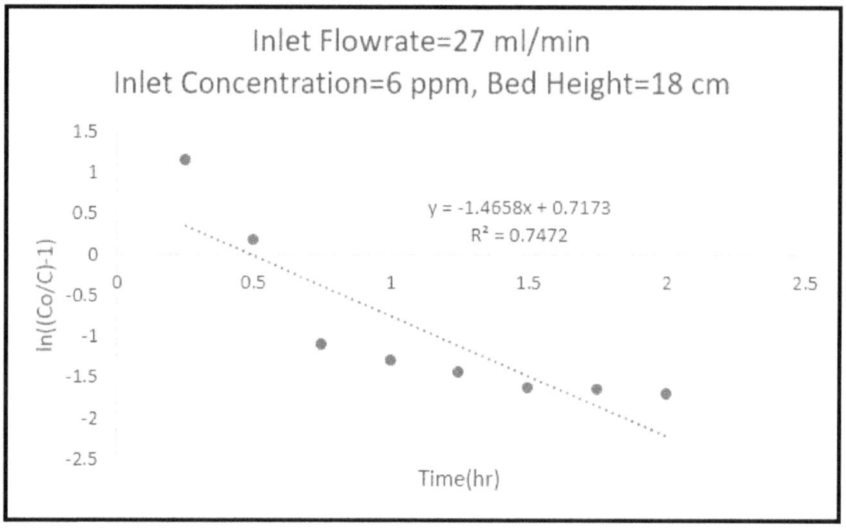

Table 7. Effect of inlet flowrate for Yoon and Nelson model

Parameters	Flow rates at Fixed Concentration and Bed Height			Units
	10 ml/min	21 ml/min	27 ml/min	
k	-2.4319	-1.5863	-1.4658	min^{-1}
T	1.2732	1.2298	0.4894	min
R^2	0.7984	0.9302	0.7472	

Figure 38. Yoon and Nelson model for inlet flowrate of 10 ml/min

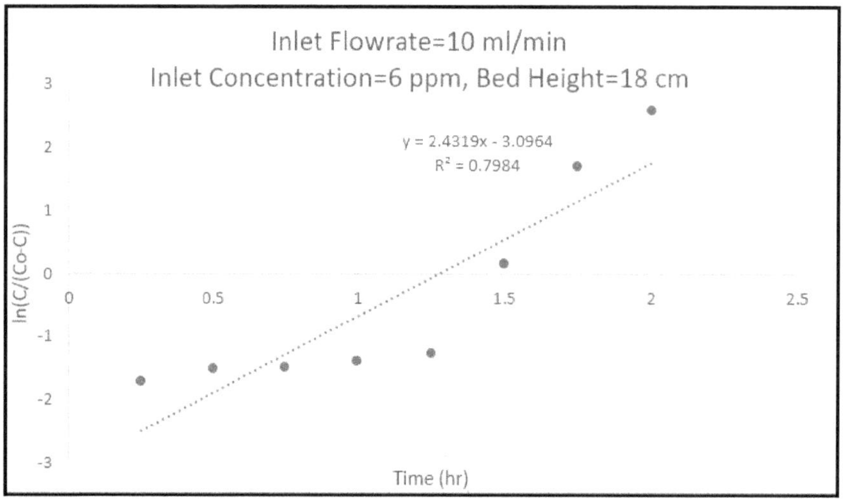

Figure 39. Yoon and Nelson model for inlet flowrate of 21 ml/min

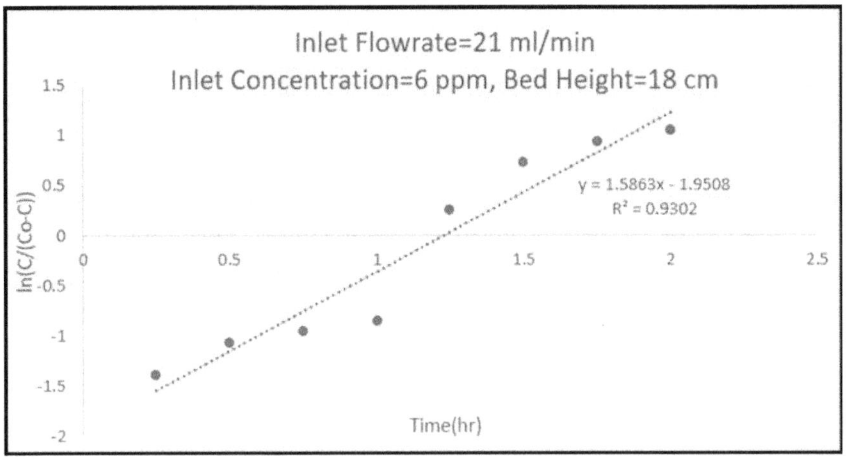

Figure 40. Yoon and Nelson model for inlet flowrate of 27 ml/min

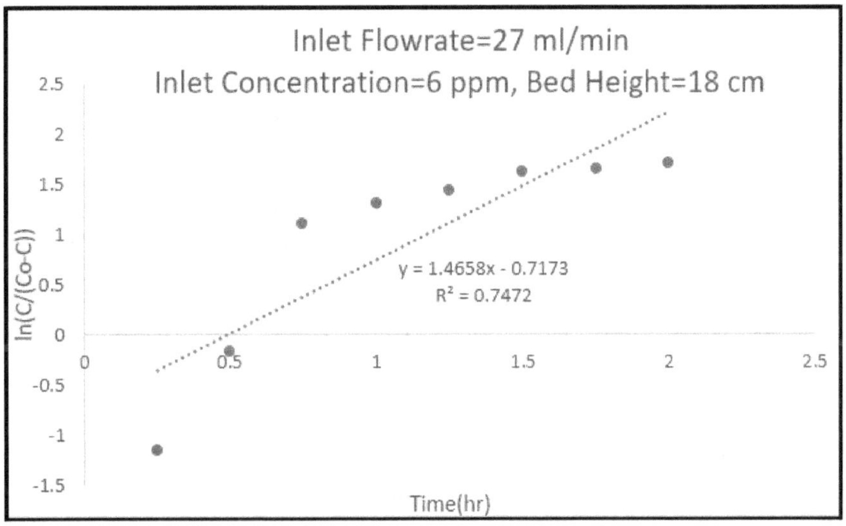

6. CONCLUSION

In this project report we have determined the unknown concentration of chromium present in the water sample. We have developed isotherms namely Freundlich Adsorption Isotherm and Langmuir Adsorption Isotherm. Different properties of adsorbent are studied. The selection of a perfect adsorbent for any process is influenced by a variety of factors such as cost, regeneration gas temperature, feed composition, and so on. The % of chromium removal increases with increasing adsorbent dosage, increasing contact time, increasing initial concentration and it changes with sample pH. The rate of adsorption is higher initially, and as time goes on, it slows down as the amount of chromium removed increases. High values of adsorption capacity are often produced by longer contact times and higher initial concentrations. With increasing pH of the chromium solution, the adsorption capacity for removing chromium ions increases.

In this adsorption process the curves obtained by performing number of experiments gives clear information about the breakthrough time. It can be determined that the breakthrough time was found earlier when the starting concentration of chromium and the bed height of the adsorbent in the column were increased, whereas the breakthrough time was found earlier when the flowrate of chromium sample was decreased. The experimental data fits well in the Thomas model and Yoon-Nelson model. The adsorption capacity increases as inlet concentration increases. The trends for each parameter were studied by plotting the graphs and breakthrough models for each parameter. Thomas model constant, maximum adsorption capacity, rate constant and time required for 50% breakthrough were calculated for each parameter.

REFERENCES

Amal, M. (2018). A General Overview on the Adsorption. *Indian Journal of Natural Science*, *9*(51), 16127–16131.

Babu, B. V., & Gupta, S. (2004). Modelling and Simulation for Dynamics of Packed Bed Adsorption. *Environmental Engineering, ChemCon'04*.

Batool, M., Khurshid, S., Qureshi, Z. M., Daoush, W. M., Hashmi, F., & Mehboob, N. (2020). Effective adsorptive removal of azodyes on synthesized copper oxide nanoparticles. *Biointerface Research in Applied Chemistry*, *10*(3), 5369–5375. doi:10.33263/BRIAC103.369375

Chu, K. H. (2020). Breakthrough curve analysis by simplistic models of fixed bed adsorption: In defense of the century-old Bohart-Adams model. *Chemical Engineering Journal*, *380*, 122513. doi:10.1016/j.cej.2019.122513

Dutta, B. K. (2009). *Principles of Mass Transfer and Separation Process*. PHI Learning Private Limited.

El-Naas, M. (2013). *Modelling of adsorption process article*. Academic Press.

Foo, K. Y., & Hameed, B. H. (2010). Insight into the modelling of adsorption isotherm systems. *Chemical Engineering Journal*, *156*(1), 2–10. doi:10.1016/j.cej.2009.09.013

Kaur, J. (2017). *Adsorption Isotherms*. Academic Press.

Kulkarni, S. J. (2020). Synthesis, Characterization and Performance of Low-Cost Unconventional Adsorbents Derived from Waste Materials. *Biointerface Research in Applied Chemistry*, *10*(6), 7243–7256. doi:10.33263/BRIAC106.72437256

Kulkarni, S. J., Mishra, A., Patil, S., & Uphale, K. (2018). Modeling of Breakthrough Curves for Removal of Chromiumfrom Waste Water. *International Journal of Advanced Research in Chemical Sciences*, *5*(6), 29–40.

Liakos E. V., Mone M., Lambropoulou D. A., Bikiaris D. N., & Kyzas G. Z. (2021). Adsorption Evaluation for the Removal of Nickel, Mercury and Barium Ions from Single-Component and Mixtures of Aqueous Solutions by Using an Optimized Biobased Chitosan Derivative. *Polymers (Basel), 13*(2), 1-20.

Malkoc, E., & Nuhoglu, Y. (2007). Potential of Tea Factory Waste for Chromium (VI) Removal from Aqueous Solu-tions: Thermodynamic and Kinetic Studies. *Separation and Purification Technology*, *54*(2), 291–298. doi:10.1016/j.seppur.2006.09.017

Mitra, S., Sarkar, A., & Sen, S. (2017). Removal of Chromium from Industrial Effluents using Nano-technology: A Review. *Nanotechnology Environmental Engineering*, *2*(11), 1–14. doi:10.100741204-017-0022-y

Samarghandi, M. R., Hadi, M., & McKay, G. (2014). Breakthrough Curve Analysis for Fixed-Bed Adsorption of Azo Dyes using Novel Pine Cone Derived Active Carbon. *Adsorption Science and Technology*, *32*(10), 791–806. doi:10.1260/0263-6174.32.10.791

Shanmugasundaram, M., & Sudalaimani, K. (2012). A study on natural adsorbents for the removal of chloride ion in water. *International Journal of Engineering Research & Technology (Ahmedabad)*, *1*(5), 1–12.

Sircar, S., Golden, T. C., & Rao, M. B. (1996). Activated carbon for gas separation and storage. *Carbon*, *34*(1), 1–12. doi:10.1016/0008-6223(95)00128-X

Treybal, R. E. (1981). *Mass Transfer Operations*. McGraw-Hill Book Company.

Unuabonah, E. I., Omorogie, M. O., & Oladoja, N. A. (2019). *Modelling in Adsorption: Fundamentals and Applications Composite Nanoadsorbents*. Micro and Nano Technologies.

Compilation of References

A Complex formula: girls and women in science, technology, engineering and mathematics in Asia. (2015). UNESCO Digital Library.

Abdussalam, O., Trochu, J., Fello, N., & Chaabane, A. (2021). Recent advances and opportunities in planning green petroleum supply chains: A model-oriented review. *International Journal of Sustainable Development and World Ecology*, *28*(6), 524–539. Advance online publication. doi:10.1080/13504509.2020.1862935

Abel, A. (2012). The history of dyes and pigments: From natural dyes to high performance pigments. In *Colour design* (pp. 557–587). Woodhead Publishing. doi:10.1016/B978-0-08-101270-3.00024-2

Abraham-Dukuma, M. C. (2021). Dirty to clean energy: Exploring 'oil and gas majors transitioning''.'. *The Extractive Industries and Society*, *8*(3), 100936. Advance online publication. doi:10.1016/j.exis.2021.100936

Accelerating Women's Advancement in STEM. (2021). *The Asia Foundation*, https://asiafoundation.org/wp-content/uploads/2021/06/Accelerating-Womens-Advancement-in-STEM_Report_update7.26.21.pdf

Acharyya, J. (2009). FDI, Growth and the Environment: Evidence from India on CO_2 Emission during the Last Two Decades. *Journal of Economic Development*, *34*(1), 43–58. doi:10.35866/caujed.2009.34.1.003

Adadzi, F. D., Godson-Amamoo, N. S., & Nunoo, J. (2022). The Ghanaian state and governance of the upstream oil and gas industry. In *Petroleum Resource Management in Africa*. Lessons from Ten Years of Oil and Gas Production in Ghana. doi:10.1007/978-3-030-83051-9_12

Adebiyi, F. M. (2022). Air quality and management in petroleum refining industry: A review. *Environmental Chemistry and Ecotoxicology*, *4*, 89–96. Advance online publication. doi:10.1016/j.enceco.2022.02.001

Adeel, S., Abrar, S., Kiran, S., Farooq, T., Gulzar, T., & Jamal, M. (2018). *Sustainable application of natural dyes in cosmetic industry. In Handbook of Renewable Materials for Coloration and Finishing*. John Wiley & Sons, Inc.

Adiono, T., Fathany, M. Y., Putra, R. V. W., Afifah, K., Santriaji, M. H., Lawu, B. L., & Fuada, S. (2016). Live demonstration: MINDS - Meshed and internet networked devices system for smart home: Track selection: Embedded systems. *Proceedings of the 2016 IEEE Asia Pacific Conference on Circuits and Systems, APCCAS*, 736–737.

Ado, A., Yahaya, H., Kwalli, A. A., & Abdulkadir, R. S. (2014). Dyeing of textiles with eco-friendly natural dyes: A review. *International Journal of Environmental Monitoring and Protection*, *1*(5), 76–81.

Adrosko, R. J. (2012). *Natural dyes and home dyeing* (Vol. 281). Courier Corporation.

Agarwal, A., Singh, P., & Kumar, A. (2021). Internet of Things (IoT) for smart cities: challenges and opportunities in India. *Journal of Ambient Intelligence*.

Ahmad, M., Ahmed, Z., Majeed, A., & Huang, B. (2021). An environmental impact assessment of economic complexity and energy consumption: Does institutional quality make a difference? *Environmental Impact Assessment Review*, *89*, 9. doi:10.1016/j.eiar.2021.106603

Ahmad, M., Khayal, M. S. H., & Tahir, A. (2022). Analysis of Factors Affecting Adoption of Volunteered Geographic Information in the Context of National Spatial Data Infrastructure. *ISPRS International Journal of Geo-Information*, *11*(2), 120. Advance online publication. doi:10.3390/ijgi11020120

Ahmad, R. W., Salah, K., Jayaraman, R., Yaqoob, I., & Omar, M. (2022). Blockchain in oil and gas industry: Applications, challenges, and future trends. *Technology in Society*, *68*, 101941. Advance online publication. doi:10.1016/j.techsoc.2022.101941

Ahmed, I., Zhang, Y., Jeon, G., Lin, W., Khosravi, M. R., & Qi, L. (2022). A blockchain- and artificial intelligence-enabled smart IoT framework for sustainable city. *International Journal of Intelligent Systems*, *37*(9), 6493–6507. Advance online publication. doi:10.1002/int.22852

Ajaps & McLellan. (2015). We don't know enough": Environmental education and pro-environmental behaviour perceptions. *Cogent Education, 2*(1), 1-17.

Akhayere, E., Kartal, M. T., Adebayo, T. S., & Kavaz, D. (2023). Role of energy consumption and trade openness towards environmental sustainability in Turkey. *Environmental Science and Pollution Research International*, *30*(8), 21156–21168. doi:10.100711356-022-23639-9 PMID:36261639

Akinyemi, F. O. (2007). Spatial data needs for poverty management. *Research and Theory in Advancing Spatial Data Infrastructure Concepts, 5.*

Akkaya, H., & Gerçek, A. (2019). The Taxation of Digital Economy in OECD and some Seleceted Countries: Inferences for Turkey. *International Journal of Public Finance*, *4*(2), 166–188. doi:10.30927/ijpf.634050

Alam, M. A., & Saiyeda, A. (2020). A Cloud Based Solution for Smart Education. *International Journal of Smart Education and Urban Society*, *11*(2), 28–37. Advance online publication. doi:10.4018/IJSEUS.2020040103

Alam, M. J., Begum, I. A., Buysse, J., Rahman, S., & Huylenbroeck, G. V. (2011). Dynamic Modeling of Causal Relationship between Energy Consumption, CO_2 Emissions and Economic Growth in India. *Renewable & Sustainable Energy Reviews*, *15*(6), 3243–3251. doi:10.1016/j.rser.2011.04.029

Albeldawi, M. (2023). Environmental impacts and mitigation measures of offshore oil and gas activities. In Developments in Petroleum Science (Vol. 78). doi:10.1016/B978-0-323-99285-5.00002-8

Aldhshan, S. R. S., Abdul Maulud, K. N., Wan Mohd Jaafar, W. S., Karim, O. A., & Pradhan, B. (2021). Energy consumption and spatial assessment of renewable energy penetration and building energy efficiency in Malaysia: A review. *Sustainability (Basel)*, *13*(16), 9244. doi:10.3390u13169244

Alhamwi, A., Medjroubi, W., Vogt, T., & Agert, C. (2017). OpenStreetMap data in modelling the urban energy infrastructure: A first assessment and analysis. *Energy Procedia*, *142*, 1968–1976. Advance online publication. doi:10.1016/j.egypro.2017.12.397

Alharthi, H. (2018). Healthcare predictive analytics: An overview with a focus on Saudi Arabia. *Journal of Infection and Public Health*, *11*(6), 749–756. doi:10.1016/j.jiph.2018.02.005 PMID:29526444

Aliyu, M. A. (2005). Foreign Direct Investment and the Environment: Pollution Haven Hypothesis Revisited. *8th Annual Conference on Global Economic Analysis*, Lubeck, Germany.

Alkraiji, A. I., Jackson, T., & Murray, I. (2013). Barriers to the Widespread Adoption of Health Data Standards: An Exploratory Qualitative Study in Tertiary Healthcare Organizations in Saudi Arabia. *Journal of Medical Systems*, *37*(2), 1–13. doi:10.100710916-012-9895-2 PMID:23321966

Allen, D. T., Chen, Q., & Dunn, J. B. (2021). Consistent Metrics Needed for Quantifying Methane Emissions from Upstream Oil and Gas Operations. *Environmental Science & Technology Letters*, *8*(4), 345–349. Advance online publication. doi:10.1021/acs.estlett.0c00907

Al-Rasheedi, G., & Khan, N. (2021). The Role of Cloud Computing in Education: Saudi Arabian Study. *International Journal of Interactive Mobile Technologies*, *15*(24), 191–200. Advance online publication. doi:10.3991/ijim.v15i24.27363

Alreshidi, E., Mourshed, M., & Rezgui, Y. (2018). Requirements for cloud-based BIM governance solutions to facilitate team collaboration in construction projects. *Requirements Engineering*, *23*(1), 1–31. Advance online publication. doi:10.100700766-016-0254-6

Al-Rwaidan, R. M., Aldossary, N., Eldahamsheh, M. M., Al-Azzam, M. K. A., Al-Quran, A. Z., & Al-Hawary, S. I. S. (2023). The impact of cloud-based solutions on digital transformation of HR practices. *International Journal of Data and Network Science*, *7*(1), 83–90. Advance online publication. doi:10.5267/j.ijdns.2022.12.003

Al-Samhan, M., Al-Fadhli, J., Al-Otaibi, A. M., Al-Attar, F., Bouresli, R., & Rana, M. S. (2022). Prospects of refinery switching from conventional to integrated: An opportunity for sustainable investment in the petrochemical industry. In Fuel (Vol. 310). doi:10.1016/j.fuel.2021.122161

Alsolami, E. (2018). Security threats and legal issues related to Cloud based solutions. *International Journal of Computer Science and Network Security, 18*(5).

Al-Turjman, F.; Malekloo, A. (2019). Smart parking in IoT-enabled cities: A survey. *Sustain. Cities Soc., 49*.

Al-Turjman, F. (2018). Internet of things for smart cities: Technologies and applications. *Intelligent Buildings International*, *10*(4), 223–239.

Amadi, S., & Beutler, J. (2019). *Open Mapping for SDGs*. https://sdgs.hotosm.org/

Amal, M. (2018). A General Overview on the Adsorption. *Indian Journal of Natural Science*, *9*(51), 16127–16131.

Amin, I., Zailani, S., & Rahman, M. K. (2020). Predicting employees' engagement in environmental behaviours with supply chain firms. *Management Research Review*, *44*(6), 825–848. Advance online publication. doi:10.1108/MRR-05-2020-0280

Andersen, A. D., & Gulbrandsen, M. (2020). The innovation and industry dynamics of technology phase-out in sustainability transitions: Insights from diversifying petroleum technology suppliers in Norway. *Energy Research & Social Science*, *64*, 101447. Advance online publication. doi:10.1016/j.erss.2020.101447

Andreão, R. V., Athayde, M., Boudy, J., Aguilar, P., de Araujo, I., & Andrade, R. (2018). Raspcare: A Telemedicine Platform for the Treatment and Monitoring of Patients with Chronic Diseases. In Assistive Technologies in Smart Cities. IntechOpen.

Anil, B. C., Dayananda, P., Nethravathi, B., & Raisinghani, M. S. (2022). Efficient local cloud-based solution for liver cancer detection using deep learning. *International Journal of Cloud Applications and Computing*, *12*(1), 1–13. Advance online publication. doi:10.4018/IJCAC.2022010109

Antweiler, W., Copeland, B. R., & Taylor, M. S. (2001). Is Free Trade Good for the Environment? *The American Economic Review*, *91*(4), 877–908. doi:10.1257/aer.91.4.877

Aponte, F. G. (2021). Innovación verde: Indicadores y principales tendencias. *Gestión I+D*, 12-48.

Araújo, S. O., Peres, R. S., Barata, J., Lidon, F., & Ramalho, J. C. (2021). Characterising the agriculture 4.0 landscape—emerging trends, challenges and opportunities. In Agronomy (Vol. 11, Issue 4). doi:10.3390/agronomy11040667

Arnott, J. C., & Lemos, M. C. (2021). Understanding knowledge use for sustainability. *Environmental Science & Policy*, *120*, 222–230. doi:10.1016/j.envsci.2021.02.016

Arogyaswamy, B. (2020). Big tech and societal sustainability: An ethical framework. *AI & Society*, *35*(4), 829–840. doi:10.100700146-020-00956-6 PMID:32218647

Arora, J., Agarwal, P., & Gupta, G. (2017). Rainbow of natural dyes on textiles using plants extracts: Sustainable and eco-friendly processes. *Green and Sustainable Chemistry*, *7*(1), 35–47. doi:10.4236/gsc.2017.71003

Asia InCH. (2023). *Natural Dye Sources*. Retrieved from https://asiainch.org/craft/natural-dye-sources/

Asteriou, D., & Hall, S. G. (2007). *Applied Econometrics: A Modern Approach Using Eviews and Microfit* (Revised Edition). Palgrave Macmillan.

Attenborough, D. (2021). Foreword. In P. Dasgupta (Ed.), *Economics of biodiversity: Dasgupta review* (pp. 1–2). HM Treasury.

Attia, A. M., Ghaithan, A. M., & Duffuaa, S. O. (2019). A multi-objective optimization model for tactical planning of upstream oil & gas supply chains. *Computers & Chemical Engineering*, *128*, 216–227. Advance online publication. doi:10.1016/j.compchemeng.2019.06.016

AWS. (2017). *OpenStreetMap Public Data Set Now Available on AWS*. https://aws.amazon.com/about-aws/whats-new/2017/06/openstreetmap-public-data-set-now-available-on-aws/

Aytun, C., & Akın, C. S. (2014). Telecommunications infrastructure and economic growth in OECD countries: Cross-Section Dependent Heterogeneous Panel Causality Analysis. *Economics. Business & Finance*, *29*(340), 69–94.

Babu, B. V., & Gupta, S. (2004). Modelling and Simulation for Dynamics of Packed Bed Adsorption. *Environmental Engineering, ChemCon'04.*

Balassa, B. (1986). Policy responses to exogenous shocks in developing countries. *The American Economic Review*, *5*, 75–78.

Baldassarre, B., Calabretta, G., Bocken, N. M. P., & Jaskiewicz, T. (2017). Bridging sustainable business model innovation and user-driven innovation: A process for sustainable value proposition design. *Journal of Cleaner Production*, *147*, 175–186. doi:10.1016/j.jclepro.2017.01.081

Balducci, F. (2021). Is OpenStreetMap a good source of information for cultural statistics? The case of Italian museums. *Environment and Planning. B, Urban Analytics and City Science*, *48*(3), 503–520. Advance online publication. doi:10.1177/2399808319876949

Balfour-Paul, J. (1999). Indigo in South and South-East Asia. *Textile History*, *30*(1), 98–112. doi:10.1179/004049699793710688 PMID:22523776

Bandyopadhyay, D., & Sen, J. (2011). Internet of things: Applications and challenges in technology and standardization. *Wireless Personal Communications*, *58*(1), 49–69. doi:10.100711277-011-0288-5

Barışık, S., & Yirmibeşcik, O. (2006). Harmonization efforts aimed at accelerating the formation process of the new economy in Turkey. *ZKU Journal of Social Sciences*, *2*(4), 39–62.

Barney, J. (1991). Firm resources and sustained competitive advantage. *Journal of Management*, *17*(1), 99–120. doi:10.1177/014920639101700108

Baruah, D. (2022). STEMming Gender Divide in India. *The British Councel*. https://education-services.britishcouncil.org/insights-blog/stemming-gender-divide-india

Basile, V., Capobianco, N., & Vona, R. (2021). The usefulness of sustainable business models: Analysis from oil and gas industry. *Corporate Social Responsibility and Environmental Management*, *28*(6), 1801–1821. Advance online publication. doi:10.1002/csr.2153

Baskarada, S., & Koronios, A. (2013). Data, Information, Knowledge, Wisdom (DIKW): A Semiotic Theoretical and Empirical Exploration of the Hierarchy and its Quality Dimension. *AJIS. Australasian Journal of Information Systems*, *18*(1), 5–24. doi:10.3127/ajis.v18i1.748

Batool, M., Khurshid, S., Qureshi, Z. M., Daoush, W. M., Hashmi, F., & Mehboob, N. (2020). Effective adsorptive removal of azodyes on synthesized copper oxide nanoparticles. *Biointerface Research in Applied Chemistry*, *10*(3), 5369–5375. doi:10.33263/BRIAC103.369375

Bazzaz, F., Ceballos, C., Davis, M., Dirzo, R., Ehrlich, P. R., Eisner, T., Levin, S., Lawton, J. H., Lubchenco, J., Matson, P. A., Mooney, H. A., Raven, P. H., Roughgarden, J. E., Sarukhan, J., Tilman, C. D., Vitousek, P., Walker, B., Wall, D. H., Wilson, E., ... Woodwell, G. M. (1998). Ecological science and the human predicament. *Science*, *282*(5390), 879. doi:10.1126cience.282.5390.879c

Bechtold, T., Turcanu, A., Ganglberger, E., & Geissler, S. (2003). Natural dyes in modern textile dyehouses—How to combine experiences of two centuries to meet the demands of the future? *Journal of Cleaner Production*, *11*(5), 499–509. doi:10.1016/S0959-6526(02)00077-X

Beck, C., Rashidbeigi, S., Roelofsen, O., & Speelman, E. (2020). *The future of oil and gas is now: How companies can decarbonize.* McKinsey & Company.

Bedely, R. (2017). *Understanding the Role of Business Analytics Applications in Healthcare Value Chain*. Academic Press.

Behera, S., Khandual, A., & Luximon, Y. (2019). An Insight In To the Ikat Technology in India: Ancient To Modern Era. *IOSR Journal of Polymer and Textile Engineering*.

Bekar, M., Tüfekci, A., Yalçın, Y., & Emirmahmutoğlu, F. (2022). Panel data analysis of the relationship between telecommunications and economic growth. *Journal of Economics. Policy & Finance Studies*, *7*(2), 366–385.

Ben-Naceur, K. (2019). How the Oil and Gas Industry Is Contributing to Sustainability. *Journal of Petroleum Technology*, *71*(03), 38–39. Advance online publication. doi:10.2118/0319-0038-JPT

Benoit, K. (2011). *Linear Regression Models with Logarithmic Transformations*. London School of Economics.

Berry, L. L., Shankar, V., Parish, J. T., Cadwallader, S., & Dotzel, T. (2006). Creating new marketsthrough service innovation. *Sloan Management Review*, *47*(2), 56–62.

Bertolotto, M., McArdle, G., & Schoen-Phelan, B. (2020). Volunteered and crowdsourced geographic information: The openstreetmap project. *Journal of Spatial Information Science*, *20*(20). Advance online publication. doi:10.5311/JOSIS.2020.20.659

Beutler, J. (2021). *OpenStreetMap for Emergency Preparedness in the US*. https://www.openstreetmap.us/news/2021/06/map-for-emergency

Bhuiyan, M. R., Islam, A., Ali, A., & Islam, M. N. (2017). Color and chemical constitution of natural dye henna (Lawsonia inermis L) and its application in the coloration of textiles. *Journal of Cleaner Production*, *167*, 14–22. doi:10.1016/j.jclepro.2017.08.142

Bigliardi, B., & Bertolini, M. (2012). Green innovation management: Theory and practice. *European Journal of Innovation Management*, *15*(4). Advance online publication. doi:10.1108/ejim.2012.22015daa.001

Binh, P. T. (2013). *Unit Root Tests, Cointegration, ECM, VECM and Causality Models*. Topics in Time Series Econometrics.

Blanco, R., Fontrodona, J., & Poveda, C. (2017). *La Industria 4.0: El estado de la cuestión*. Academic Press.

Blanco, F., Castro, J. M., Gayoso, R. A., & Santana, W. (2020). *Las claves de la cuarta revolución industrial : ¿cómo afectará a los negocios y a las personas?* ECOE Ediciones.

Boca, G. D., & Sinan, S. (2019). Environmental education and student's perception for sustainability. *Sustainability (Basel)*, *11*(6), 1553. doi:10.3390u11061553

Bocken, Short, Rana, & Evans. (2014). A literature and practice review to develop sustainable business model archetypes. *Clean. Prod.*, *65*, 42-56. U6 doi:10.1016/j.jclepro.2013.11.039

Bogdan, R., & Biklen, S. K. (2007). Investigación Cualitativa en Educación: Una introducción a teoría y métodos (4ta. ed.). Alfaomega.

Bokpin, G. A. (2017). Foreign direct investment and environmental sustainability in Africa: The role of institutions and governance. *Research in International Business and Finance*, *39*, 239–247. doi:10.1016/j.ribaf.2016.07.038

Bos, K., & Gupta, J. (2019). Stranded assets and stranded resources: Implications for climate change mitigation and global sustainable development. In Energy Research and Social Science (Vol. 56). doi:10.1016/j.erss.2019.05.025

Boulatoff, C., & Jenkins, M. (2010). Long-term Nexus Between Openness, Income, and Environmental Quality. *International Advances in Economic Research*, *16*(4), 410–418. doi:10.100711294-010-9283-y

Bradu, P., Biswas, A., Nair, C., Sreevalsakumar, S., Patil, M., Kannampuzha, S., Mukherjee, A. G., Wanjari, U. R., Renu, K., Vellingiri, B., & Gopalakrishnan, A. V. (2022). Recent advances in green technology and Industrial Revolution 4.0 for a sustainable future. *Environmental Science and Pollution Research International*, 1–32. doi:10.100711356-022-20024-4 PMID:35397034

Brandon, G., & Lewis, A. (1999). Reducing household energy consumption: A qualitative and quantitative field study. *Journal of Environmental Psychology*, *19*(1), 75–85. doi:10.1006/jevp.1998.0105

Brandt, D., Alnyme, O., & Heldt, T. (2022). Building a spatial decision support system for tourism and infrastructure planning: Technical solution and data integration challenges. *European Journal of Geography*, *13*(1), 94–108. Advance online publication. doi:10.48088/ejg.d.bra.13.1.094.108

Buchholz, H., Eberle, T., Klevesath, M., Jürgens, A., Beal, D., Baic, A., & Radeke, J. (2020). Forward Thinking for Sustainable Business Value: A New Method for Impact Valuation. *Sustainability (Basel)*, *12*(20), 8420. doi:10.3390u12208420

Buchner, B., Falconer, A., Hervé-Mignucci, M., Trabacchi, C., & Brinkman, M. (2011). *The landscape of climate finance*. Climate Policy Initiative.

Bukth, R., & Heeks, R. (2017). *Defining, conceptualising and measuring the digital economy.* Working Paper Series, Paper No. 68.

Burrows, L. C., Haeri, F., Cvetic, P., Sanguinito, S., Shi, F., Tapriyal, D., Goodman, A., & Enick, R. M. (2020). A Literature Review of CO2, Natural Gas, and Water-Based Fluids for Enhanced Oil Recovery in Unconventional Reservoirs. *Energy & Fuels*, *34*(5), 5331–5380. Advance online publication. doi:10.1021/acs.energyfuels.9b03658

Buslaev, G., Morenov, V., Konyaev, Y., & Kraslawski, A. (2021). Reduction of carbon footprint of the production and field transport of high-viscosity oils in the Arctic region. *Chemical Engineering and Processing*, *159*, 108189. Advance online publication. doi:10.1016/j.cep.2020.108189

Cabrera, J., Irfan, M., Dai, Y., Zhang, P., Zong, Y., & Liu, X. (2021). Bioelectrochemical system as an innovative technology for treatment of produced water from oil and gas industry: A review. *Chemosphere*, *285*, 131428. Advance online publication. doi:10.1016/j.chemosphere.2021.131428 PMID:34237499

Cai, Y. J., & Choi, T. M. (2020). A United Nations' Sustainable Development Goals perspective for sustainable textile and apparel supply chain management. *Transportation Research Part E, Logistics and Transportation Review*, *141*, 102010. doi:10.1016/j.tre.2020.102010 PMID:32834741

Camilleri, M.A. & Falzon, L. (2020). Understanding motivations to use online streaming services: Integrating the technology acceptance model(TAM) and the uses and gratifications theory (UGT). *Spanish Journal of Marketing*. doi:10.1108/SJME-04-2020-0074

Carlson, B. (2004). The digital economy: What is new and what is not? *Structural Change and Economic Dynamics*, *15*(3), 245–264. doi:10.1016/j.strueco.2004.02.001

Cash, D. W., Clark, W. C., Alcock, F., Dickson, N. M., Selin, N., Guston, D., Jager, J., & Mitchell, R. (2003). Knowledge systems for sustainable development. *Proceedings of the National Academy of Sciences of the United States of America*, *100*(14), 8086–8091. doi:10.1073/pnas.1231332100 PMID:12777623

Castellacci, F., & Mee, C. (2017). A taxonomy of green innovators: Empirical evidence from South Korea. *Journal of Cleaner Production*, *143*, 1036–1047. doi:10.1016/j.jclepro.2016.12.016

Castro, R., Tierra, A., & Luna, M. (2019). Assessing the horizontal positional accuracy in openstreetmap: A big data approach. *Advances in Intelligent Systems and Computing*, *931*, 513–523. Advance online publication. doi:10.1007/978-3-030-16184-2_49

Cesaratto, A., Leona, M., & Pozzi, F. (2019). Recent advances on the analysis of polychrome works of art: SERS of synthetic colorants and their mixtures with natural dyes. *Frontiers in Chemistry*, *7*, 105. doi:10.3389/fchem.2019.00105 PMID:30886843

Chakraborty, D., & Mukherjee, S. (2013). How do Trade and Investment Flows affect Environmental Sustainability? Evidence from Panel Data. *Environmental Development*, *6*, 34–47. doi:10.1016/j.envdev.2013.02.005

Chamalwa, H. A., & Bakari, H. R. (2016). A Vector Autoregressive (VAR) Cointegration and Vector Error Correction Model (VECM) Approach for Financial Deepening Indicators and Economic Growth in Nigeria. *American Journal of Mathematical Analysis*, *4*(1), 1–6.

Chandia, K. E., Gul, I., Aziz, S., Sarwar, B., & Zulfiqar, S. (2018). An analysis of the association among carbon dioxide emissions, energy consumption and economic performance: An econometric model. [CrossRef]. *Carbon Management*, *9*(3), 227–241. doi:10.1080/17583004.2018.1457930

Chaula, J. (2019). Opportunities and Challenges for Integrating Statistical and Spatial Data in East African Countries. *American Scientific Research Journal for Engineering, 59*(1).

Chavan, R. B. (2013). *Health and environmental hazards of synthetic dyes*. Garment and Fashion Design at the University of Bahir Dar.

Chavula, H. K. (2013). Telecommunications development and economic growth in Africa. *Information Technology for Development*, *19*(1), 5–23. doi:10.1080/02681102.2012.694794

Chen, K. C., Kulkarni, P., McMullen, P., Biddick, D., Nader, L., & Sellers, C. (2019). High reliability protector-less artificial lift technology for electrical submersible pumps/compressors. *Society of Petroleum Engineers - SPE Gulf Coast Section Electric Submersible Pumps Symposium 2019, ESP 2019*. 10.2118/194391-MS

Chen, X., Wang, M., Wang, B., Hao, H., Shi, H., Wu, Z., Chen, J., Gai, L., Tao, H., Zhu, B., & Wang, B. (2023). Energy Consumption Reduction and Sustainable Development for Oil & Gas Transport and Storage Engineering. In Energies (Vol. 16, Issue 4). doi:10.3390/en16041775

Chen, B., Tu, Y., Song, Y., Theobald, D. M., Zhang, T., Ren, Z., Li, X., Yang, J., Wang, J., Wang, X., Gong, P., Bai, Y., & Xu, B. (2021). Mapping essential urban land use categories with open big data: Results for five metropolitan areas in the United States of America. *ISPRS Journal of Photogrammetry and Remote Sensing*, *178*, 203–218. Advance online publication. doi:10.1016/j.isprsjprs.2021.06.010

Cheng, C. Y., Chien, M. S., & Lee, C. C. (2021). ICT diffusion, financial development, and economic growth: An international cross-country analysis. *Economic Modelling*, *94*, 662–671. doi:10.1016/j.econmod.2020.02.008

Chen, Q., Dunn, J. B., & Allen, D. T. (2019). Aggregation and Allocation of Greenhouse Gas Emissions in Oil and Gas Production: Implications for Life-Cycle Greenhouse Gas Burdens. *ACS Sustainable Chemistry & Engineering*, *7*(20), 17065–17073. Advance online publication. doi:10.1021/acssuschemeng.9b03136

Chen, X., Shao, N., Zhang, H., Fang, W., Xu, X., Xu, T., Lv, H., & Li, Y. (2022). A Green and Low-Carbon Transformation of Oil and Gas Companies. *Advances in Transdisciplinary Engineering*, *23*. Advance online publication. doi:10.3233/ATDE220317

Cherepovitsyn, A., Rutenko, E., & Solovyova, V. (2021). Sustainable development of oil and gas resources: A system of environmental, socio-economic, and innovation indicators. *Journal of Marine Science and Engineering*, *9*(11), 1307. Advance online publication. doi:10.3390/jmse9111307

Chien, F., Paramaiah, C., Thuy, N. T. T., Phan, T. T. H., & Ngo, T. Q. (2023). The impact of eco-innovation, trade openness, financial development, green energy and government governance on sustainable development in ASEAN countries. *Renewable Energy*, *211*, 259–268. doi:10.1016/j.renene.2023.04.109

Choi, E., Heshmati, A., & Cho, Y. (2010). An Empirical Study of the Relationships between CO_2 Emissions, Economic Growth and Openness. *SSRN*, (5304), 1–27. doi:10.2139srn.1708750

Choughri, R., & Soubjaki, M. (2017). *The Future of Business Analytics for SMEs in MENA Region*. Academic Press.

Chu, K. H. (2020). Breakthrough curve analysis by simplistic models of fixed bed adsorption: In defense of the century-old Bohart-Adams model. *Chemical Engineering Journal*, *380*, 122513. doi:10.1016/j.cej.2019.122513

Chu, W. (2001). Dye removal from textile dye wastewater using recycled alum sludge. *Water Research*, *35*(13), 3147–3152. doi:10.1016/S0043-1354(01)00015-X PMID:11487111

Chu, Z., Wang, L., & Lai, F. (2018). Customer pressure and green innovations at third party logistics providers in China: The moderation effect of organizational culture. *International Journal of Logistics Management*, *30*(1), 57–75. doi:10.1108/IJLM-11-2017-0294

Cillo, V., Messeni Petruzzelli, A., Ardito, L., & Del Giudice, M. (2019). Understanding sustainable innovation: A systematic literature review. *Corporate Social Responsibility and Environmental Management*, *26*(5), 1012–10. doi:10.1002/csr.1783

Ciobanu, A. M. (2021). The impact of FDI on economic growth in case of Romania. *International Journal of Economics and Finance*, *12*(12), 1–81.

Cole, M. A., & Elliott, R. J. R. (2003). Determining the trade–environment composition effect: The role of capital, labor and environmental regulations. *Journal of Environmental Economics and Management*, *46*(3), 363–383. doi:10.1016/S0095-0696(03)00021-4

Colotla, I., Fæste, A., Heidemann, A., Winther, A., Andersen, P., Duvold, T., & Hansen, M. (2016). *Winning the Industry 4.0 race. How ready are Danish manufacturers?* The Boston Consulting Group.

Comisión Económica para América Latina y el Caribe. (2021). *Innovación para el Desarrollo*. Obtenido de Conferencia de Ciencia, Innovación y Tecnologías de la Información y las Comunicaciones de CEPAL: https://innovalac.cepal.org/3/sites/innovalac3/files/c2100805_web.pdf

Concha, J. M., & Monsalvo, M. G. (2020). *Cuarta Revolución Industrial y el Trabajo en México*. UNAM. Ciudad de México: Dirección General de Bibliotecas. Tesis Digitales. Obtenido de http://132.248.9.195/ptd2020/enero/0800339/Index.html

Cooper, C. M., McCall, J., Stokes, S. C., McKay, C., Bentley, M. J., Rosenblum, J. S., Blewett, T. A., Huang, Z., Miara, A., Talmadge, M., Evans, A., Sitterley, K. A., Kurup, P., Stokes-Draut, J. R., Macknick, J., Borch, T., Cath, T. Y., & Katz, L. E. (2022). Oil and Gas Produced Water Reuse: Opportunities, Treatment Needs, and Challenges. In ACS ES and T Engineering (Vol. 2, Issue 3). doi:10.1021/acsestengg.1c00248

Corbett, D. C., & Rostow, W. W. (1960). The Stages of Economic Growth: A Non-Communist Manifesto. *International Journal (Toronto, Ont.)*, *16*(1), 83. Advance online publication. doi:10.2307/40198523

Cox, D. R., Newton, A. M. W., & Huuse, M. (2020). An introduction to seismic reflection data: Acquisition, processing and interpretation. In Regional Geology and Tectonics: Volume 1: Principles of Geologic Analysis. doi:10.1016/B978-0-444-64134-2.00020-1

Cramwinckel, J. (2019). *The role of global open geospatial data in measuring SDG indicator 11.7. 1: Public open spaces*. Wageningen University.

Creswell, J. W., & Poth, C. N. (2017). *Qualitative Inquiry and Research Design: Choosing Among Five Approaches*. Sage Publications.

Cuevas, I. (2016). *El uso de las Tecnologías Verdes y sus efectos en el desempeño del sector agroindustrial en México*. Obtenido de Repositorio Instituto Politécnico Nacional: https://tesis.ipn.mx/handle/123456789/22783

Dasgupta, P. (2021). *Economics of biodiversity: Dasgupta review*. H M Treasury.

David, O. O. (2019). Nexus between telecommunication infrastructures, economic growth and development in Africa: Panel vector autoregression (PVAR) analysis. *Telecommunications Policy*, *43*(8), 1–17. doi:10.1016/j.telpol.2019.03.005

De La Torre-Díez, I., Lopez-Coronado, M., Garcia-Zapirain Soto, B., & Mendez-Zorrilla, A. (2015). Secure cloud-based solutions for different eHealth services in spanish rural health centers. *Journal of Medical Internet Research*, *17*(7), e157. Advance online publication. doi:10.2196/jmir.4422 PMID:26215155

Del Río, P. (2003). La adopción y difusión de tecnologías limpias. *Economía Industrial*, 47-58. Obtenido de Economía Industrial: https://digital.csic.es/handle/10261/10419

del Valle Inclán, D. (2016). La globalización y su incidencia en el desarrollo sostenible. In A. Hérnandez, H. González, & G. Tamez (Eds.), *Desarrollo sustentable: de la teoría a la práctica (págs* (pp. 37–56). Universidad Auntónoma de Nuevo León.

Delefosse, M., Jacobsen, M. W., Balle, J. D., Hansen, B. K., Middelboe, A. L., Nielsen, E. E., & Teilmann, J. (2020). Marine mammal biodiversity around oil and gas platforms - Challenges and successes of long-term monitoring. *Society of Petroleum Engineers - SPE International Conference and Exhibition on Health, Safety, Environment, and Sustainability 2020, HSE and Sustainability 2020.* 10.2118/199479-MS

Depken, D., & Zeman, C. (2018). Small business challenges and the triple bottom line, TBL: Needs assessment in a Midwest State, USA. *Technological Forecasting and Social Change, 135*, 44–50. doi:10.1016/j.techfore.2017.05.032

Desai, J. N., Pandian, S., & Vij, R. K. (2021). Big data analytics in upstream oil and gas industries for sustainable exploration and development: A review. In Environmental Technology and Innovation (Vol. 21). doi:10.1016/j.eti.2020.101186

Dhaoui, I. (2019). *Healthcare system efficiency and its determinants: A two-stage Data Envelopment Analysis (DEA) from MENA countries.* Academic Press.

Diaz, C. D. P., Hoyos, C. C. M., Páez, E. G. M., & Martin, C. A. G. (2022). Wind Potencial as an Oportunity for Energy Transition in Oil and Gas Industry: Colombian Caribbean Offshore Case of Study. *Proceedings of the Annual Offshore Technology Conference.* 10.4043/31911-MS

Dickey, D. A., & Fuller, W. A. (1979). Distribution of the Estimators for Autoregressive Time Series with a Unit Root. *Journal of the American Statistical Association, 74*(366), 427–431. doi:10.2307/2286348

Dickey, D. A., & Fuller, W. A. (1981). Likelihood ratio statistics for autoregressive time series with a unit root. *Econometrica, 49*(4), 1057–1072. doi:10.2307/1912517

Dickinson, S., Plisson-Saune, S., & Mani, P. (2020). Establishing public biodiversity commitments within an oil & gas company. *Society of Petroleum Engineers - SPE International Conference and Exhibition on Health, Safety, Environment, and Sustainability 2020, HSE and Sustainability 2020.* 10.2118/199508-MS

Didenko, N., Skripnuk, D., Merkulov, V., Kikkas, K. N., & Skripniuk, K. (2023). Methodology for the Formation of a Digital Model of the Life Cycle of an Offshore Oil and Gas Platform. *Resources, 12*(8), 86. Advance online publication. doi:10.3390/resources12080086

Diggle, P. (2005). Applied Spatial Statistics for Public Health Data. *Journal of the American Statistical Association, 100*(470), 702–703. Advance online publication. doi:10.1198/jasa.2005.s15

Dikshit, R., & Tallapragada, P. (2018). Comparative study of natural and artificial flavoring agents and dyes. In *Natural and artificial flavoring agents and food dyes* (pp. 83–111). Academic Press. doi:10.1016/B978-0-12-811518-3.00003-X

Dinana, H. O. (2019). The impact of information & communication technologies (ICT) on sustainable development goals (SDGs) progression in the era of digital economy. *Sustainable Development Conference*, 1-13.

Donaghy, T. Q., Healy, N., Jiang, C. Y., & Battle, C. P. (2023). Fossil fuel racism in the United States: How phasing out coal, oil, and gas can protect communities. In Energy Research and Social Science (Vol. 100). doi:10.1016/j.erss.2023.103104

Dossi, S. (2022). How is the current STEM gender bias? *Imperial London College*, https://www.imperial.ac.uk/news/233901/how-current-stem-gender-bias/

Dovers, S. (2001). *Institutions for sustainability.* Australian Conservation Foundation.

Dowd, M., Dixon, A., & Kinsella, B. (2020). Optimizing waste management collection routes in Urban Haiti: A collaboration between DataKind and SOIL. *ArXiv, 2012*.

Doytch, N. (2020). The impact of foreign direct investment on the ecological footprints of nations. *Environmental and Sustainability Indicators*, *8*, 100085. doi:10.1016/j.indic.2020.100085

Dupre, D. (2020). Urban and socio-economic correlates of property prices in dublin's area. *Proceedings - 2020 IEEE 7th International Conference on Data Science and Advanced Analytics, DSAA 2020*. 10.1109/DSAA49011.2020.00070

Dutrénit, G. (2020). In R. Álavrez (Ed.), *Nuevas tecnologías, industria 4.0 y sociedad* (pp. 6–7). Ciencia.

Dutta, J., Chowdhury, C., Roy, S., Middya, A. I., & Gazi, F. (2017). Towards smart city: Sensing air quality in city based on opportunistic crowd-sensing. *ACM Int. Conf. Proc. Ser.*

Dutta, A. (2001). Telecommunications and economic activity: An analysis of Granger causality. *Journal of Management Information Systems*, *17*(4), 71–95. doi:10.1080/07421222.2001.11045658

Dutta, B. K. (2009). *Principles of Mass Transfer and Separation Process*. PHI Learning Private Limited.

Dutta, P., Chavhan, R., Gowtham, P., & Singh, A. (2023). The individual and integrated impact of Blockchain and IoT on sustainable supply chains: A systematic review, Supply Chain Forum. *International Journal (Toronto, Ont.)*, *24*(1), 103–126. doi:10.1080/16258312.2022.2082851

Eadie, L., & Ghosh, T. K. (2011). Biomimicry in textiles: Past, present and potential. An overview. *Journal of the Royal Society, Interface*, *8*(59), 761–775. doi:10.1098/rsif.2010.0487 PMID:21325320

Ebrahim, S. A., Mosaad, M. M., Othman, H., & Hassabo, A. G. (2022). A valuable observation of eco-friendly natural dyes for valuable utilisation in the textile industry. *Journal of Textiles. Coloration and Polymer Science*, *19*(1), 25–37.

Edensor, T. (2004). Automobility and national identity: Representation, geography and driving practice. *Theory, Culture & Society*, *21*(4-5), 101–120. doi:10.1177/0263276404046063

Egan, J., & Salmon, S. (2022). Strategies and progress in synthetic textile fiber biodegradability. *SN Applied Sciences*, *4*(1), 1–36. doi:10.100742452-021-04851-7

Egorov, V. V. (2017). Dozy-chaos end of the human civilization. [Section B]. *Journal of Ultra Scientist of Physical Sciences*, *29*(4), 87–96. doi:10.22147/jusps-B/290402

Ekardt, F., Bärenwaldt, M., & Heyl, K. (2022). The Paris Target, Human Rights, and IPCC Weaknesses: Legal Arguments in Favour of Smaller Carbon Budgets. *Environments, 9*(9). doi:10.3390/environments9090112

Ekins, P., Simon, S., Lisa, D., Folke, C., & De Groot, R. (2003). A framework for the practical application of the concepts of critical natural capital and strong sustainability. *Ecological Economics*, *44*(2-3), 165–185. doi:10.1016/S0921-8009(02)00272-0

Ekin, Z. (2005). Resurgence of safflower (Carthamus tinctorius L.) utilization: A global view. *Journal of Agronomy*, *4*(2), 83–87. doi:10.3923/ja.2005.83.87

El Baz, J., & Laguir, I. (2017). Third-party logistics providers (TPLs) and environmental sustainability practices in developing countries: The case of Morocco. *International Journal of Operations & Production Management*, *37*(10), 1451–1474. doi:10.1108/IJOPM-07-2015-0405

Elgarahy, A. M., Elwakeel, K. Z., Mohammad, S. H., & Elshoubaky, G. A. (2021). A critical review of biosorption of dyes, heavy metals and metalloids from wastewater as an efficient and green process. *Cleaner Engineering and Technology*, *4*, 100209. doi:10.1016/j.clet.2021.100209

Elijah, O., Ling, P. A., Rahim, S. K. A., Geok, T. K., Arsad, A., Kadir, E. A., Abdurrahman, M., Junin, R., Agi, A., & Abdulfatah, M. Y. (2021). *A Survey on Industry 4.0 for the Oil and Gas Industry: Upstream Sector*. IEEE. doi:10.1109/ACCESS.2021.3121302

Ellingsen, S. (2014). 'Seismic Shifts: Platforms, Content Creators andSpreadable Media'. *Media International Australia, Incorporating Culture & Policy*, *150*(1), 106–113. doi:10.1177/1329878X1415000121

El-Naas, M. (2013). *Modelling of adsorption process article*. Academic Press.

El-Sayed, H., Sankar, S., Prasad, M., Puthal, D., Gupta, A., Mohanty, M., & Lin, C.T. (2017). Edge of Things: The Big Picture on the Integration of Edge, IoT and the Cloud in a Distributed Computing Environment. *IEEE Access*, *6*, 1706–1717.

Engle, R., & Granger, C. W. J. (1987). Cointegration and Error Correction: Representation, Estimation and Testing. *Econometrica*, *55*(2), 251–276. doi:10.2307/1913236

Environmental Accounting and Financial Performance of Oil and Gas Companies in Nigeria. (2019). *Research Journal of Finance and Accounting*. doi:10.7176/RJFA/10-10-21

Evans, S., Vladimirova, D., Holgado, M., Van Fossen, K., Yang, M., Silva, E., & Barlow, C. (2017). *Business model innovation for sustainability: towards a unified perspective for creation of sustainable business models*. Bus. Strat. Environ. doi:10.1002/bse.1939

Ewing, B., Moore, D., Goldfinger, S., Oursler, A., Reed, A., & Wackenagel, M. (2010). *Ecological footprint atlas 2010*. Oakland.

Faiyaad, C. M., & Sadiki, A. B. (2022). *How healthcare industry in Arabs can use data science for sustainable healthcare practices*. Business & IT. doi:10.14311/bit.2022.01.22

Falk, M. (2016). A gravity model of foreign direct investment in the hospitality industry. *Tourism Management*, *55*, 225–237. doi:10.1016/j.tourman.2016.02.012

Farag, S. G. (2019). Application of Smart Structural System for Smart Sustainable Cities. In *Proceedings of the 2019 4th MEC International Conference on Big*. 10.1109/ICBDSC.2019.8645582

Farooq, M. S., Riaz, S., Abid, A., Abid, K., & Naeem, M. A. (2019). A Survey on the Role of IoT in Agriculture for the Implementation of Smart Farming. In IEEE Access (Vol. 7). doi:10.1109/ACCESS.2019.2949703

Felício, S., Hora, J., Ferreira, M. C., Abrantes, D., Costa, P. D., Dangelo, C., Silva, J., & Galvão, T. (2022). Handling OpenStreetMap georeferenced data for route planning. *Transportation Research Procedia*, *62*, 189–196. Advance online publication. doi:10.1016/j.trpro.2022.02.024

Fernández de Sevilla, M., Gutiérrez, Ó., Gómez, J., Tayebi, A., Álvarez, Á., & Sáez de Adana, F. (2021). On the application of radio planning tools in open environments for the improvement of autoguidance systems used in precision agriculture. *Computers and Electronics in Agriculture*, *187*, 106258. Advance online publication. doi:10.1016/j.compag.2021.106258

Ferreira, E. S., Hulme, A. N., McNab, H., & Quye, A. (2004). The natural constituents of historical textile dyes. *Chemical Society Reviews*, *33*(6), 329–336. doi:10.1039/b305697j PMID:15280965

Fischer, J., Dyball, R., Faezy, I., Gross, C., Dovers, S., Ehrlich, P. R., Bulle, R. J., Christensen, C., & Borden, R. (2012). Human behaviour and sustainability. *Frontiers in Ecology and the Environment*, *10*(3), 153–160. doi:10.1890/110079

Fonte, C. C., Patriarca, J. A., Minghini, M., Antoniou, V., See, L., & Brovelli, M. A. (2019). Using OpenStreetMap to Create Land Use and Land Cover Maps: Development of an Application. In *Geospatial Intelligence* (Vol. 2). Concepts, Methodologies, Tools, and Applications. doi:10.4018/978-1-5225-8054-6.ch047

Foo, K. Y., & Hameed, B. H. (2010). Insight into the modelling of adsorption isotherm systems. *Chemical Engineering Journal, 156*(1), 2–10. doi:10.1016/j.cej.2009.09.013

Forbes, D. (2016). How Can We Define 'Innovation'? *Entrepreneur & Innovation Exchange, 1-4*. Advance online publication. doi:10.17919/X9DW2X

Foster, G. (2020). Circular economy strategies for adaptive reuse of cultural heritage buildings to reduce environmental impacts. *Resources, Conservation and Recycling, 152*, 104507. Advance online publication. doi:10.1016/j.resconrec.2019.104507

Fox, T. A., Barchyn, T. E., Risk, D., Ravikumar, A. P., & Hugenholtz, C. H. (2019). A review of close-range and screening technologies for mitigating fugitive methane emissions in upstream oil and gas. In Environmental Research Letters (Vol. 14, Issue 5). doi:10.1088/1748-9326/ab0cc3

Frankel, J. A., & Romer, D. (1999). Does trade cause growth? *The American Economic Review, 89*(3), 379–399. doi:10.1257/aer.89.3.379

Freeman, R. E., Martin, K. E., & Parmar, B. L. (2020). *The Power of and: Responsible Business without Trade-Offs*. Columbia University Press.

Fuchs, C. (2006). Social Informatics: An Information Society for All? In Remembrance of Rob Kling. In B. T., M. I. Numinen, & T. Impagliazzo (Eds), IFIP International Federation for Information Processing (pp. 219-230). Springer.

Galicia, R. (2015). Innovación Tecnológica. *Ecorfan Journal*. Obtenido de https://www.ecorfan.org/proceedings/CTI_I/3.pdf

Galleguillos, L., & Santelices, I. (2012). Cambio tecnológico y desarrollo sostenible. *Revista Ingeniería Industrial, 1*(1), 59-66. Obtenido de https://dialnet.unirioja.es/descarga/articulo/3995770.pdf

Gandhi, S., Mangla, S. K., Kumar, P., & Kumar, D. (2015). Evaluating factors in implementation of successful green supply chain management using DEMATEL: A case study. *International Strategic Management Review, 3*(1-2), 96-109.

García, M., & Oliveira, A. (2023). Industria 4.0: ¿una nueva revolución tecnológica? (R. Álvarez, Ed.) *Ciencia. Nuevas tecnologías, industria 4.0 y sociedad, 74*(2). Obtenido de https://www.revistaciencia.amc.edu.mx/images/revista/74_2/PDF/Ciencia74-2.pdf

Garcia-Gonzales, D. A., Shonkoff, S. B. C., Hays, J., & Jerrett, M. (2019). Hazardous Air Pollutants Associated with Upstream Oil and Natural Gas Development: A Critical Synthesis of Current Peer-Reviewed Literature. In Annual Review of Public Health (Vol. 40). doi:10.1146/annurev-publhealth-040218-043715

Gardner, Z., Mooney, P., De Sabbata, S., & Dowthwaite, L. (2020). Quantifying gendered participation in OpenStreetMap: Responding to theories of female (under) representation in crowdsourced mapping. *GeoJournal, 85*(6), 1603–1620. Advance online publication. doi:10.100710708-019-10035-z

Gassner, F. (2015). *Comparative study between Buddhism and Christanity on peace and war [Masters degree dissertation]*. Portuguese Catholic University.

Gavito, M., van der Wal, H., Aldasoro, M., Ayala, B., Bullén, A., & Cach, M. (n.d.). *Ecología, tecnología e innovación para la sustentabilidad: retos y perspectivas en México*. Academic Press.

Ghasem, N. (2020). CO2 removal from natural gas. In *Advances in Carbon Capture*. Methods, Technologies and Applications. doi:10.1016/B978-0-12-819657-1.00021-9

Ghosh, A., & Chakraborty, N. (2016). Design of smart grid in an University Campus using ZigBee mesh networks. *Proceedings of the 1st IEEE International Conference on Power Electronics, Intelligent Control and Energy Systems*, 1–6. 10.1109/ICPEICES.2016.7853432

Gibbard, P. L., & Walker, M. J. C. (2013). Term 'Anthropocene' in the context of formal geological classification. *Special Publication - Geological Society of London*, *395*(1), 29–37. doi:10.1144/SP395.1

Gil, J. (2015). Building a multimodal urban network model using OpenStreetMap data for the analysis of sustainable accessibility. *OpenStreetMap in GIScience: Experiences, Research, and Applications*, 229–251.

Gill, K. S. (1996). Knowledge and the Post-industrial Society. In Information Society. Human-centred Systems (pp. 3-29). Springer. doi:10.1007/978-1-4471-3249-3_1

Gómez, J., Cuevas, I., & Soto, M. d. (2022). El cambio tecnológico y su incidencia en la incorporación de estrategias de sustentabilidad en la Industria de los Detergentes en México. *Transformación digital como propuesta de valor para la competitividad, 16*(16), 89-106. Obtenido de https://www.riico.net/index.php/riico/article/view/2102/1918

Gonzalez, D. J. X., Francis, C. K., Shaw, G. M., Cullen, M. R., Baiocchi, M., & Burke, M. (2022). Upstream oil and gas production and ambient air pollution in California. *The Science of the Total Environment*, *806*, 150298. Advance online publication. doi:10.1016/j.scitotenv.2021.150298 PMID:34844318

Gozgor, G. (2011). *Munich Personal RePEc Archive Panel unit root tests of purchasing power parity hypothesis: Evidence from Turkey Panel Unit Root Tests of Purchasing Power Parity Hypothesis: Evidence from Turkey*. Academic Press.

Gözüküçük, M. F. (2020). *Digital transformation and economic growth* [Unpublished Master Thesis]. Istanbul Commerce University Institute of Social Sciences.

Green, J., Hadden, J., Hale, T., & Mahdavi, P. (2022). Transition, hedge, or resist? Understanding political and economic behavior toward decarbonization in the oil and gas industry. *Review of International Political Economy*, *29*(6), 2036–2063. Advance online publication. doi:10.1080/09692290.2021.1946708

Grinberger, A. Y., Minghini, M., Juhász, L., Yeboah, G., & Mooney, P. (2022). OSM Science—The Academic Study of the OpenStreetMap Project, Data, Contributors, Community, and Applications. In ISPRS International Journal of Geo-Information (Vol. 11, Issue 4). doi:10.3390/ijgi11040230

Grippa, T., Georganos, S., Zarougui, S., Bognounou, P., Diboulo, E., Forget, Y., Lennert, M., Vanhuysse, S., Mboga, N., & Wolff, E. (2018). Mapping urban land use at street block level using OpenStreetMap, remote sensing data, and spatial metrics. *ISPRS International Journal of Geo-Information*, *7*(7), 246. Advance online publication. doi:10.3390/ijgi7070246

Groshek, J., & Krongard, S. (2016). Netflix and engage? Implications fo rstreaming television on political participation during the 2016 US presidential campaign. *Social Sciences (Basel, Switzerland)*, *5*(4), 65–83. doi:10.3390ocsci5040065

Gubbi, J., Buyya, R., Marusic, S., & Palaniswami, M. (2013). Internet of Things (IoT): A vision, architectural elements, and future directions. *Future Generation Computer Systems*, *29*(7), 1645–1660. doi:10.1016/j.future.2013.01.010

Guerrero-Martin, C. A., Ortega-Ramírez, A. T., Rodríguez, P. A. P., López, S. J. R., Guerrero-Martin, L. E., Salinas-Silva, R., & Camacho-Galindo, S. (2023). Analysis of Environmental Sustainability through a Weighting Matrix in the Oil and Gas Industry. *Sustainability (Basel)*, *15*(11), 9063. Advance online publication. doi:10.3390u15119063

Gujarati, D. N. (2012). *Econometrics By Example* (2nd ed.). Palgrave Macmillan.

Güney, T. (2016). *Governance and sustainable development: How effective is governance?* doi:10.1080/09638199.2016.1249391

Guoping, L., Yun, H., & Aizhi, W. (2017). Fourth Industrial Revolution: Technological Drivers. *Impacts and Coping Methods.*, *27*(4), 626–637.

Gurgul, H., & Lach, Ł. (2014). Globalization and economic growth: Evidence from two decades of transition in CEE. *Economic Modelling*, *36*, 99–107. doi:10.1016/j.econmod.2013.09.022

Gürses, A., Açıkyıldız, M., Güneş, K., Gürses, M. S., Gürses, A., Açıkyıldız, M., ... Gürses, M. S. (2016). Historical development of colorants. *Dyes and Pigments*, 1–12.

Habib, B. (2015). Sustainability is not enough: A call for regeneration. *Ethos (Berkeley, Calif.)*, *23*(2 (Term 2)), 7–10.

Haksoon, K. (2010). Political Stability and Foreign Direct Investment. *International Journal of Economics and Finance*, *2*(3). Advance online publication. doi:10.5539/ijef.v2n3p59

Hamdaqa, M., & Tahvildari, L. (2012). Cloud Computing Uncovered: A Research Landscape. In Advances in Computers (Vol. 86). doi:10.1016/B978-0-12-396535-6.00002-8

Hamel, G., & Prahalad, C. K. (1989). Strategic intent. *Harvard Business Review*, (May–June), 63–76. PMID:10303477

Hamelink, M., & Opdenakker, R. (2019). How business model innovation affects firm performance in the energy storage market. *Renewable Energy*, *131*, 120–127. doi:10.1016/j.renene.2018.07.051

Handayani, W., Kristijanto, A. I., & Hunga, A. I. R. (2018). Are natural dyes eco-friendly? A case study on water usage and wastewater characteristics of batik production by natural dyes application. *Sustainable Water Resources Management*, *4*(4), 1011–1021. doi:10.100740899-018-0217-9

Haq, R. (2023). Over A Million Pakistani University Students Enrolled in STEM Fields. *Haq's Musings*, http://www.riazhaq.com/2023/02/over-million-pakistani-university.html

Harris, Wise, Gallagher, & Goodwin. (Eds.). (2001). A Survey of Sustainable Development: Social and Economic Dimensions. Island Press.

Hartmann, J., Inkpen, A. C., & Ramaswamy, K. (2021). Different shades of green: Global oil and gas companies and renewable energy. *Journal of International Business Studies*, *52*(5), 879–903. Advance online publication. doi:10.105741267-020-00326-w

Hashim, H. H., Abdul Talip, N. A., Muhamad Pikri, M. H., Abu Bakar, N. A., & Wan Zakaria, A. (2022). Sustainable Methane Emissions Management. *IOP Conference Series. Materials Science and Engineering*, *1257*(1), 012013. Advance online publication. doi:10.1088/1757-899X/1257/1/012013

Hassan, A., Mahmoud, M., Bageri, B. S., Aljawad, M. S., Kamal, M. S., Barri, A. A., & Hussein, I. A. (2020). Applications of chelating agents in the upstream oil and gas industry: A review. In Energy and Fuels (Vol. 34, Issue 12). doi:10.1021/acs.energyfuels.0c03279

Haverkort, B. R., & Zimmermann, A. (2017). Smart Industry: How ICT Will Change the Game! *IEEE Internet Computing*, *21*(1), 8–10. doi:10.1109/MIC.2017.22

Heeren, A. J., Singh, A. S., Zwickle, A., Koontz, T. M., Slagle, K. M., & McCreery, A. C. (2016). Is sustainability knowledge half the battle?: An examination of sustainability knowledge, attitudes, norms, and efficacy to understand sustainable behaviours. *International Journal of Sustainability in Higher Education*, *17*(5), 613–632. doi:10.1108/IJSHE-02-2015-0014

Heidari, H., Akbari, M., Souhankar, A., & Hafezi, R. (2022). Review of global energy trends towards 2040 and recommendations for Iran oil and gas sector. *International Journal of Environmental Science and Technology*, *19*(8), 8007–8018. Advance online publication. doi:10.100713762-022-03963-w

Heider, K., Lopez, J. M. R., & Scheffran, J. (2018). The potential of volunteered geographic information to investigate peri-urbanization in the conservation zone of Mexico City. *Environmental Monitoring and Assessment*, *190*(4), 219. Advance online publication. doi:10.100710661-018-6597-3 PMID:29541867

Hennig, S. (2017). OpenStreetMap used in protected area management. The example of the recreational infrastructure in Berchtesgaden National Park. *Eco.Mont*, *9*(2), 30–41. Advance online publication. doi:10.1553/eco.mont-9-2s30

Herbert, W., Nwaorgu, I., Onyilo, F., & Iormbagah, J. (2020). Sustainability Reporting and Performance of Listed Upstream Oil and Gas Firms in Nigeria: A Content Evaluation Approach. *International Journal of Applied Economics. Finance and Accounting*, *8*(1). Advance online publication. doi:10.33094/8.2017.2020.81.46.61

Hitam, M. B., & Borhan, H. B. (2012). FDI, Growth and the Environment: Impact on Quality of Life in Malaysia. *Procedia: Social and Behavioral Sciences*, *50*, 333–342. doi:10.1016/j.sbspro.2012.08.038

Ho, C. Y., Wang, W., & Yu, J. (2013). Growth spillover through trade: A spatial dynamic panel data approach. *Economics Letters*, *120*(3), 450–453. doi:10.1016/j.econlet.2013.05.027

Holfelder. (2019). Towards a sustainable future with education? *Sustainability Science, 14*, 943-952. https://doi.org/https://doi.org/10.1007/s11625-019-00682-z

Hopper, J. R., & Nielsen, J. M. (1991). Recycling as altruistic behavior: Normative and behavioral strategies to expand participation in a community recycling program. *Environment and Behavior*, *23*(2), 195–220. doi:10.1177/0013916591232004

HOTOSM. (2021). *Somaliland Wajaale Farmyards*. https://tasks.hotosm.org/projects/10301

HOTOSM. (2023a). *Hot's Response to the Turkey & Syria Earthquake*. https://www.hotosm.org/projects/join-the-turkey-and-syria-earthquake-response/

HOTOSM. (2023b). *Open Cities: Mexico*. https://www.hotosm.org/open-cities-mexico.html

Hottenrot, H., Rexhausser, S., & Veugelers, R. (2016). Organisational change and the productivity effects of green technology adoption. *Resource and Energy Economics*, *43*, 172–194. doi:10.1016/j.reseneeco.2016.01.004

Hsu, K. F. (2020). Reframing public spaces: From recreational amenity to civic good in Resilient Cities. COVID-19 offers an opportunity to re-evaluate neighbourhood spaces to ensure they are available to all. *Journal of Public Space*, *5*(3), 67–82. doi:10.32891/jps.v5i3.1416

Hu, X., Gong, J., Renard, E. G., & Parashar, M. (2019). Two-Stage Framework for Big Spatial Data Analytics to Support Disaster Response. *Proceedings - 2019 IEEE International Conference on Big Data, Big Data 2019*. 10.1109/BigData47090.2019.9005613

Huang, J. A. J.-L. (2020). Historical comparison of gender inequality in scientific careers across countries and disciplines. *PNAS*, https://www.pnas.org/doi/10.1073/pnas.1914221117

Huang, Y., Dang, Z., Choi, Y., Andrade, J., & Bar-ilan, A. (2018). High-precision smart system on accelerometers and inclinometers for Structural Health Monitoring: Development and applications. *Proceedings of the 2018 12th France-Japan and 10th Europe-Asia Congress on Mechatronics*, 52–57.

Huawei Technologies Co. Ltd. (2023). Introduction to Cloud Computing Computing. In Cloud Computing Technology. doi:10.1007/978-981-19-3026-3_1

Hunt, J. D., Nascimento, A., Nascimento, N., Vieira, L. W., & Romero, O. J. (2022). Possible pathways for oil and gas companies in a sustainable future: From the perspective of a hydrogen economy. *Renewable & Sustainable Energy Reviews*, *160*, 112291. Advance online publication. doi:10.1016/j.rser.2022.112291

Hu, Y., Fan, H., Chen, Y., Chang, J., Zhan, X., Wu, H., Zhang, B., Wang, M., Zhang, W., Yang, L., Hou, X., Shen, X., Pan, T., Wu, W., Li, J., Hu, H., & Wei, F. (2021). Spatial patterns and conservation of genetic and phylogenetic diversity of wildlife in China. *Science Advances*, *7*(4), eabd5725. Advance online publication. doi:10.1126ciadv.abd5725 PMID:33523945

India IoT Market Overview 2020-2025. (2020). *MarketsandMarkets*. www.marketsandmarkets.com/Market-Reports/india-iotmarket-277424354.html

Innovation and Technology for Gender Equality. (2022). *UNICEF*. https://www.unicef.org/eap/innovation-and-technology-gender-equality

Instituto Nacional de Contadores Públicos. (2022). *Las 10 tecnologías verdes que pueden ayudar a preservar el medio ambiente*. Obtenido de https://incp.org.co/las-10-tecnologias-verdes-que-pueden-ayudar-a-preservar-el-medio-ambiente/#:~:text=Estas%20tecnolog%C3%ADas%20incluyen%20bio%2Dremediaci%C3%B3n,si%20mismo%20con%20el%20tiempo

Ionescu, L. (2021). Transitioning to a low-carbon economy: Green financial behavior, climate change mitigation, and environmental energy sustainability. *Geopolitics, History, and International Relations*, *13*(1), 86. Advance online publication. doi:10.22381/GHIR13120218

IPCC. (1996). Second Assessment on Climate Change. International Panel on Climate Change. Cambridge, UK: Cambridge University Press.

Islam, M. R., & Hossain, M. E. (2020). Drilling Engineering: Towards achieving total sustainability. In Drilling Engineering: Towards Achieving Total Sustainability. doi:10.1016/C2019-0-00943-0

Jia, C. (2020). Development challenges and future scientific and technological researches in China's petroleum industry upstream. *Shiyou Xuebao. Acta Petrolei Sinica*, *41*(12). Advance online publication. doi:10.7623yxb202012001

Jitt-Aer, K., Wall, G., Jones, D., & Teeuw, R. (2022). Use of GIS and dasymetric mapping for estimating tsunami-affected population to facilitate humanitarian relief logistics: A case study from Phuket, Thailand. *Natural Hazards*, *113*(1), 185–211. Advance online publication. doi:10.100711069-022-05295-x

Jo, A., Lee, S. K., & Kim, J. (2020). Gender gaps in the use of urban space in seoul: Analyzing spatial patterns of temporary populations using mobile phone data. *Sustainability (Basel)*, *12*(16), 6481. Advance online publication. doi:10.3390u12166481

Johansen, S. (1988). Statistical Analysis of Cointegration Vectors. *Journal of Economic Dynamics & Control*, *12*(2), 231–254. doi:10.1016/0165-1889(88)90041-3

Johansen, S., & Juselius, K. (1992). Testing Structural Hypothesis in a Multivariate Cointegration Analysis of the PPP and the UIP for UK. *Journal of Econometrics*, *53*(1-3), 211–244. doi:10.1016/0304-4076(92)90086-7

Johnson, D., & Turner, C. (2010). *International Business: Themes and Issues in the Modern Global Economy* (2nd ed.). Routledge, Taylor & Francis Group. doi:10.4324/9780203861615

Johnson, L., Callaghan, C., Balasubramanian, M., Haq, H., & Spallek, H. (2019). Cost Comparison of an On-Premise IT Solution with a Cloud-Based Solution for Electronic Health Records in a Dental School Clinic. *Journal of Dental Education*, *83*(8), 895–903. Advance online publication. doi:10.21815/JDE.019.089 PMID:31010892

Johnston, J. E., Lim, E., & Roh, H. (2019). Impact of upstream oil extraction and environmental public health: A review of the evidence. In Science of the Total Environment (Vol. 657). doi:10.1016/j.scitotenv.2018.11.483

Jorgenson, A. K., Dick, C., & Mahutga, M. C. (2007). Foreign Investment Dependence and the Environment: An Eco-structural Approach. Society for the Study of Social Problems, 54(3), 371-394.

Jung, S., Yuan, Y., Del Carpio, M. S., Pawlik, T., Hartmann, S., Estrella, N., Oteros, J., Traidl-Hoffmann, C., Damialis, A., Buters, J., & Menzel, A. (2022). Impact of Local Grasslands on Wild Grass Pollen Emission in Bavaria, Germany. *Land (Basel)*, *11*(2), 306. Advance online publication. doi:10.3390/land11020306

Juré, G., & Fittipaldi, I. (2017). Del origen y tratamiento de la cuestión ambiental a la propuesta del desarrollo sustentable. In Introducción al desarrollo sustentable (pp. 63-84). Universidad Nacional Arturo Jauretche.

Ju, Y., Dronova, I., & Delclòs-Alió, X. (2022). A 10 m resolution urban green space map for major Latin American cities from Sentinel-2 remote sensing images and OpenStreetMap. *Scientific Data*, *9*(1), 586. doi:10.103841597-022-01701-y PMID:36153342

Kadolph, S. (2008). Natural dyes: A traditional craft experiencing new attention. *Delta Kappa Gamma Bulletin*, *75*(1), 14.

Kaitkowski, D., Philips, P. C. B., Schmidt, P., & Shin, Y. (1992). Testing the null hypothesis of stationarity against the alternative of a unit root. *Journal of Econometrics*, *54*(3), 159–178. doi:10.1016/0304-4076(92)90104-Y

Kaloxylos, A., Groumas, A., Sarris, V., Katsikas, L., Magdalinos, P., Antoniou, E., Politopoulou, Z., Wolfert, S., Brewster, C., Eigenmann, R., & Maestre Terol, C. (2014). A cloud-based farm management system: Architecture and implementation. *Computers and Electronics in Agriculture*, *100*, 168–179. Advance online publication. doi:10.1016/j.compag.2013.11.014

Kalyani, Y., & Collier, R. (2021). A systematic survey on the role of cloud, fog, and edge computing combination in smart agriculture. In Sensors (Vol. 21, Issue 17). doi:10.339021175922

Kamanga, C. T., Bugingo, E., Badibanga, S. N., & Mukendi, E. M. (2023). A multi-criteria decision making heuristic for workflow scheduling in cloud computing environment. *The Journal of Supercomputing*, *79*(1), 243–264. Advance online publication. doi:10.100711227-022-04677-z

Kanchana, R., Fernandes, A., Bhat, B., Budkule, S., Dessai, S., & Mohan, R. (2013). Dyeing of textiles with natural dyes-an eco-friendly approach. *International Journal of Chemtech Research*, *5*(5), 2102–2109.

Kang, L., Poslad, S., Wang, W., Li, X., Zhang, Y., & Wang, C. (2016). A public transport bus as a flexible mobile smart environment sensing platform for IoT. *Proceedings of the 12th International Conference on Intelligent Environments*, 1–8. 10.1109/IE.2016.10

Karaçor, Z., Güvenek, B., & Karadoğan, E. (2020). Ekonominin değişen maskesi: Dijital ekonomi. *International Journal of Social. Humanities and Administrative Sciences*, *6*(34), 2140–2149.

Kardos, M. (2014). The relevance of Foreign Direct Investment for sustainable development. Empirical evidence from European Union. *Procedia Economics and Finance*, *15*, 1349–1354. doi:10.1016/S2212-5671(14)00598-X

Karimi, S., Sayyadi, H., & Shahabaldini, Z. (2021). Green innovation: A systematic literature review. *Journal of Cleaner Production*, *279*, 122474. Advance online publication. doi:10.1016/j.jclepro.2020.122474

Kasiri, M. B., & Safapour, S. (2014). Natural dyes and antimicrobials for green treatment of textiles. *Environmental Chemistry Letters*, *12*(1), 1–13. doi:10.100710311-013-0426-2

Kaur, J. (2017). *Adsorption Isotherms*. Academic Press.

Keane, P. A., & Topol, E. J. (2018). With an eye to AI and autonomous diagnosis. *NPJ Digital Medicine*, *1*(1), 10–12. doi:10.103841746-018-0048-y PMID:31304321

Khan, H., Weili, L., & Khan, I. (2021). Environmental innovation, trade openness and quality institutions: An integrated investigation about environmental sustainability. *Environment, Development and Sustainability*, 1–31.

Khan, Z., Anjum, A., Soomro, K., & Tahir, M. A. (2015). Towards cloud based big data analytics for smart future cities. *Journal of Cloud Computing (Heidelberg, Germany)*, *4*(1), 4. doi:10.118613677-015-0026-8

Kihumbe. (2019). *Kihumbe- Survey Digitization and Mapping for HIV Monitoring*. Academic Press.

Kırışkan, I. & Yılmaz, D. H. (2020). Telecommunications infrastructure and economic growth in Turkey. *Bilig / Journal of Turkish World Social Sciences, 92*, 55 – 84.

KLL. (2015). *Agriculture Resource Mapping*. http://kathmandulivinglabs.org/our-work/agriculture-resource-mapping

Kong, Q., Peng, D., Ni, Y., Jiang, X., & Wang, Z. (2021). Trade openness and economic growth quality of China: Empirical analysis using ARDL model. *Finance Research Letters*, *38*, 101488. doi:10.1016/j.frl.2020.101488

Koroteev, D., & Tekic, Z. (2021). Artificial intelligence in oil and gas upstream: Trends, challenges, and scenarios for the future. *Energy and AI*, *3*, 100041. Advance online publication. doi:10.1016/j.egyai.2020.100041

Koubaa, A., Aldawood, A., Saeed, B., Hadid, A., Ahmed, M., Saad, A., Alkhouja, H., Ammar, A., & Alkanhal, M. (2020). Smart palm: An IoT framework for red palm weevil early detection. *Agronomy (Basel)*, *10*(7), 987. doi:10.3390/agronomy10070987

Kristensen, A., Yang, J. K., Bozhevolnyi, S. I., Link, S., Nordlander, P., Halas, N. J., & Mortensen, N. A. (2016). Plasmonic colour generation. *Nature Reviews. Materials*, *2*(1), 1–14. doi:10.1038/natrevmats.2016.88

Křížová, H. (2015). *Natural dyes: their past, present, future and sustainability. In Recent Developments in Fibrous Material Science*. Kosmas Publishing.

Kulanov, A., Issakhova, A., Koshkina, O., Issakhova, P., & Karshalova, A. (2020). Venture financing and the fuel and energy complex: Investing in alternative energy. *International Journal of Energy Economics and Policy*, *10*(5), 531–538. Advance online publication. doi:10.32479/ijeep.9963

Kulkarni, S. J. (2020). Synthesis, Characterization and Performance of Low-Cost Unconventional Adsorbents Derived from Waste Materials. *Biointerface Research in Applied Chemistry*, *10*(6), 7243–7256. doi:10.33263/BRIAC106.72437256

Kulkarni, S. J., Mishra, A., Patil, S., & Uphale, K. (2018). Modeling of Breakthrough Curves for Removal of Chromiumfrom Waste Water. *International Journal of Advanced Research in Chemical Sciences*, *5*(6), 29–40.

Kulkarni, S. S., Bodake, U. M., & Pathade, G. R. (2011). Extraction of Natural Dye from Chili (Capsicum Annum) for Textile Coloration. *Universal Journal of Environmental Research and Technology*, *1*(1).

Kumar, R., & Chandrakar, R. (2012). Overview of green supply chain management: Operation and environmental impact at different stages of the supply chain. *International Journal of Engineering and Advanced Technology*, *1*(3), 1–6.

Kuzlu, M., & Pipattanasomporn, M. (2013). Assessment of communication technologies and network requirements for different smart grid applications. *Proceedings of the 2013 IEEE PES Innovative Smart Grid*, 1–6. 10.1109/ISGT.2013.6497873

Kwarto, F., Nurafiah, N., Suharman, H., & Dahlan, M. (2022). *The potential bias for sustainability reporting of global upstream oil and gas companies: a systematic literature review of the evidence*. Management Review Quarterly. doi:10.100711301-022-00292-7

Kwarto, F., Nurafiah, N., Suharman, H., & Dahlan, M. (2023). Sustainability Reporting Reliability: An Industry Worker's Perspective In Indonesia. *International Journal of Business*, *28*(3). Advance online publication. doi:10.55802/IJB.028(3).002

Landes, D. (1979). *Proceso Tecnológico y Revolución Industrial*. Tecnos.

Lantano, F., Petruzzelli, A. M., & Panniello, U. (2022). Business model innovation in video-game consoles to face the threats of mobile gaming: Evidence from the case of Sony PlayStation. *Technological Forecasting and Social Change*, *174*, 121210. doi:10.1016/j.techfore.2021.121210

Latif, S., Islam, K. M. R., Khan, M. M. I., & Ahmed, S. I. (2011). OpenStreetMap for the disaster management in Bangladesh. *2011 IEEE Conference on Open Systems, ICOS 2011*, 435–439. 10.1109/ICOS.2011.6079240

Laturkar, K., & Laturkar, K. (2023). *Environmental Sustainability Measures in the Oil and Gas Industry*. doi:10.4018/978-1-6684-8223-0.ch008

Lee, J., Sorensen, C., Lemery, J., Workman, C. F., Linstadt, H., & Bazilian, M. D. (2022). Managing upstream oil and gas emissions: A public health oriented approach. In Journal of Environmental Management (Vol. 310). doi:10.1016/j.jenvman.2022.114766

Lee, K. H., & Wu, Y. (2014). Integrating sustainability performance measurement into logistics and supply networks: A multi-methodological approach. *The British Accounting Review*, *46*(4), 361–378. doi:10.1016/j.bar.2014.10.005

Lee, S. H., Levendis, J., & Gutierrez, L. (2012). Telecommunications and economic growth: An empirical analysis of Sub-Saharan Africa. *Applied Economics*, *44*(4), 461–469. doi:10.1080/00036846.2010.508730

Lele, A. (2019). Internet of things (IoT). *Smart Innov. Syst. Technol.*, *132*, 187–195. doi:10.1007/978-981-13-3384-2_11

Lema, R., Fu, X., & Rabellotti, R. (2020). Green windows of opportunity: Latecomer development in the age of transformation toward sustainability. *Industrial and Corporate Change*, 1193–1209.

Le, T. H., Chang, Y., & Park, D. (2016). Trade openness and environmental quality: International evidence. *Energy Policy*, *92*, 45–55. doi:10.1016/j.enpol.2016.01.030

Levin, N., Lechner, A. M., & Brown, G. (2017). An evaluation of crowdsourced information for assessing the visitation and perceived importance of protected areas. *Applied Geography (Sevenoaks, England)*, *79*, 115–126. Advance online publication. doi:10.1016/j.apgeog.2016.12.009

Lewis, S. L., & Maslin, M. A. (2015). Defining the Anthropocene. *Nature*, *519*(7542), 171–180. doi:10.1038/nature14258 PMID:25762280

Liakos E. V., Mone M., Lambropoulou D. A., Bikiaris D. N., & Kyzas G. Z. (2021). Adsorption Evaluation for the Removal of Nickel, Mercury and Barium Ions from Single-Component and Mixtures of Aqueous Solutions by Using an Optimized Biobased Chitosan Derivative. *Polymers (Basel)*, *13*(2), 1-20.

Li, H., Zech, J., Ludwig, C., Fendrich, S., Shapiro, A., Schultz, M., & Zipf, A. (2021). Automatic mapping of national surface water with OpenStreetMap and Sentinel-2 MSI data using deep learning. *International Journal of Applied Earth Observation and Geoinformation*, *104*, 102571. Advance online publication. doi:10.1016/j.jag.2021.102571

Li, M., Wiedmann, T., & Hadjikakou, M. (2019). Enabling Full Supply Chain Corporate Responsibility: Scope 3 Emissions Targets for Ambitious Climate Change Mitigation. *Environmental Science & Technology*. Advance online publication. doi:10.1021/acs.est.9b05245 PMID:31766851

Linchant, J., Lisein, J., Semeki, J., Lejeune, P., & Vermeulen, C. (2015). Are unmanned aircraft systems (UAS s) the future of wildlife monitoring? A review of accomplishments and challenges. *Mammal Review*, *45*(4), 239–252. doi:10.1111/mam.12046

Linh, H. T. (2009). Natural dyes in Eastern Asia (Vietnam and neighbouring countries). Handbook of Natural Colorants, 65-72.

Litvinenko, V., Bowbrick, I., Naumov, I., & Zaitseva, Z. (2022). Global guidelines and requirements for professional competencies of natural resource extraction engineers: Implications for ESG principles and sustainable development goals. In Journal of Cleaner Production (Vol. 338). doi:10.1016/j.jclepro.2022.130530

Liu, S., & Gao, B. (2022). Econometric Analysis of Leisure Agriculture and Rural Tourism Scenery Based on Spatial Data Analysis. *Computational Intelligence and Neuroscience*, *2022*, 1–8. Advance online publication. doi:10.1155/2022/2286803 PMID:35898780

Liu, Z. (2021). Identifying urban land use social functional units: A case study using OSM data. *International Journal of Digital Earth*, *14*(12), 1798–1817. Advance online publication. doi:10.1080/17538947.2021.1988161

Lucas, R. E. Jr. (1990). Liquidity and interest rates. *Journal of Economic Theory*, *50*(2), 237–264. doi:10.1016/0022-0531(90)90001-Z

Ludwig, C., Hecht, R., Lautenbach, S., Schorcht, M., & Zipf, A. (2021). Mapping public urban green spaces based on OpenStreetMap and Sentinel-2 imagery using belief functions. *ISPRS International Journal of Geo-Information*, *10*(4), 251. doi:10.3390/ijgi10040251

Luqun Xie, J. Z. (2020, February). Gender diversity in R&D teams and innovation efficiency: Role of the innovation context. *Research Policy*, *49*(1), 103885. Advance online publication. doi:10.1016/j.respol.2019.103885

Ma, D., Sandberg, M., & Jiang, B. (2015). Characterizing the heterogeneity of the openstreetmap data and community. *ISPRS International Journal of Geo-Information*, *4*(2), 535–550. Advance online publication. doi:10.3390/ijgi4020535

Malik, N. A., & Shaikh, M. A. (2019). Spatial distribution and accessibility to public sector tertiary care teaching hospitals: Case study from Pakistan. *Eastern Mediterranean Health Journal*, *25*(6), 431–434. Advance online publication. doi:10.26719/emhj.18.049 PMID:31469163

Malkoc, E., & Nuhoglu, Y. (2007). Potential of Tea Factory Waste for Chromium (VI) Removal from Aqueous Solutions: Thermodynamic and Kinetic Studies. *Separation and Purification Technology*, *54*(2), 291–298. doi:10.1016/j.seppur.2006.09.017

Maniatis, K., Chiaramonti, D., & van den Heuvel, E. (2021). Post covid-19 recovery and 2050 climate change targets: Changing the emphasis from promotion of renewables to mandated curtailment of fossil fuels in the eu policies. *Energies*, *14*(5), 1347. Advance online publication. doi:10.3390/en14051347

Mariotti, N., Bonomo, M., Fagiolari, L., Barbero, N., Gerbaldi, C., Bella, F., & Barolo, C. (2020). Recent advances in eco-friendly and cost-effective materials towards sustainable dye-sensitized solar cells. *Green Chemistry*, *22*(21), 7168–7218. doi:10.1039/D0GC01148G

Markides, C. (2006). Disruptive Innovation: In Need of Better Theory*. *Journal of Product Innovation Management*, *23*(1), 19–25. doi:10.1111/j.1540-5885.2005.00177.x

Marshall, C., & Rossman, G. B. (2016). *Designing qualitative research*. Sage Publications.

Martin & Evans. (2018). Sustainable business model innovation: A review. *Journal of Cleaner Production, 198*, 401-416. https://www.sciencedirect.com/science/article/pii/S0959652618318961 doi:10.1016/j.jclepro.2018.06.240

Martín, M. (2018). ¿Por qué Gran Bretaña?: Revolución. *Universidad de Valladolid. Facultad de Ciencias Económicas y Empresariales. Grado en Marketing e Investigación de mercados.*

Martin, J. L., Maris, V., & Simberloff, D. S. (2016). Need to respect nature and its limits challenges society and conservarion science. *Proceedings of the National Academy of Sciences of the United States of America*, *113*(22), 6105–6112. doi:10.1073/pnas.1525003113 PMID:27185943

Massa, L., Tucci, C., & Afuah, A. (2017). A critical assessment of business model research Acad. *The Academy of Management Annals*, *11*(1), 73–104. doi:10.5465/annals.2014.0072

Mathew, K. S. (2011). Professional associations and academic librarianship in Kerala. In Library 2.0 and Information Management: essays in honour of Dr. Jalaja (pp. 133-140). Atlantic Publishers.

Mauser, W., Klepper, G., Rice, M., Schmalzbauer, B. S., Hackmann, H., Leemans, R., & Moore, H. (2013). Transdisciplinary global change research: Co-creation of knowledge for sustainability. *Current Opinion in Environmental Sustainability*, *5*(3-4), 420–431. doi:10.1016/j.cosust.2013.07.001

Maximo, B. (2017). *Impacto industrial y laboral.* CCOO de Industria.

Maxwell, N. (2012). How universities can help humanity learn How to resolve crises of our times - From Knowledge to wisdom: University College London experience. In G. Heam, T. Katlelle, & D. Rooney (Eds.), *Handbook on the knowledge economy* (pp. 158–179). Edward Elgar. doi:10.4337/9781781005132.00019

Maxwell, R. (2012). *Textiles of Southeast Asia: Trade, tradition and transformation.* Tuttle Publishing.

Melane-Lavado, A. (2018). Green foreign direct investments and the deepening of capabilities for sustainable innovation in multinationals: Insights from renewable energy. *Journal of Cleaner Production.*

Melville, N. P. (2010). Information systems innovation for environmental sustainability. *Management Information Systems Quarterly*, *34*(1), 1–21. doi:10.2307/20721412

Mikulecky, P. (2015). Cloud – Based solutions for outdoor ambient intelligence support. *Studies in Computational Intelligence*, *598*, 229–238. Advance online publication. doi:10.1007/978-3-319-16211-9_24

Miller, T. R., Munoz-Erickson, T., & Redman, C. L. (2011). Transforming knowledge for sustainability: Towards adaptive academic institutions. *International Journal of Sustainability in Higher Education*, *12*(2), 177–192. doi:10.1108/14676371111118228

Minatogawa, V. L. F., Franco, M. M. V., Rampasso, I. S., Anholon, R., Quadros, R., Durán, O., & Batocchio, A. (2020). Operationalizing Business Model Innovation through Big Data Analytics for Sustainable Organizations. *Sustainability (Basel)*, *12*(1), 277. doi:10.3390u12010277

Minghini, M., Delucchi, L., Sarretta, A., Lupia, F., Napolitano, M., & Palmas, A. (2017). Collaborative mapping response to disasters through openstreetmap: The case of the 2016 Italian earthquake. In Geoingegneria Ambientale e Mineraria (Vol. 151, Issue 2). Academic Press.

Min, J., Yan, G., Abed, A. M., Elattar, S., Amine Khadimallah, M., Jan, A., & Elhosiny Ali, H. (2022). The effect of carbon dioxide emissions on the building energy efficiency. *Fuel*, *326*, 124842. Advance online publication. doi:10.1016/j.fuel.2022.124842

Mitra, S., Sarkar, A., & Sen, S. (2017). Removal of Chromium from Industrial Effluents using Nanotechnology: A Review. *Nanotechnology Environmental Engineering*, *2*(11), 1–14. doi:10.100741204-017-0022-y

Mohamed, M. B. H., Skima, I., Saafi, S., & Farhat, A. (2014). Price modeling: Analysis with a Vector Error Correction Model. *International Conference on Innovation & Engineering Management, Proceedings*, 99-102.

Mollahoseini Ardakani, M. R., Hashemi, S. M., & Razzazi, M. (2019). A Cloud-based solution/reference architecture for establishing collaborative networked organizations. *Journal of Intelligent Manufacturing*, *30*(5), 2273–2289. Advance online publication. doi:10.100710845-017-1387-2

Montabon, F., Pagell, M., & Wu, Z. (2016). Making sustainability sustainable. *The Journal of Supply Chain Management*, *52*(2), 11–27. doi:10.1111/jscm.12103

Mooney, P., Corcoran, P., & Ciepluch, B. (2013). The potential for using volunteered geographic information in pervasive health computing applications. *Journal of Ambient Intelligence and Humanized Computing*, *4*(6), 731–745. doi:10.100712652-012-0149-4

Moore, S. (2020). *OpenStreetMap Tourist Attractions for Europe*. https://hub.arcgis.com/datasets/f3dcc26dd868456da288baca83b8a2cd_0/about

Mora, H. (2022). *Diccionario Etimológico*. Obtenido de Tecnologías: http://etimologias.dechile.net/?tecnologi.a

Morandín-Ahuerma, I., Contreras-Hernández, A., Ayala-Ortiz, D. A., & Pérez-Maqueo, O. (2019). Socio–ecosystemic sustainability. *Sustainability*, 11.

Mufutau Opeyemi, B. (2021). Path to sustainable energy consumption: The possibility of substituting renewable energy for non-renewable energy. *Energy*, *228*, 120519. Advance online publication. doi:10.1016/j.energy.2021.120519

Muhammad Shahbaz, S. K. (2013). The dynamic links between energy consumption, economic growth, financial development and trade in China: Fresh evidence from multivariate framework analysis. *Energy Economics*, *40*, 8–21. doi:10.1016/j.eneco.2013.06.006

Mukherjee, S. (2015). Environmental and social impact of fashion: Towards an eco-friendly, ethical fashion. *International Journal of Interdisciplinary and Multidisciplinary Studies*, *2*(3), 22–35.

Munawar, H. S., Mojtahedi, M., Hammad, A. W. A., Kouzani, A., & Mahmud, M. A. P. (2022). Disruptive technologies as a solution for disaster risk management: A review. In Science of the Total Environment (Vol. 806). doi:10.1016/j.scitotenv.2021.151351

Mungmachon, M. R. (2012). Knowledge and local wisdom: Community treasure. *International Journal of Humanities and Social Science*, *2*(13), 174–181.

Munir, A., Asmat, A., & Malik Sikander, H. K. (2016a). Potential of Volunteered Geographic Information for Adaptation of Climate Change Effects in Pakistan. *4th International Conference on Energy, Environment and Sustainable Development.*

Munir, A., Asmat, A., & Malik Sikander, H. K. (2016b). Promoting Heritage Tourism in Pakistan using Volunteered Geographic Information: An Investigative Outlook. *24th International Conference of Historians of Asia.*

Munir, K., Ghafoor, M., Khafagy, M., & Ihshaish, H. (2022). AgroSupportAnalytics: A Cloud-based Complaints Management and Decision Support System for Sustainable Farming in Egypt. *Egyptian Informatics Journal*, *23*(1), 73–82. Advance online publication. doi:10.1016/j.eij.2021.06.002

Murphy, W. (2019). Advice for Men Who Are Nervous About Mentoring Women. *Harvard Business Review*. https://hbr.org/2019/03/advice-for-men-who-are-nervous-about-mentoring-women

Mussak, R. A., & Bechtold, T. (2009). Natural colorants in textile dyeing. Handbook of natural colorants, 315-335.

Muthu, S. S. (2014). *Roadmap to sustainable textiles and clothing: Eco-friendly raw materials, technologies, and processing methods* (1st ed.). Springer.

Muzaffar, H. M., Tahir, A., Ali, A., Ahmad, M., & McArdle, G. (2017). Quality assessment of volunteered geographic information for educational planning. In *Volunteered Geographic Information and the Future of Geospatial Data*. IGI Global. doi:10.4018/978-1-5225-2446-5.ch005

Naciones Unidas. (2019). *Science, technology & innovation capacity development course.* Obtenido de United Nations Conference on Trade and Development: https://unctad.org/system/files/official-document/dtlstictinf2019d1_en.pdf

Naji, G. M. A., Isha, A. S. N., Alazzani, A., Brough, P., Saleem, M. S., Mohyaldinn, M. E., & Alzoraiki, M. (2022). Do Leadership, Organizational Communication, and Work Environment Impact Employees' Psychosocial Hazards in the Oil and Gas Industry? *International Journal of Environmental Research and Public Health, 19*(8), 4432. Advance online publication. doi:10.3390/ijerph19084432 PMID:35457300

Neubauer, N., Over, M., Schilling, A., & Zipf, A. (2009). Virtual Cities 2.0: Generating web-based 3D city models and landscapes based on free and user generated data (OpenStreetMap). GeoViz2009.

Neumayer, E. (2012). Weak versus strong sustainability: Exploring the limits of two opposing paradigms. *Journal of Human Development and Capabilities, 13*(4), 561–579. doi:10.1080/19452829.2012.693067

New report on diversity trends in STEM workforce and education. (2023). *National Science Foundation.* https://phys.org/news/2023-01-diversity-trends-stem-workforce.html

Nikolai, S. (2009). *Green Technologies: Opportunities for South-South Trade*. International Trade Forum.

Norouzi, N. (2021). Post-COVID-19 and globalization of oil and natural gas trade: Challenges, opportunities, lessons, regulations, and strategies. In International Journal of Energy Research (Vol. 45, Issue 10). doi:10.1002/er.6762

O'Connor, A. E., Gallagher, W. M., & Byrne, A. T. (2009). Porphyrin and nonporphyrin photosensitizers in oncology: Preclinical and clinical advances in photodynamic therapy. *Photochemistry and Photobiology, 85*(5), 1053–1074. doi:10.1111/j.1751-1097.2009.00585.x PMID:19682322

O'Grady, M., Langton, D., & O'Hare, G. (2019). Edge computing: A tractable model for smart agriculture? *Artif. Intell. Agric., 3*, 42–51. doi:10.1016/j.aiia.2019.12.001

OECD. (2005). *Manual de OSLO.* OECD.

OECD. (2019). Tax challenges arising from digitalisation-interim report 2019 (Inclusive Framework on BEPS, OECD/G20 Base Erosion and Profit Shifting Project). OECD.

Ogolo, N. A., Anih, O. C., & Onyekonwu, M. O. (2022). Sources and effects of environmental pollution from oil and gas industrial operations. *Arabian Journal of Chemical and Environmental Research, 9*(1).

Ogunbiyi, O., Al-Rewaily, R., Saththasivam, J., Lawler, J., & Liu, Z. (2023). Oil spill management to prevent desalination plant shutdown from the perspectives of offshore cleanup, seawater intake and onshore pretreatment. In Desalination (Vol. 564). doi:10.1016/j.desal.2023.116780

Ojuederie, O. B., & Babalola, O. O. (2017). Microbial and plant-assisted bioremediation of heavy metal polluted environments: A review. *International Journal of Environmental Research and Public Health, 14*(12), 1504. doi:10.3390/ijerph14121504 PMID:29207531

Okere, C. J., & Sheng, J. J. (2023). Review on clean hydrogen generation from petroleum reservoirs: Fundamentals, mechanisms, and field applications. *International Journal of Hydrogen Energy*. Advance online publication. doi:10.1016/j.ijhydene.2023.06.135

Olas, B., Białecki, J., Urbańska, K., & Bryś, M. (2021). The effects of natural and synthetic blue dyes on human health: A review of current knowledge and therapeutic perspectives. *Advances in Nutrition, 12*(6), 2301–2311. doi:10.1093/advances/nmab081 PMID:34245145

ONU. (2023). *Objetivos del Desarrollo Sostenible*. Obtenido de https://www.un.org/sustainabledevelopment/es/objetivos-de-desarrollo-sostenible/

ONU. (2023). *Sostenibilidad.* Obtenido de https://www.un.org/es/impacto-acad%C3%A9mico/sostenibilidad#:~:text=En%201987%2C%20la%20Comisi%C3%B3n%20Brundtland,mundo%20que%20buscan%20formas%20de

ONUDI. (2020). *Informe sobre el Desarrollo Industrial 2020. La industrialización en la era digital.* Viena.: ONUDI.

Opricovic, S., & Tzeng, G. H. (2003). Defuzzification within a multicriteria decision model. *International Journal of Uncertainty, Fuzziness and Knowledge-based Systems, 11*(05), 635–652. doi:10.1142/S0218488503002387

Organización Mundial de la Propiedad Intelectual. (2022). *Índice Mundial de Innovación.* Obtenido de https://www.wipo.int/global_innovation_index/es/#:~:text=El%20%C3%8Dndice%20Mundial%20de%20Innovaci%C3%B3n%202022%20toma%20el%20pulso%20de,de%20suministro%20y%20la%20energ%C3%ADa

Osinubi, T. T., Ajide, F. M., & Osinubi, O. B. (2022). Governance, FDI and tourism in top 10 tourist destinations in Africa. *Transnational Corporations Review, 14*(2), 140–152. doi:10.1080/19186444.2022.2040345

Otero, A. S., & de Grossi, M. (1998). *La llamada revolución industrial.* Universidad Catolica Andres. Obtenido de https://books.google.es/books?id=YmbEneoFEI0C&dq=revoluci%C3%B3n+industrial+definicion&lr=&hl=es&source=gbs_navlinks_s

Oxford Learner's Dictionary. (2022). *Societal.* Oxford Learner's Dictionary. Retrieved august 23, 2022, from https://www.oxfordlearnersdictionaries.com/definition/english/societal?q=societal

Oxford Learner's Dictionary. (2022). *Sustainability.* Oxford Learner's Dictionary. Retrieved august 23, 2022, from https://www.oxfordlearnersdictionaries.com/definition/english/sustainability?q=sustainability

Özcan, M. P. (2016). Problems encountered in taxation of the digital economy: An evaluation within the scope of BEPS action plan 1. *Electronic Journal of Vocational Colleges, 6*(2), 73–82.

Öztürk, L. (2002). The global dimensions of the digital divide. *Ege Academic Review, 2*(1), 1–10.

Palmer, M. A., Bernhardt, E. S., Chornesky, E. A., Collins, S. L., Dobson, A. P., Duke, C. S., Gold, B. D., Jacobson, R. B., Kingsland, S. E., Kranz, R. H., Mappin, M. J., Martinez, M. L., Micheli, F., Morse, J. L., Pace, M. L., Pascual, M., Palumbi, S. S., Reichman, O. J., Townsend, A. R., & Turner, M. G. (2005). Ecological science and sustainability for the 21st century. *Frontiers in Ecology and the Environment, 3*(1), 4–11. doi:10.1890/1540-9295(2005)003[0004:ESASFT]2.0.CO;2

Pardini, K., Rodrigues, J. J., Kozlov, S. A., Kumar, N., & Furtado, V. (2019). IoT-based solid waste management solutions: A survey. *J. Sens. Actuator Netw., 8*(1), 5. doi:10.3390/jsan8010005

Paredes, C. A., & Tigrera, C. R. (2006). Gerencia del proceso de innovación: Un aporte para el desarrollo de tecnologías verdes en el Parque Tecnológico Universitario. *Revista de Ciencias Sociales*, 6–23.

Parvin, F., Islam, S., Akm, S. I., Urmy, Z., & Ahmed, S. (2020). A study on the solutions of environment pollutions and worker's health problems caused by textile manufacturing operations. *Biomedical Journal of Scientific & Technical Research, 28*, 21831–21844.

Pattanayak, S. (2021). *Homo sapiens: do the species making themselves extinct?* Academia Letters. doi:10.20935/AL1617

Pazarlıoğlu, M., & Kiren Gürler, Ö. (2007). Telecommunications investments and economic growth: Panel data approach. *Finance. Political & Economic Comments, 44*(508), 35–43.

Pechmann, J., & de los Reyes, C. (2023, March 17). *Using OSM Data in the Turkey and Syria Earthquake Response.* https://www.hotosm.org/updates/using-osm-data-for-the-turkey-and-syria-earthquake-response/

Peña, C., Ortega, L., & San Juan, L. (2020). Desarrollo sustentable: un detonante para el crecimiento. *Boletín Científico INVESTIGIUM de la Escuela Superior de Tizayuca, 6*(11), 15-23. Obtenido de https://repository.uaeh.edu.mx/revistas/index.php/investigium/issue/archive

Penchev, G. (2016). Using Space Syntax for Estimation of Potential Disaster Indirect Economic Losses. *Comparative Economic Research, 19*(5), 125–142. Advance online publication. doi:10.1515/cer-2016-0041

Phillips, C.B. P. & Perron, P. (1988). Testing for a unit root in time series regressions. *Biometrika, 75*(2), 335-346.

Pickl, M. J. (2019). The renewable energy strategies of oil majors – From oil to energy? *Energy Strategy Reviews, 26*, 100370. Advance online publication. doi:10.1016/j.esr.2019.100370

Pieroni, M. P., McAloone, T. C., & Pigosso, D. C. A. (2019). D.C. Business model innovation for circular economy and sustainability: A review of approaches. *Journal of Cleaner Production, 215*, 198–216. doi:10.1016/j.jclepro.2019.01.036

Poiani, T. H., Rocha, R. dos S., Degrossi, L. C., & de Albuquerque, J. P. (2016). Potential of Collaborative Mapping for Disaster Relief : A Case Study of OpenStreetMap in the Nepal Earthquake 2015. *Hawaii International Conference on System Sciences*, 1–10. 10.1109/HICSS.2016.31

Prakash, P., Balasundaram, J., Al-Enizi, A. M., Ubaidullah, M., & Pandit, B. (2022). Effect of photovoltaic performance of plant-based cocktail DSSCs and adsorption of nano TiO2 onto the solvent-influenced dye sensitizers. *Optical Materials, 133*, 113031. doi:10.1016/j.optmat.2022.113031

Preiser, R., Swilling, M., & Nnoli-Edozien, N. (2021). *Towards new narratives of hope for fostering transformative African futures.* Club of Rome. Retrieved November 05, 2021, from https://www.clubofrome.org/publication/towards-new-narratives-of-hope-for-fostering-transformative-african-futures/

Presenza, A., Messeni Petruzzelli, A., & Natalicchio, A. (2019). Business model innovation for sustainability. Highlights from the tourism and hospitality industry. *Sustainability (Basel), 11*(1), 2–5. doi:10.3390u11010212

Pruthviraja, D., Anil, B. C., & Sowmyarani, C. N (2021). Efficient local cloud-based solution for diabetic retinopathy detection. *International Journal of Web-Based Learning and Teaching Technologies, 16*(3), 39–46. Advance online publication. doi:10.4018/IJWLTT.20210501.oa3

Pudasaini, N. (2016). Mapping a sustainable future. *Geospatial World, 7*(1).

PwC. (2019). *A period of video streaming peace: Consumers have found their video consumption groove: PwC.* PwC. Available at: https://www.pwc.com/us/en/services/consulting/library/consumer-intelligence-series/consumer-video-streaming-behavior

Qamar, M., Noor, M., Ali, W., & Obaid, M. (2020). Green Technology and its implications Worlwide. *The inquisitive Meridian*. Obtenido de https://www.researchgate.net/publication/350443477_Green_Technology_and_its_Implications_Worldwide

Rabiei-Dastjerdi, H., McArdle, G., Matthews, S. A., & Keenan, P. (2021). Gap analysis in decision support systems for real-estate in the era of the digital earth. *International Journal of Digital Earth*, *14*(1), 121–138. Advance online publication. doi:10.1080/17538947.2020.1808719

Radnejad, A. B., Osiyevskyy, O., & Vredenburg, H. (2020). Barriers to radical process innovation: A case of environmental technology in the oil industry. *Journal of Strategy and Management*, *13*(4), 453–476. Advance online publication. doi:10.1108/JSMA-11-2019-0206

Radu, M. (2015). Political stability-a condition for sustainable growth in Romania? *Procedia Economics and Finance*, *30*, 751–757. doi:10.1016/S2212-5671(15)01324-6

Rafiee, J., Sarma, P., Gutierrez, F., Hilliard, R., Calad, C. M., Angulo, O., & Boyer, B. (2022). Energy Transition Meets Digital Transformation: Design and Implementation of a Comprehensive Carbon Emissions Estimation and Forecasting Platform. *Proceedings of the Annual Offshore Technology Conference*. 10.4043/31747-MS

Rahman, A., Hossain, M. S., Alrajeh, N. A., & Guizani, N. (2020). B5G and Explainable Deep Learning Assisted Healthcare Vertical at the Edge: COVID-19 Perspective. *IEEE Network*, *34*(4), 98–105. doi:10.1109/MNET.011.2000353

Ralston, P. (2014). *Supply chain collaboration: A literature review and empirical analysis to investigate uncertainty and collaborative benefits in regards to their practical impact on collaboration and performance*. Academic Press.

Ramadhanis, Z. (2023). *Promoting Indonesia's historical places through digital mapping with GLAM*. https://openstreetmap.or.id/en/2023/05/22/promoting-indonesias-historical-places-through-digital-mapping-with-glam/

Ramírez, J. (2020). *La globalización y el desarrollo sustentable*. Universidad del Valle de México.

Ramos, D. (2020). *Innovación en tecnología verde y las restricciones estructurales para su desarrollo en los países de periferia*. Universidad Nacional Autónoma de México. Obtenido de http://ru.iiec.unam.mx/5251/1/2-096-Ramos.pdf

Randall, A. (2022). Driving with eyes on the rear-view mirror—Why weak sustainability is not enough. *Sustainability (Basel)*, *14*(16), 13. doi:10.3390u141610203

Ranjbari, M., Morales-Alonso, G., Esfandabadi, Z. S., & Carrasco-Gallego, R. (2019). Sustainability and the sharing economy: Modelling the interconnections. *Direccion y Organizacion*, *68*, 33–40. Advance online publication. doi:10.37610/dyo.v0i68.549

Ranjith, V. K. (2016). Business Models and Competitive Advantage. *Procedia Economics and Finance, 37*, 203-207.

Rayan, Z., Alfonse, M., & Salem, A. B. M. (2018). Machine Learning Approaches in Smart Health. *Procedia Computer Science*, *154*, 361–368. doi:10.1016/j.procs.2019.06.052

Real Academia Española. (2023). *Real Academia Española*. Obtenido de https://dle.rae.es/revoluci%C3%B3n

RealHomes. (2023). https://openstreetmap.realhomes.io/

Report, F. (2019). Women in STEM – A Baseline Study. *PBCWE*. https://pbcwe.com.ph/2019/11/29/full-report-women-in-stem-a-baseline-study/

Resetting tech culture. (2020). *Accenture*. https://www.accenture.com/content/dam/accenture/final/a-com-migration/pdf/pdf-134/accenture-a4-gwc-report-final1.pdf#zoom=50

Rezny, L., White, J. B., & Maresova, P. (2019). The knowledge economy: Key to sustainable development? *Structural Change and Economic Dynamics*, *51*(C), 291–300. doi:10.1016/j.strueco.2019.02.003

Riandita, A. (2020). *Inter-firm relationships for sustainability: Incumbent firms and sustainability ventures*. Academic Press.

Richardson, J. (2008). The business model: An integrative framework for strategy execution. *Strategic Change*, *17*(5-6), 133–144. doi:10.1002/jsc.821

RIDAA. (1996). La innovación tecnológica: definiciones y elementos de base. *Redes, 6*(3), 131-175. Obtenido de https://ridaa.unq.edu.ar/bitstream/handle/20.500.11807/500/07R1996v3n6.pdf?sequence=1&isAllowed=y

Robertson, J. O., & Chilingar, G. V. (2017). Environmental Aspects of Oil and Gas Production. In Environmental Aspects of Oil and Gas Production. doi:10.1002/9781119117421

Roberts, R., & Flin, R. (2020). Unlocking the Potential: Understanding the Psychological Factors That Influence Technology Adoption in the Upstream Oil and Gas Industry. *SPE Journal*, *25*(1), 515–528. Advance online publication. doi:10.2118/198903-PA

Roco, M. C., & Bainbridge, W. S. (2013). The new world of discovery, invention, and innovation: Convergence of knowledge, technology, and society. *Journal of Nanoparticle Research*, *15*(9), 15. doi:10.100711051-013-1946-1

Rojek, I., & Studzinski, J. (2019). Detection and localization of water leaks in water nets supported by an ICT system with artificial intelligence methods as away forward for smart cities. *Sustainability (Basel)*, *11*(2), 518. doi:10.3390u11020518

Roller, L. H., & Waverman, L. (2001). Telecommunications infrastructure and economic development: A simultaneous approach. *The American Economic Review*, *91*(4), 909–923. doi:10.1257/aer.91.4.909

Roy Choudhury, A. K. (2014). Environmental impacts of the textile industry and its assessment through life cycle assessment. *Roadmap to sustainable textiles and clothing: environmental and social aspects of textiles and clothing supply chain*, 1-39.

Ruessmann, M., Lorenz, M., Gerbert, P., Waldner, M., Engel, P., Harnisch, M., & Justus, J. (2015). *Industria 4.0: El futuro de la productividad y el crecimiento en las industrias manufactureras*. Obtenido de The Boston Consulting Group.: https://www.bcg.com/publications/2015/engineered_products_project_business_industry_4_future_productivity_growth_manufacturing_industries

Rukhaya, S., Yadav, S., Rose, N. M., Grover, A., & Bisht, D. (2021). Sustainable approach to counter the environmental impact of fast fashion. *The Pharma Innovation Journal*, *10*(8S), 517–523.

Rymbai, H., Sharma, R. R., & Srivastav, M. (2011). Bio-colorants and its implications in health and food industry–a review. *International Journal of Pharmacological Research*, *3*(4), 2228–2244.

Saha, S. S., Schweitzer, E., Scaglione, A., & Johnson, N. G. (2019). A Framework for Generating Synthetic Distribution Feeders using OpenStreetMap. *51st North American Power Symposium, NAPS 2019*. 10.1109/NAPS46351.2019.9000187

Saini, A., & Malik, A. (2017). Routing in internet of things: A survey. *Commun. Comput. Syst.*, *I*, 855–860.

Sakhardande, P., Hanagal, S., & Kulkarni, S. (2016). Design of disaster management system using IoT based interconnected network with smart city monitoring. *Proceedings of the 2016 International Conference on Internet of Things and Applications*, 185–190. 10.1109/IOTA.2016.7562719

Saleh, S., Alameddine, M., & El-Jardali, F. (2009). The case for developing publicly-accessible datasets for health services research in the Middle East and North Africa (MENA) region. *BMC Health Services Research*, *9*(1), 197–197. doi:10.1186/1472-6963-9-197 PMID:19874590

Samanta, A. K., & Konar, A. (2011). Dyeing of textiles with natural dyes. *Natural Dyes, 3*, 30-56.

Samanta, P. (2020). A review on application of natural dyes on textile fabrics and its revival strategy. *Chemistry and technology of natural and synthetic dyes and pigments*, 1-25.

Samarghandi, M. R., Hadi, M., & McKay, G. (2014). Breakthrough Curve Analysis for Fixed-Bed Adsorption of Azo Dyes using Novel Pine Cone Derived Active Carbon. *Adsorption Science and Technology, 32*(10), 791–806. doi:10.1260/0263-6174.32.10.791

Samsami, S., Mohamadizaniani, M., Sarrafzadeh, M. H., Rene, E. R., & Firoozbahr, M. (2020). Recent advances in the treatment of dye-containing wastewater from textile industries: Overview and perspectives. *Process Safety and Environmental Protection, 143*, 138–163. doi:10.1016/j.psep.2020.05.034

Sánchez, I. (2019). La Cuarta Revolución Industrial en México. *OBSERVATORIO DEL DESARROLLO, 8*(24), 58-64. Obtenido de https://estudiosdeldesarrollo.mx/observatoriodeldesarrollo/wp-content/uploads/2021/02/OD245.pdf

Sandra Sancier-Sultan, S. S. (2022). Why bringing women into tech roles is good for society. *World Economic Forum*. https://www.weforum.org/agenda/2022/02/women-gender-technology-economy-representation-equality/

Santibanez-Borda, E., Korre, A., Nie, Z., & Durucan, S. (2021). A multi-objective optimisation model to reduce greenhouse gas emissions and costs in offshore natural gas upstream chains. *Journal of Cleaner Production, 297*, 126625. Advance online publication. doi:10.1016/j.jclepro.2021.126625

Sarkar, S., Banerjee, A., Halder, U., Biswas, R., & Bandopadhyay, R. (2017). Degradation of synthetic azo dyes of textile industry: A sustainable approach using microbial enzymes. *Water Conservation Science and Engineering, 2*(4), 121–131. doi:10.100741101-017-0031-5

Sass, L., & Carboni, G. (2020). Acesso a tecnologias verdes e propriedade intelectual. *JOTA Desenvolvimiento Sustentável*, 1-10. Obtenido de https://www.jota.info/opiniao-e-analise/artigos/acesso-a-tecnologias-verdes-e-propriedade-intelectual-03082020

Saxena, S., & Raja, A. S. M. (2014). Natural dyes: sources, chemistry, application and sustainability issues. *Roadmap to sustainable textiles and clothing: eco-friendly raw materials, technologies, and processing methods*, 37-80.

Schaltegger, S., Freund, F. L., & Hansen, E. G. (2012). Business cases for sustainability: The role of business model innovation for corporate sustainability. *International Journal of Innovation and Sustainable Development, 6*(2), 95–119. doi:10.1504/IJISD.2012.046944

Schiefelbein, J., Rudnick, J., Scholl, A., Remmen, P., Fuchs, M., & Müller, D. (2019). Automated urban energy system modeling and thermal building simulation based on OpenStreetMap data sets. *Building and Environment, 149*, 630–639. Advance online publication. doi:10.1016/j.buildenv.2018.12.025

Schilling, J., & Tränckner, J. (2020). Estimation of wastewater discharges by means of OpenStreetMap data. *Water (Basel), 12*(3), 628. doi:10.3390/w12030628

Scholz, S., Knight, P., Eckle, M., Marx, S., & Zipf, A. (2018). Volunteered geographic information for disaster risk reduction—The missing maps approach and its potential within the red cross and red crescent movement. *Remote Sensing (Basel), 10*(8), 1239. doi:10.3390/rs10081239

Schwab, K. (2016). *La Cuarta Revolución Industrial*. Debate.

Schweizer, J., Poliziani, C., Rupi, F., Morgano, D., & Magi, M. (2021). Building a large-scale micro-simulation transport scenario using big data. *ISPRS International Journal of Geo-Information, 10*(3), 165. Advance online publication. doi:10.3390/ijgi10030165

Scotter, M. J. (2011). Emerging and persistent issues with artificial food colours: Natural colour additives as alternatives to synthetic colours in food and drink. *Quality Assurance and Safety of Crops & Foods*, *3*(1), 28–39. doi:10.1111/j.1757-837X.2010.00087.x

Scott, G., & Rajabifard, A. (2017). Sustainable development and geospatial information: A strategic framework for integrating a global policy agenda into national geospatial capabilities. *Geo-Spatial Information Science*, *20*(2), 59–76. doi:10.1080/10095020.2017.1325594

Seker, F., Ertugrul, H. M., & Cetin, M. (2015). The Impact of Foreign Direct Investment on Environmental Quality: A Bounds Testing and Causality Analysis for Turkey. *Renewable & Sustainable Energy Reviews*, *52*, 347–356. doi:10.1016/j.rser.2015.07.118

Selvaraj, V., Karthika, T. S., Mansiya, C., & Alagar, M. (2021). An over review on recently developed techniques, mechanisms and intermediate involved in the advanced azo dye degradation for industrial applications. *Journal of Molecular Structure*, *1224*, 129195. doi:10.1016/j.molstruc.2020.129195

Seuring, S., & Müller, M. (2008). From a literature review to a conceptual framework for sustainable supply chain management. *Journal of Cleaner Production*, *16*(15), 1699–1710. doi:10.1016/j.jclepro.2008.04.020

Sezer, A., Deniz, M., & Topuz, M. (2018). Analysis of Accessibility of Schools in Usak City via Geographical Information Systems (GIS). *Tarih Kultur Ve Sanat Arastirmalari Dergisi-Journal of History Culture and Art Research, 7*(5).

Shahbaz, M., Gao, C., Zhai, L., Shahzad, F., & Arshad, M.R. (2020). Moderating Effects of Gender and Resistance to Change on the Adoption of Big Data Analytics in Healthcare. *Complex., 2020*, 2173765:1-2173765:13.

Shahbaz, M., Gao, C., Zhai, L., Shahzad, F., & Hu, Y. (2019). Investigating the adoption of big data analytics in healthcare: The moderating role of resistance to change. *Journal of Big Data*, *6*(1), 6. doi:10.118640537-019-0170-y

Shahbaz, M., Nasreen, S., Abbas, F., & Anis, O. (2015). Does Foreign Direct Investment Impede Environmental Quality in High, Middle and Low-income Countries? *Energy Economics*, *51*, 1–35. doi:10.1016/j.eneco.2015.06.014

Shahbaz, M., Nasreen, S., Ahmed, K., & Hammoudeh, S. (2017). Trade openness–carbon emissions nexus: The importance of turning points of trade openness for country panels. *Energy Economics*, *61*, 221–232. doi:10.1016/j.eneco.2016.11.008

Shahid, M., & Mohammad, F. (2013). Recent advancements in natural dye applications: A review. *Journal of Cleaner Production*, *53*, 310–331. doi:10.1016/j.jclepro.2013.03.031

Shang, C., Wu, T., Huang, G., & Ganlin, J. (2019). Weak sustainability is not sustainable: Socioeconomic and environmental assessment of Inner Mongolia for the past three decades. *Resources, Conservation and Recycling*, *141*, 243–252. doi:10.1016/j.resconrec.2018.10.032

Shanmugasundaram, M., & Sudalaimani, K. (2012). A study on natural adsorbents for the removal of chloride ion in water. *International Journal of Engineering Research & Technology (Ahmedabad)*, *1*(5), 1–12.

Sharma, N., Solanki, V. K., & Davim, J. P. (2019). Basics of the Internet of Things (IoT) and Its Future. In Handbook of IoT and Big Data. CRC Press.

Sheikh, M., Malik, M., & Masood, R. (2021). Dynamic linkage between trade openness and sustainable development: Evidence from the BRICS countries. *BRICS Journal of Economics*, *2*(1), 23–39. doi:10.38050/2712-7508-2021-27

Shiau, S. J. H., Huang, C. Y., Yang, C. L., & Juang, J. N. (2018). *A derivation of factors influencing the innovation diffusion of the OpenStreetMap in STEM education.* Sustainability. doi:10.3390u10103447

Sigurdson, G. T., Tang, P., & Giusti, M. M. (2017). Natural colorants: Food colorants from natural sources. *Annual Review of Food Science and Technology*, *8*(1), 261–280. doi:10.1146/annurev-food-030216-025923 PMID:28125346

Silva, J. (2011). *Rol de las ecotecnologías, tecnologías limpias y de tratamiento, en el control de la contaminación generada por las curtiembres de Trujillo-Perú.* Obtenido de Tesis doctoral: https://dspace.unitru.edu.pe/handle/UNITRU/8191

Sims, C. A. (1980). Macroeconomics and Reality. *Econometrica*, *48*(1), 1–48. doi:10.2307/1912017

Singh, A., Mehrotra, R., Rajput, V. D., Dmitriev, P., Singh, A. K., Kumar, P., Tomar, R. S., Singh, O., & Singh, A. K. (2022). Geoinformatics, artificial intelligence, sensor technology, big data: emerging modern tools for sustainable agriculture. *Sustainable Agriculture Systems and Technologies*, 295–313.

Singh, H., & Jun, K. W. (1995). Some new evidence on determinants of foreign direct investment in developing countries (No. 1531). World Bank Publications.

Singh, A. (2019). Remote sensing and GIS applications for municipal waste management. *Journal of Environmental Management*, *243*, 22–29. Advance online publication. doi:10.1016/j.jenvman.2019.05.017 PMID:31077867

Sircar, S., Golden, T. C., & Rao, M. B. (1996). Activated carbon for gas separation and storage. *Carbon*, *34*(1), 1–12. doi:10.1016/0008-6223(95)00128-X

Skarzynski, P., & Gibson, R. (2013). *Innovation to the Core: a Blueprint for Transforming the Way Your Company Innovates.* Harvard Business Press.

Slama, H. B., Chenari Bouket, A., Pourhassan, Z., Alenezi, F. N., Silini, A., Cherif-Silini, H., Oszako, T., Luptakova, L., Golińska, P., & Belbahri, L. (2021). Diversity of synthetic dyes from textile industries, discharge impacts and treatment methods. *Applied Sciences (Basel, Switzerland)*, *11*(14), 6255. doi:10.3390/app11146255

Smajlović, S., Umihanić, B., & Turulja, L. (2019). The Interplay of Technological Innovation and Business Model Innovation toward Company Performance. *Management*, *24*(2), 63–79. doi:10.30924/mjcmi.24.2.5

Soeder, D. J. (2021a). Balancing Energy, Environment, and Economics. In Fracking and the Environment. doi:10.1007/978-3-030-59121-2_11

Soeder, D. J. (2021b). Fracking and the Environment. In *Fracking and the Environment.* Springer International Publishing. doi:10.1007/978-3-030-59121-2_6

Spangenberg, J. H. (2005). Will the information society be sustainable? Towards criteria and indicators for a sustainable knowledge society. *International Journal of Innovation and Sustainable Development*, *1*(1/2), 85–102. doi:10.1504/IJISD.2005.008082

Stasiak, N., Kukula-Koch, W., & Glowniak, K. (2014). Modern industrial and pharmacological applications of indigo dye and its derivatives-a review. *Acta Poloniae Pharmaceutica*, *71*(71), 215–221. PMID:25272640

Stock, J. H., & Watson, M. W. (2007). *Introduction to Econometrics* (2nd ed.). Pearson Education.

Suasana, I. G., & Ekawati, N. (2018). Environmental commitment and green innovation reaching success new products of creative industry in Bali. *Journal of Business and Retail Management Research, 12*(4), 246-250. Obtenido de https://www.jbrmr.com/cdn/article_file/content_90455_18-07-04-02-02-40.pdf

Subramaniam Ananthram, D. B. (2021). The gender gap in STEM is not due to women's lack of confidence. *World Economic Forum.* https://www.weforum.org/agenda/2021/03/gender-gap-in-stem-not-due-to-womens-lack-of-confidence/

Sun, Z., Wang, D., & Zhong, G. (2018). Extraction of farmland geographic information using OpenStreetMap data. *2018 7th International Conference on Agro-Geoinformatics, Agro-Geoinformatics 2018*. 10.1109/Agro-Geoinformatics.2018.8476088

Sun, Y., Yang, Y., Huang, N., & Zou, X. (2020). The impacts of climate change risks on financial performance of mining industry: Evidence from listed companies in China. *Resources Policy*, *69*, 101828. Advance online publication. doi:10.1016/j.resourpol.2020.101828

Tahmad & Adow. (2018). The impact of trade openness on foreign direct investment in Sudan by sector in the 1990-2017 period: An empirical analysis. *Economic Analysis, 7*(8), 14–21.

Talari, S., Shafie-Khah, M., Siano, P., Loia, V., Tommasetti, A., & Catalão, J. P. (2017). A review of smart cities based on the internet of things concept. *Energies*, *10*(4), 421. doi:10.3390/en10040421

Tamburini, D., Breitung, E., Mori, C., Kotajima, T., Clarke, M. L., & McCarthy, B. (2020). Exploring the transition from natural to synthetic dyes in the production of 19th-century Central Asian ikat textiles. *Heritage Science*, *8*(1), 1–27. doi:10.118640494-020-00441-9

Tang, M., Walsh, G., Lerner, D., Fitza, M. A., & Li, Q. (2017). Green Innovation, Managerial Concern and Firm Performance: An Empirical Study. *Business Strategy and the Environment.*

Tank, B., & Gandhi, V. (2023). A Comparative Study on Cloud Computing, Edge Computing and Fog Computing. *Advances in Transdisciplinary Engineering*, *32*. Advance online publication. doi:10.3233/ATDE221329

Tao, F., Cheng, J., & Qi, Q. (2018). IIHub: An industrial internet-of-things hub toward smart manufacturing based on cyber-physical system. *IEEE Transactions on Industrial Informatics*, *14*(5), 2271–2280. doi:10.1109/TII.2017.2759178

Tapscott, D. (1995). *The Digital Economy: Promise and Peril in the Age of Networked Intelligence*. McGraw-Hill Publisher.

Taymaz, E. (2018). *Digital Technologies and Economic Growth: Turkey's Position, Opportunities and Options in Digital Technology Sectors*. TÜSİAD.

TED. (2023). *Mapping the Future: Colombian OpenStreetMappers Impress ITDP Sustainable Transport Expert*. https://www.trufi-association.org/mapping-the-future-colombian-openstreetmappers-impress-itdp-sustainable-transport-expert/

Tefertiller, A. (2018). Media substitution in cable cord-cutting: The adoption ofweb-streaming television. *Journal of Broadcasting & Electronic Media*, *62*(3), 390–407. doi:10.1080/08838151.2018.1451868

Tesfaye, T., Ayele, M., Gibril, M., Ferede, E., Limeneh, D. Y., & Kong, F. (2022). Beneficiation of avocado processing industry by-product: A review on future prospect. *Current Research in Green and Sustainable Chemistry*, *5*, 100253. doi:10.1016/j.crgsc.2021.100253

Thakur, A., & Koul, B. (2022). Impact of oil exploration and spillage on marine environments. In Advances in Oil-Water Separation. doi:10.1016/B978-0-323-89978-9.00018-5

The Future is equal: Gender Equality in the Technology Industry. (2021). *ESCAP*. https://www.unescap.org/kp/2021/report-gender-equality-technology-industry

Toda, H. Y., & Yamamoto, T. (1995). Statistical inference in vector auto regressions with possibly integrated processes. *Journal of Econometrics*, *66*(1-2), 225–250. doi:10.1016/0304-4076(94)01616-8

Tohănean, D., Buzatu, A. I., Baba, C.-A., & Georgescu, B. (2020). Business model innovation through the use of digital technologies: Managing risks and creating sustainability. *Amfiteatru Economic*, *22*(55), 758–774. doi:10.24818/EA/2020/55/758

Tontiwachwuthikul, P., Chan, C. W., Zeng, F. B., Liang, Z. H., Sema, T., & Min, C. (2020). Recent progress and new developments of applications of artificial intelligence (AI), knowledge-based systems (KBS), and Machine Learning (ML) in the petroleum industry. *Petroleum*, *6*(4), 319–320. Advance online publication. doi:10.1016/j.petlm.2020.08.001

Topçu, H., & Gedik, A. (2021). The effect of telecommunication investments on economic growth in Turkey. *Journal of Academic Value Studies*, *7*(3), 316–343.

Trakadas, P., Simoens, P., Gkonis, P., Sarakis, L., Angelopoulos, A., Ramallo-González, A. P., Skarmeta, A., Trochoutsos, C., Calvo, D., Pariente, T., & ... . (2020). An artificial intelligence-based collaboration approach in industrial iot manufacturing: Key concepts, architectural extensions and potential applications. *Sensors (Basel)*, *20*(19), 5480. doi:10.339020195480 PMID:32987911

Trencher, G., & Karvonen, A. (2019). Stretching "smart": Advancing health and wellbeing through the smart city agenda. *Local Environment*, *24*(7), 610–627. doi:10.1080/13549839.2017.1360264

Treybal, R. E. (1981). *Mass Transfer Operations*. McGraw-Hill Book Company.

Tzavella, K., Skopeliti, A., & Fekete, A. (2022). Volunteered geographic information use in crisis, emergency and disaster management: A scoping review and a web atlas. *Geo-Spatial Information Science*, 1–32. Advance online publication. doi:10.1080/10095020.2022.2139642

Ugwuegbula, L. (2020). TEM Gender Bias Cripples Asia-Pacific Region. *The Diplomat*. https://thediplomat.com/2020/02/stem-gender-bias-cripples-asia-pacific-region/

UNCTAD. (2023). *Informe sobre Tecnología e Innovación 2023*. UNCTAD.

UNDP. (2023). *Sustainable Development Goals*. https://www.undp.org/sustainable-development-goals

UN-GGIM. (2012). *Monitoring sustainable development: Contribution of geospatial information to the rio+ 20 processes*. United Nations New York.

United Nations. (2019, March 28). *Meetings coverage: Seventy third session, High level meeting on climate and sustainable development (am and pm)*. United Nations. Retrieved August 15, 2021, from https://www.un.org/press/en/2019/ga12131.doc.htm

United Nations. (2023). *Sustainable development goals*. https://www.un.org/sustainabledevelopment/

Unuabonah, E. I., Omorogie, M. O., & Oladoja, N. A. (2019). *Modelling in Adsorption: Fundamentals and Applications Composite Nanoadsorbents*. Micro and Nano Technologies.

Useche, A. J., Juarez, F., & Ramirez, A. (2022). *Tecnologías de la cuarta revolución industrial y su aplicación en la Armada Nacional de Colombia*. Universidad del Rosario. Obtenido de., doi:10.12804/urosario9789587848786

Van den Berg, R., & De Langen, P. W. (2017). Environmental sustainability in container transport: The attitudes of shippers and forwarders. *International Journal of Logistics*, *20*(2), 146–162. doi:10.1080/13675567.2016.1164838

Van Tonder, C., Schachtebeck, C., Nieuwenhuizen, C., & Bossink, B. (2020). A framework for digital transformation and business model innovation. *Management*, *25*(2), 111–132. doi:10.30924/mjcmi.25.2.6

Vardhan, K. H., Kumar, P. S., & Panda, R. C. (2019). A review on heavy metal pollution, toxicity and remedial measures: Current trends and future perspectives. *Journal of Molecular Liquids*, *290*, 111197. doi:10.1016/j.molliq.2019.111197

Varsei, M., Soosay, C., Fahimnia, B., & Sarkis, J. (2014). Framing sustainability performance of supply chains with multidimensional indicators. *Supply Chain Management*, *19*(3), 242–257. doi:10.1108/SCM-12-2013-0436

Velu, C. (2015). Business model innovation and third-party alliance on the survival of new firms. *Technovation, 35*, 1–11. doi:10.1016/j.technovation.2014.09.007

Venil, C. K., Velmurugan, P., Dufossé, L., Renuka Devi, P., & Veera Ravi, A. (2020). Fungal pigments: Potential coloring compounds for wide ranging applications in textile dyeing. *Journal of Fungi (Basel, Switzerland), 6*(2), 68. doi:10.3390/jof6020068 PMID:32443916

Vidgen, R. T., Shaw, S., & Grant, D. B. (2017). Management challenges in creating value from business analytics. *European Journal of Operational Research, 261*(2), 626–639. doi:10.1016/j.ejor.2017.02.023

Villa de Prado, R. (2019). Consecuencias económicas y sociales de la cuarta revolución industrial y estrategias pensadas para la adaptación de la actividad económica. *Revista Aportes de la Comunicación y la Cultura*. Obtenido de http://www.scielo.org.bo/scielo.php?script=sci_arttext&pid=S2306-86712019000100010&lng=es&tlng=es

Villagaray, S., & Bautista, E. (2011). Sistemas agroforestales con tecnología limpia en los suelos del VRAEM, Perú. *Acta Nova, 2*(5), 289-311. Obtenido de http://scielo.iics.una.py/scielo.php?script=sci_arttext&pid=S2414-89382020000100097#B63

Vivanco, E. (2020). *Energías renovables y no renovables*. Obtenido de Biblioteca del Congreso Nacional de Chile: https://www.bcn.cl/obtienearchivo?id=repositorio/10221/29102/1/BCN_Energia_renovable_y_no_renovable_ventajas_y_desventajas_final.pdf

Wadanambi, R. T., Wandana, L. S., Chathumini, K. K. G. L., Dassanayake, N. P., Preethika, D. D. P., & Arachchige, U. S. P. R. (2020). The effects of industrialization on climate change. *Journal of Research Technology and Engineering, 1*(4).

Wagner, M., Doherty, G., & van Zanten, B. (2019). *Leveraging OpenStreetMap to improve disaster risk management in the Seychelles*. https://www.preventionweb.net/news/leveraging-openstreetmap-improve-disaster-risk-management-seychelles

Wagner, D. L., Grames, E. M., Forister, M. L., Berenbaum, M. R., & Stopak, D. (2021). Insect decline in the Anthropocene: Death by a thousand cuts. *Proceedings of the National Academy of Sciences of the United States of America, 118*(2), 1–10. doi:10.1073/pnas.2023989118 PMID:33431573

Walsh, K. B., & Rose, J. (2022). A Review of Restoration Techniques and Outcomes for Rangelands Affected by Oil and Gas Production in North America. *Ecological Restoration, 40*(4), 259–269. Advance online publication. doi:10.3368/er.40.4.259

Wan, J., Yang, J., Wang, Z., & Hua, Q. (2018). Artificial intelligence for cloud-assisted smart factory. *IEEE Access, 6*, 55419–55430.

Wang, Q., Su, M., & Li, R. (2018). Toward to economic growth without emission growth: the role of urbanization and industrialization in China and India. *J Clean Prod, 205*, 499-511. https://doi.org/. doi:10.1016/j.jclepro.2018.09.034

Wang, Y., Ram, S., Currim, F., Dantas, E., & Sabóia, L. A. (2016). A big data approach for smart transportation management on bus network. *Proceedings of the IEEE 2nd International Smart Cities Conference: Improving the Citizens Quality of Life*, 1–6. 10.1109/ISC2.2016.7580839

Wang, C., Li, Y., & Shi, X. (2019). Information mining for urban building energy models (UBEMs) from two data sources: OpenStreetMap and baidu map. *Building Simulation Conference Proceedings,* 5. 10.26868/25222708.2019.210545

Wang, Q., & Zhang, F. (2020). The effects of trade openness on decoupling carbon emissions from economic growtheevidence from 182 countries. *Journal of Cleaner Production, 279*, 123838. doi:10.1016/j.jclepro.2020.123838 PMID:32863606

Wang, W., Rehman, M. A., & Fahad, S. (2022). The dynamic influence of renewable energy, trade openness, and industrialization on the sustainable environment in G-7 economies. *Renewable Energy*, *198*, 484–491. doi:10.1016/j.renene.2022.08.067

Wang, Y., & Byrd, T. A. (2017). Business analytics-enabled decision-making effectiveness through knowledge absorptive capacity in health care. *Journal of Knowledge Management*, *21*(3), 517–539. doi:10.1108/JKM-08-2015-0301

Wan, T., Lu, H., Lu, Q., & Luo, N. (2017). Classification of High-Resolution Remote-Sensing Image Using OpenStreetMap Information. *IEEE Geoscience and Remote Sensing Letters*, *14*(12), 2305–2309. Advance online publication. doi:10.1109/LGRS.2017.2762466

Ward, M. J., Marsolo, K. A., & Froehle, C. M. (2014). Applications of Business Analytics in Healthcare. *Business Horizons*, *57*(5), 571–582. doi:10.1016/j.bushor.2014.06.003 PMID:25429161

Weiss, E. B. (1992). Environment and trade as partners in sustainable development: A commentary. *The American Journal of International Law*, *86*(4), 728–735. doi:10.2307/2203789

Weitekamp, C. A., Stevens, T., Stewart, M. J., Bhave, P., & Gilmour, M. I. (2020). Health effects from freshly emitted versus oxidatively or photochemically aged air pollutants. In Science of the Total Environment (Vol. 704). doi:10.1016/j.scitotenv.2019.135772

Whitmarsh, L. (2011). Scepticism and uncertainty about climate change: Dimensions, determinants and change over time. *Global Environmental Change*, *21*(2), 690–700. doi:10.1016/j.gloenvcha.2011.01.016

Wichmann, J. R. K., Scholdra, T. P., & Reinartz, W. J. (2023). Propelling International Marketing Research with Geospatial Data. *Journal of International Marketing*, *31*(2), 82–102. doi:10.1177/1069031X221149951

Widiawati, D., Sn, S., Rosandini, M., & Ds, S. (2012). Natural Dyes on Indonesian Traditional Textiles-A Case Study: Geringsing Woven Fabric, In Tenganan Pegeringsingan Village Bali. *The Research Journal of the Costume Culture*, *20*(1), 111–120. doi:10.7741/rjcc.2012.20.1.111

Wieland, T. (2017). Market area analysis for retail and service locations with MCI. *The R Journal*, *9*(1), 298. Advance online publication. doi:10.32614/RJ-2017-020

Wiki. (2019). *Wildlife sanctuaries of India.* https://wiki.openstreetmap.org/wiki/Wildlife_sanctuaries_of_India

Wiki. (2023). *Niskala Institute: Preservation of Ciamis Regency Cultural Heritage Through OSM Data.* https://wiki.openstreetmap.org/wiki/Niskala_Institute:_Preservation_of_Ciamis_Regency_Cultural_Heritage_Through_OSM_Data

Williams, B. (2021). *Saving us from ourselves: Can we repair 50 years of ecological overshoot.* https://poemsforparliament.uk

Wilson, I., Saini, S., Sreenivasan, H., Sahu, C., Krishna, S., & Gupta, P. (2023). Review and Perspectives of Energy-Efficient Methane Production from Natural Gas Hydrate Reservoirs Using Carbon Dioxide Exchange Technology. In Energy and Fuels. doi:10.1021/acs.energyfuels.3c00715

Winterfeldt, D. (2013). Bridging the gap between science and decision making. *Proceedings of the National Academy of Sciences of the United States of America*, *110*(Suppl. 3), 14055–14061. doi:10.1073/pnas.1213532110 PMID:23940310

Wollin, K. M., Damm, G., Foth, H., Freyberger, A., Gebel, T., Mangerich, A., Gundert-Remy, U., Partosch, F., Röhl, C., Schupp, T., & Hengstler, J. G. (2020). Critical evaluation of human health risks due to hydraulic fracturing in natural gas and petroleum production. In Archives of Toxicology (Vol. 94, Issue 4). doi:10.100700204-020-02758-7 women and digital revolution. (2021). *UNESCO science report*. https://www.unesco.org/reports/science/2021/en/women-digital-revolution

Women in Science. (2020). *UNESCO Fact Sheet No. 60*. https://uis.unesco.org/sites/default/files/documents/fs60-women-in-science-2020-en.pdf

Women Role in STEM Education in Pakistan. (2022). *Stem Education*. https://stemeducation.org.pk/2022/05/09/women-role-in-stem-education/

Wood, A., & Berge, K. (1997). Exporting manufactures: Human resources, natural resources,and trade policy. *The Journal of Development Studies*, *34*(1), 35–59. doi:10.1080/00220389708422502

Wooldridge, J. M. (2013). *Introductory Econometrics: A Modern Approach* (5th ed.). South-Western Cengage Learning.

World Bank Group. (Ed.). (2016). *World Development Indicators 2016*. World Bank Publications.

World Commission on Environment and Development. (1987). *Our Common Future*. Oxford University Press.

World Economic Forum (WEF). (2018). *The future of jobs report 2018*. https://www3.weforum.org/docs/WEF_Future_of_Jobs_2018.pdf

World Economic Situation and Prospects. (2020). https://www.un.org/development/desa/dpad/wpcontent/uploads/sites/45/WESP2020_Annex.pdf

World Health Organization. (2022). *9 out of 10 people worldwide breathe polluted air, but more countries are taking action*. Retrieved October 14, 2022, from https://www.who.int/news/item/02-05-2018-9-out-of-10-people-worldwide-breathe-polluted-air-but-more-countries-are-taking
-action

World Soil Information. (2020). *SoilGrids*. SoilGrids250m.

Wossen, T., Muller, A., & Nichterlein, K. (2019). Eco-efficiency and agricultural innovation systems in developing countries: Evidence from macro-level analysis. *PLoS One*, *14*(4), e0214115. Advance online publication. doi:10.1371/journal.pone.0214115 PMID:30951548

Wrolstad, R. E., & Culver, C. A. (2012). Alternatives to those artificial FD&C food colorants. *Annual Review of Food Science and Technology*, *3*(1), 59–77. doi:10.1146/annurev-food-022811-101118 PMID:22385164

Wu, J., & Strezov, V. (2023). *Green technologies and sustainability: A new trend*. Green Technologies and Sustainability.

Xepapadeas, A. (2005). Economic Growth and the Environment. Handbook of Environmental Economics, 1(2), 1219-1271.

Yan, Y., Qian, Y., & Sharif, H. (2011). A secure data aggregation and dispatch scheme for home area networks in smart grid. *Proceedings of the GLOBECOM—IEEE Global Telecommunications Conference*, 1–6. 10.1109/GLOCOM.2011.6133655

Yang, C., Huang, Q., Li, Z., Liu, K., & Hu, F. (2017). Big Data and cloud computing: innovation opportunities and challenges. In International Journal of Digital Earth (Vol. 10, Issue 1). doi:10.1080/17538947.2016.1239771

Yang, D. (2019). Mapping Regional Landscape by Using OpenStreetMap (OSM): A Case Study to Understand Forest Patterns in Maya Zone, Mexico. In Environmental Information Systems: Concepts, Methodologies, Tools, and Applications (pp. 771–790). IGI Global.

Yang, D., Fu, C. S., Smith, A. C., & Yu, Q. (2017). Open land-use map: A regional land-use mapping strategy for incorporating OpenStreetMap with earth observations. *Geo-Spatial Information Science*, *20*(3), 269–281. Advance online publication. doi:10.1080/10095020.2017.1371385

Yapraklı, S., & Sağlam, T. (2010). Information and communication technologies and economic growth in Turkey: An econometric analysis (1980-2008). *Ege Academic Review*, *10*(2), 575–596. doi:10.21121/eab.2010219637

Yasukawa, A., Chida, A., Kato, Y., & Kasai, M. (2017). Dyeing silk and cotton fabrics using natural blackcurrants. *Textile Research Journal*, *87*(19), 2379–2387. doi:10.1177/0040517516671125

Yıldız, F. (2012). The impact of telecommunications investments on economic growth: An empirical study on OECD countries. *Süleyman Demirel University Journal of the Faculty of FEAS*, *17*(3), 233–258.

Yin, C. (2021). International law regulation of offshore oil and gas exploitation. *Environmental Impact Assessment Review*, *88*, 106551. Advance online publication. doi:10.1016/j.eiar.2021.106551

Yousefpour, A., Fung, C., Nguyen, T., Kadiyala, K., Jalali, F., Niakanlahiji, A., Kong, J., & Jue, J. P. (2019). All one needs to know about fog computing and related edge computing paradigms: A complete survey. *Journal of Systems Architecture*, *98*, 289–330. doi:10.1016/j.sysarc.2019.02.009

Yousuf, N., Olayiwola, O., Guo, B., & Liu, N. (2021). A comprehensive review on the loss of wellbore integrity due to cement failure and available remedial methods. In Journal of Petroleum Science and Engineering (Vol. 207). doi:10.1016/j.petrol.2021.109123

Yüce, M., & Akbulut, N. (2018). Measures taken regarding the taxation of the digital economy within the scope of the BEPS action. *Academic Perspective International Refereed Journal of Social Sciences*, (68), 105–123.

Yusianto, R., Marimin, S., & Hardjomidjojo, H. (2020). Intelligent spatial decision support system concept in the potato agro-industry supply chain. *2020 International Conference on Computer Science and Its Application in Agriculture, ICOSICA 2020.* 10.1109/ICOSICA49951.2020.9243233

Yusoff, M. B., & Nuh, R. (2015). Foreign direct investment, trade openness and economic growth: Empirical evidence from Thailand. *Foreign Trade Review*, *50*(2), 73–84. doi:10.1177/0015732515572055

Yusuf, M. (2019). Synthetic dyes: A threat to the environment and water ecosystem. *Textiles and clothing*, 11-26.

Zabulis, X., Meghini, C., Partarakis, N., Beisswenger, C., Dubois, A., Fasoula, M., Nitti, V., Ntoa, S., Adami, I., Chatziantoniou, A., Bartalesi, V., Metilli, D., Stivaktakis, N., Patsiouras, N., Doulgeraki, P., Karuzaki, E., Stefanidi, E., Qammaz, A., Kaplanidi, D., ... Galanakis, G. (2020). Representation and preservation of heritage crafts. *Sustainability (Basel)*, *12*(4), 1461. Advance online publication. doi:10.3390u12041461

Zaman, M., Pinglu, C., Hussain, S. I., Ullah, A., & Qian, N. (2021). Does regional integration matter for sustainable economic growth? Fostering the role of FDI, trade openness, IT exports, and capital formation in BRI countries. *Heliyon*, *7*(12), e08559. doi:10.1016/j.heliyon.2021.e08559 PMID:34917822

Zarkogianni, M., Mikropoulou, E., Varella, E., & Tsatsaroni, E. (2011). Colour and fastness of natural dyes: Revival of traditional dyeing techniques. *Coloration Technology*, *127*(1), 18–27. doi:10.1111/j.1478-4408.2010.00273.x

Zhang, C., & Zhou, X. (2016). Does foreign direct investment lead to lower CO2 emissions? Evidence from a regional analysis in China. *Renewable & Sustainable Energy Reviews*, *58*, 943–951. doi:10.1016/j.rser.2015.12.226

Zimmerle, D., Vaughn, T., Bell, C., Bennett, K., Deshmukh, P., & Thoma, E. (2020). Detection Limits of Optical Gas Imaging for Natural Gas Leak Detection in Realistic Controlled Conditions. *Environmental Science & Technology*, *54*(18), 11506–11514. Advance online publication. doi:10.1021/acs.est.0c01285 PMID:32786569

Zubairu, A., & Mshelia, Y. M. (2015). Effects of selected mordants on the application of natural dye from onion skin (Allium cepa). *Science and Technology*, *5*(2), 26–32.

About the Contributors

Supriya Lamba Sahdev is currently an Assistant Professor for Amity International Business School at Amity University in Uttar Pradesh.

Chitra Krishnan is currently an Associate Professor at the Symbiosis Centre for Management Studies in Noida.

Ahdi Hassan has been Associate or Consulting Editor of numerous journals and also served the editorial review board from 2013- to till now. He has a number of publications and research papers published in various domains. He has given contribution with the major roles such as using modern and scientific techniques to work with sounds and meanings of words, studying the relationship between the written and spoken formats of various Asian/European languages, developing the artificial languages in coherence with modern English language, and scientifically approaching the various ancient written material to trace its origin. He teaches topics connected but not limited to communication such as English for Young Learners, English for Academic Purposes, English for Science, Technology and Engineering, English for Business and Entrepreneurship, Business Intensive Course, Applied Linguistics, interpersonal communication, verbal and nonverbal communication, cross cultural competence, language and humor, intercultural communication, culture and humor, language acquisition and language in use.

* * *

Vernika Agarwal is currently working as an Associate Professor in operations management at Apeejay School of Management, Dwarka, New Delhi, India. Dr Vernika Agarwal is an active reviewer in several international journals and has edited international books on Smart Cities & Lean-Green & Society 4.0 in Wiley and Springer publishers. She works in Green Supply Chain Management, Sustainable Supply Chain Management, Multi-Criteria Decision Making, Third Party Logistic Provider, Sustainable Lean Manufacturing, Circular Economy and Sustainability, cross-disciplinary research in supply and operations management, Optimisation, and Reverse Logistics.

Munir Ahmad is a Ph.D. in Computer Science. over 24 years of extensive experience in spatial data development, management, processing, visualization, and quality control. He is dedicated expertise in open data, crowdsourced data, volunteered geographic information and spatial data infrastructure. A seasoned professional with extensive knowledge in the field, having served as a trainer for the latest spatial technologies. With a passion for research and over 30 publications in the same field. In 2022, he got PhD degree in Computer Science from Preston University Rawalpindi, Pakistan. He is dedicated to advancing the industry and spreading knowledge through my expertise and experience. #SpatialData #GIS #GeoTech.

Asmat Ali is PhD in Remote Sensing & GIS. He is Director of Cartography and GIS at Survey of Pakistan. In, 1998 he earned Professional Master Degree in Geoinformatics with specialization in Cartography from Faculty of Geo-Information Science and Earth Observation (ITC), University of Twente, Enschede The Netherlands. Later on, he got MSc Degree in Geo-information Science and Earth Observation, with specialization in Geo-Information Management from the same university. In 2022, he got PhD degree in Remote Sensing and GIS from PMAS-Arid Agriculture University Rawalpindi, Pakistan. He has 34 years of experience in geospatial information production and management discipline as practitioner, trainer and educator. He has served on a number of operational, administrative and instructional appointments. He was in-charge of the team which started GIS mapping and digital cartography at Survey of Pakistan in 1999. His more than 25 research papers including two book chapters and articles on Spatial Data Infrastructure (SDI), GIS, Remote Sensing, E-governance, as well as Land Administration have been published in various international conferences and renowned journals. In 2008, SDI Asia-Pacific identified him as focal point for SDI development in Pakistan. GSDI in 2016, acknowledged and awarded him as SDI implementer from Pakistan. He is on the visiting faculty of Bahria University Islamabad and PMAS- Arid Agriculture University Rawalpindi, Pakistan.

Pragya Arya is currently working as a faculty of Operations Management with Apeejay School of Management, Dwarka, New Delhi. She is a Fellow from Management Development Institute, Gurugram with an M.B.A. from IBS Hyderabad and B.E. in Electrical Engineering from Nagpur University. Dr. Pragya is a passionate academician with a keen interest in Sustainability of Business Operations. In her academic career of 13 years, she has taught a vast expanse of subjects – Production and Operations Management, Supply Chain Management, Operations Strategy, Project Management, Quality Management, Business Statistics, Operations Research and Operations Analytics. She has delivered overseas assignments at partner institutes representing Apeejay School of Management. She has publications in journals indexed in UGC Care and ABDC list. Apart from this she has teaching cases published in Ivey Cases Collection.

Mona Chaudhary is associated with Amity University, Uttar Pradesh, India as faculty of Business Strategy. She is involved in student engagement and career aspirations of management programmes. She has attended conference of institutes of high repute and published research articles, business cases in international journals, books, conference proceedings.

Ingrid Cuevas Zuñiga is a specialist in innovation and sustainability; her research focuses on the incorporation of the premise of sustainable development into the business strategy in organizations, specifically through innovation. She is head of the Research Department and teacher of the ESCA Santo Tomás Unit of the National Polytechnic Institute in Mexico City. And it was awarded the Presencia Lázaro Cárdenas, the highest distinction granted by the Institute to the most outstanding members of its community.

Seema Garg is Associate Professor at Amity International Business School, Amity University. She has done her M.Sc, M.Phil (Operational Research) from Faculty of Mathematical sciences Delhi University, Delhi and Ph.D from Jamia Millia Islamia University, Delhi. Her area of specialization is Decision Sciences. She has over 19 years experience in academics and teaching. Her publications are in the areas of Operations management, Modelling and Efficiency Optimization using Data Envelopment Analysis application of DEA. She has pursued research in truly inter- disciplinary areas. She has to her credits a number of reputed national and international journals. She is the Merit certificate holder from Central Board of secondary education and also recipient of National Scholarship for academic performance. She has also received the Best teacher award from MLA of Delhi.

Sunil J. Kulkarni has completed his PhD from Sant Gadge baba Amravati University and Masters in Chemical Engineering from Tatyasaheb Kore Institute of Engineering and Technology, Warananagar. He has worked as Assistant Professor in Chemical Engineering Department of Datta Meghe College of Engineering, Airoli, Navi Mumbai, India for 13 years. He has 25 years of teaching and research experience. Dr. Sunil is involved with different academic activities across the universities in Maharashtra. His area of investigation in doctoral studies is phenol and heavy metal removal from waste water. He has also worked on kinetics and isotherms of various sorption operations. His area of research includes adsorption, clean technology and environmental biotechnology. He is on the reviewer board of many international journals and reviewed many international papers. He is also governing body member, promotional editor and executive member of many reputed international journals. Currently he is working as Associate Professor in chemical engineering at GIT, Lavel, India.

Kasturi Laturkar is currently working as a Validation Engineer for Validation Associates LLC. She graduated with a M.S. in Chemical Engineering from Syracuse University and a B.Tech in Chemical Engineering from Guru Gobind Singh Indraprastha University, Delhi, India. She has more than 4 years of experience working in commissioning, qualification and validation of upstream and downstream bioprocessing equipment and critical utilities.

Kaustubh Laturkar is currently working as an engineer at the Facility for Rare Isotope Beams which is a US-DOE project in Michigan after completing his M.S in chemical engineering from University of Florida and B.E. in Chemical Engineering from Panjab University, Chandigarh, India. He has more than 10 years of experience working in the field of process engineering, refinery operations, utility systems design and operation with a special focus on design and commissioning of engineering systems.

Rutuja Santosh Mulukh is a final-year Chemical Engineering student at Gharda Institute of Technology. She has completed her schooling with distinctions. She has published papers on solid waste, biofuels from waste feedstocks, and Temperature Swing Adsorption at SSRN, IJARCS, and JSMPM. Rutuja is looking forward to exploring more in the field of Chemical engineering as she is passionate about it. She believes in hard work, persistence & continuous learning.

Gautami Jagdish Narhari is a final year Chemical engineering student at Gharda Institute of Technology, Lavel. She has completed her studies with Good grades. She has membership of IIChe. She has been actively taken part in Various paper presentations via online mode which are published. Her interests include Content and project-based learning, Critical and Creative thinking. Her area of interest is waste water treatment.

Sathish Pachiyappan is currently serving as Assistant Professor at CHRIST (Deemed to be University), Bannerghatta Road Campus, Bengaluru. He has eight years teaching and three years research experience. He has done his doctoral degree from VIT (Deemed to be University), Vellore. He is specialized in Finance and Accounting, doing his research in the same area. He has completed MBA from Anna University and B,Com from SRM University, Chennai. Currently, as a part of his Research work, he had published articles in peer reviewed journals which includes Scopus Indexed Journals; Web of Science Indexed Journals, Australian Business Deans Council (ABDC) listed Journals and EBSCO host Journals. Also, published book chapter in Elsevier, Springer and IGI global. Presented research papers in national and international conferences and also attended Faculty development programmes and workshops which are related to SPSS, AMOS and Econometrics. He went as resource person for various FDP, Guest Lecture, Session Chair for Conference and Workshop to various reputed colleges. He is well versed in handling SPSS, R STATCRAFT, E-Views (Econometrics) for financial data analysis in research area. He is gold medalist and earned first rank in B.Com at SRM University, Chennai and also earned Class topper in MBA and 39th Rank Holder in Anna University. He received Research Award-2017 for publications in VIT University, Vellore. Recently, received award for "Best Researcher-2020" conducted by Pondicherry Research Society in Pondicherry.

Namrata Pancholi is an MBA, Ph.D. in Knowledge Management Strategy from M.L.S. University, Udaipur. She has 18 years of teaching experience in Management. Presently working as Assistant Professor in the area of Strategy, International Business and Marketing, Amity University Uttar Pradesh, Noida. She has a number of publications and research papers published in various domains .She teaches topics connected to General Management, Strategy, International Business and Consultancy. She has taught in a number of Management Institutions.The most prominent among them are Birla Institute of Technology &Science, Pilani, Jaipur National University, Jaipur and Pacific University, Udaipur.

Saima Munawar Parkar is a final year chemical engineering student at Gharda Institute of Technology. She has completed her schooling with distinctions. She has experience of various paper presentations which were published online. She has published papers related to solid waste, biofuels from waste feedstocks and temperature swing adsorption online at SSRN, IJARCS and JSMPM. Saima is looking forward to transitioning into professional world of engineering. She believes in continuous learning and evolving as an engineer.

Prabhat Ranjan is presently working as a college librarian in University of Allahabad (India). He also served Rajarshi Shahu College of Engineering, Pune as a librarian and Shri Ramswaroop Memorial University, Barabanki (Uttar Pradesh), India as a deputy librarian. He has obtained his post graduate degree in Library and Information Science from Banaras Hindu University, Varanasi. He completed his Ph.D. in the same discipline from Savitribai Phule Pune University (formerly known as University of Pune), Pune. He has thirty publications as research papers and book chapters. He is actively engaged in research on several interdisciplinary topics including information seeking behavior and information society. He is also keen on environmental studies and social sciences.

Rupayan Roy, an Assistant Professor at the National Institute of Fashion Technology in Kannur, India, specializes in fibrous structures, nonwoven structures, fibre orientation, and fibrous composites. He obtained his Ph.D. in Textile and Fibre Engineering from IIT Delhi, where a wide range of fibrous structures for various applications was explored. His doctoral research primarily focused on the development of highly efficient air filter media utilizing needle-punched nonwovens. With 3 years of experience in academics, Dr. Roy has authored more than 25 research papers and delivered presentations at over 10 conferences of International repute. His primary interest involves advancing the field of fibrous material design for high-end applications. His current research interest includes technical textiles with natural materials.

Bhupendra Kumar Singh is working as a university librarian in University of Allahabad (India) since 2016. Earlier he served as an assistant librarian in V. M. Open University, Kota and deputy librarian in University of Allahabad. He has obtained his academic qualifications i.e. B.A., M.A. and professional qualifications of B.L.I.Sc., M.L.I.Sc. and Ph.D. in Library and Information Science from Banaras Hindu University, Varanasi (Uttar Pradesh), India. He has more than sixty (60) articles and research papers published in national and international journals. Four books are in his credit. His area of specialization is library management and use of latest tools and techniques in library services. He has delivered many lectures on various topics including research methods, referencing, intellectual property rights in several educational institutions. He is University Coordinator of National Digital Library of India (NDLI) and Shodhganga Project of University Grants Commission (UGC), New Delhi in his institution.

Ma. del Rocío Soto Flores is a PhD in Business Administration and Management from the Polytechnic University of Catalonia. She is a specialist in the economics of technological change and competitiveness and industrial innovation and focuses her research on the need for innovation that companies have as a pillar of the growth and economic development of any country. She is a Professor at the National Polytechnic Institute in Mexico City. And he is a founding member of the Research Network in Teaching and Technological Innovation. In addition, he was a member of the Network of Centers in support of Innovation, funded by the CyTED of Spain. She served as Director for Mexico of the Latin-American Association of Technology Management.

Index

www.ingramcontent.com/pod-product-compliance
Lightning Source LLC
LaVergne TN
LVHW080644011026
849160LV00014B/1317